The Cremona Group and Its Subgroups

Mathematical
Surveys
and
Monographs

Volume 252

The Cremona Group and Its Subgroups

Julie Déserti

AMERICAN MATHEMATICAL SOCIETY
Providence, Rhode Island

2020 *Mathematics Subject Classification.* Primary 14E07, 14E05, 20-02, 32M05.

The author was partially supported by the ANR grant Fatou ANR-17-CE40-0002-01 and the ANR grant Foliage ANR-16-CE40-0008-01.

For additional information and updates on this book, visit
www.ams.org/bookpages/surv-252

Library of Congress Cataloging-in-Publication Data

Names: Déserti, Julie, author.
Title: The Cremona group and its subgroups / Julie Deserti.
Description: Providence, Rhode Island : American Mathematical Society, [2021] | Series: Mathematical surveys and monographs, 0076-5376 ; volume 252 | Includes bibliographical references and index.
Identifiers: LCCN 2020045532 | ISBN 9781470460129 (paperback) | 978-1-4704-6452-3 (ebook)
Subjects: LCSH: Geometry, Algebraic. | AMS: Algebraic geometry – Birational geometry – Birational automorphisms, Cremona group and generalizations. | Algebraic geometry – Birational geometry – Rational and birational maps. | Group theory and generalizations – Research exposition (monographs, survey articles). | Several complex variables and analytic spaces {For infinite-dimensional holomorphy, see 46G20, 58B12} – Complex spaces with a group of automorphisms – Complex Lie groups, automorphism groups acting
Classification: LCC QA564 .D47 2021 | DDC 516.3/5–dc23
LC record available at https://lccn.loc.gov/2020045532

∞ The paper used in this book is acid-free and falls within the guidelines
established to ensure permanence and durability.
Visit the AMS home page at https://www.ams.org/

10 9 8 7 6 5 4 3 2 1   26 25 24 23 22 21

To Benoît

Contents

Preface

The main purpose of the present treatise is to draw a portrait of the n-dimensional Cremona group $\mathrm{Bir}(\mathbb{P}^n_{\mathbb{C}})$. The study of this group started in the 19th century; the subject has known a lot of developments since the beginning of the 21st century. Old and new results are discussed; unfortunately we will not be exhaustive. The Cremona group is approached through the study of its subgroups: algebraic, finite, normal, nilpotent, simple, torsion subgroups are evoked but also centralizers of elements, representation of lattices, subgroups of automorphisms of positive entropy, etc.

Let us introduce birational self maps of the plane and the plane Cremona group from a geometrical point of view.

A plane collineation is a one-to-one map from $\mathbb{P}^2_{\mathbb{C}}$ to itself such that the images of collinear points are themselves collinear. Such maps leave the projective properties of curves unaltered. In advancing beyond such properties let us introduce other maps of the plane to itself that establish relations between curves of different orders and possessing different sets of singularities. The most general rational map of the plane is defined by equations of the form

$$\phi\colon (z_0 : z_1 : z_2) \dashrightarrow \big(\phi_0(z_0, z_1, z_2) : \phi_1(z_0, z_1, z_2) : \phi_2(z_0, z_1, z_2)\big)$$

where ϕ_0, ϕ_1 and ϕ_2 are homogeneous polynomials of degree n without common factors of positive degree. Such a map corresponds to a point p with coordinates $(p_0 : p_1 : p_2)$, a point $\phi(p) = q$ with coordinates $(q_0 : q_1 : q_2)$, where

$$\delta q_0 = \phi_0(p_0, p_1, p_2), \qquad \delta q_1 = \phi_1(p_0, p_1, p_2), \qquad \delta q_2 = \phi_2(p_0, p_1, p_2) \tag{0.1}$$

with δ in $\mathbb{C}^*$.

Consider the net of curves Λ_ϕ defined by the equation

$$\alpha\phi_0 + \beta\phi_1 + \gamma\phi_2 = 0$$

where α, β, and γ are arbitrary parameters. As p describes a line in $\mathbb{P}^2_{\mathbb{C}}$, then $q = \phi(p)$ describes a curve $\mathcal{C}$ of Λ_ϕ. The curves of the net Λ_ϕ are thus correlated by ϕ with the lines of the plane. Conversely, given any net Λ of curves such as Λ_ϕ a linear representation of the curves of Λ on the lines of the plane is equivalent to a rational map of the plane.

The curves of Λ_ϕ may have base-points p_i common to them all. Each such point is a common zero of ϕ_0, ϕ_1, and ϕ_2, so the equations (0.1) to determine its corresponding point are illusory. Conversely each point, termed a *base-point* of ϕ, which renders equation (0.1) illusory is a base-point of Λ_ϕ. In other words:

THEOREM. *The base-points of any rational map are the base-points of the associated net of curves.*

Any two general curves $\mathcal{C}$ and $\mathcal{C}'$ of Λ_ϕ define a pencil of curves $\mathcal{C}+\alpha\mathcal{C}'$ of the net. Denote by n the number of free intersections of $\mathcal{C}$ and $\mathcal{C}'$ not occurring at the base-points p_i of Λ_ϕ; denote by $r_1, r_2, \ldots, r_n$ these points. The integer n is called the *grade* of Λ_ϕ.

To curves of the arbitrary pencil $\mathcal{C}+\alpha\mathcal{C}'$ there correspond by the map ϕ lines of a pencil $L+\alpha L'$. Furthermore if the base-point of the latter pencil is q, then clearly every point r_i corresponds to q. Conversely if any two points of the plane have the same preimage q, then they belong to the same free intersection set of some pencil in Λ_ϕ.

THEOREM. *Let ϕ be a rational self map of the plane. Let Λ_ϕ be its associated net and let n be the grade of Λ_ϕ. An arbitrary point q is the transform of n points $r_1, r_2, \ldots, r_n$ which together form the free intersection set of a pencil of curves of Λ_ϕ.*

In other words the general rational map of the plane is an $(n,1)$ correspondence between the points p and q. And this means that, when the ratios of q_0, q_1, q_2 are given, the equations (0.1) have, in general, n distinct solutions for the ratios of p_0, p_1, and p_2. If $n=1$, *i.e.*, if these equations have only one solution, $(p_0:p_1:p_2)$ are rational functions of $(q_0:q_1:q_2)$. In this case the equations of the reverse map will be of the form

$$\alpha p_0=\psi_0(q_0,q_1,q_2) \qquad \alpha p_1=\psi_1(q_0,q_1,q_2) \qquad \alpha p_2=\psi_2(q_0,q_1,q_2)$$

where ψ_0, ψ_1, and ψ_2 are homogeneous polynomials of degree n'. A *Cremona map* is a rational map whose reverse is also rational; we also speak about the birational self map of the plane. The *plane Cremona group* is the group of birational self maps of the plane.

A *homaloidal net* of curves in the plane is one whose grade is 1.

Equations (0.1) define a birational map ϕ if and only if the associated net Λ_ϕ is homaloidal. Conversely, from any given homaloidal net we can derive many birational self maps of the plane; if $\overline{\phi_0}$, $\overline{\phi_1}$, and $\overline{\phi_2}$ are three independent linear combinations of ϕ_0, ϕ_1, and ϕ_2, the net

$$\alpha\,\phi_0+\beta\,\phi_1+\gamma\,\phi_2=0$$

can also be expressed in the form

$$\alpha'\,\overline{\phi_0}+\beta'\,\overline{\phi_1}+\gamma'\,\overline{\phi_2}=0$$

and the map defined by

$$(z_0:z_1:z_2)\dashrightarrow\big(\overline{\phi_0}(z_0,z_1,z_2):\overline{\phi_1}(z_0,z_1,z_2):\overline{\phi_2}(z_0,z_1,z_2)\big)$$

is based on the same net. Moreover

THEOREM. *To any birational self map of the plane there is associated a homaloidal net of curves.*

Conversely, any homaloidal net of curves generates an infinity of birational self maps of the plane, any of which is the product of any other by a plane collineation.

A collineation is the simplest kind of birational self map of the plane whose homaloidal net is composed of the lines of the plane.

The *degree* of a birational self map of the plane is the degree of the curves of its generating homaloidal net.

Let ϕ be a birational self map of the plane of degree n. Denote by n' the degree of its inverse ϕ^{-1}. If the number of intersections of two curves $\mathcal{C}$ and $\mathcal{C}'$ is denoted by $\mathcal{C} \cdot \mathcal{C}'$ and if L and L' are lines, then

$$n = L \cdot \Lambda_\phi = \phi(L) \cdot \phi(\Lambda_\phi) = \Lambda_{\phi^{-1}} \cdot L' = n'.$$

Hence

THEOREM. *A birational self map of the plane and its inverse have the same degree.*

Let us finish this introduction by pointing out that this statement is not true in higher dimensions:

$$\mathbb{P}^3_\mathbb{C} \dashrightarrow \mathbb{P}^3_\mathbb{C} \qquad (z_0 : z_1 : z_2 : z_3) \dashrightarrow (z_0^2 : z_0z_1 : z_1z_2 : z_0z_3 - z_1^2)$$

is a birational self map of $\mathbb{P}^3_\mathbb{C}$ of degree 2 whose inverse

$$\mathbb{P}^3_\mathbb{C} \dashrightarrow \mathbb{P}^3_\mathbb{C} \qquad (z_0 : z_1 : z_2 : z_3) \dashrightarrow \left(z_0^2z_1 : z_0z_1^2 : z_0^2z_2 : z_1(z_0z_3 + z_1^2)\right)$$

has degree 3. As we will see there are many other differences between the plane Cremona group and the n-dimensional Cremona group, $n \geq 3$.

Note that the study of $\mathrm{Bir}(\mathbb{P}^2_\mathbb{C})$ is central: if S is a complex rational surface, then its group of birational self maps is isomorphic to $\mathrm{Bir}(\mathbb{P}^2_\mathbb{C})$.

We now deal with the content of the book. Chapter 1 contains introductory examples and the very basic techniques used to study birational maps of the projective plane. This chapter explains in particular the importance of divisors and linear systems in the study of the plane Cremona groups.

Chapter 2 builds upon Chapter 1 by explaining how to blow-up all points in $\mathbb{P}^2_\mathbb{C}$ and subsequent blown-up surfaces. It gives rise to an infinite hyperbolic space on which the Cremona group acts. This space plays a fundamental role in the study of Cremona groups, as it allows us to apply tools from geometric group theory to study subgroups of the Cremona group, as well as degree growth and dynamical behaviours of birational maps.

Chapter 3 presents two natural topologies on the Cremona group and their properties, and the notion of algebraic subgroups of the Cremona groups. The construction of one of the topologies - the Zariski topology - is defined via the concept of morphisms. It links to the concept of an algebraic group acting on a variety, which is discussed in this chapter as well.

Chapter 4 addresses a very basic and classical interests while dealing with a group: finding a "nice" generating set and "nice" structures of the group, such as an amalgamated structure. This is quite an important topic in research on Cremona groups because for the plane Cremona group there are "nice" generating sets, and many statements are proven by using them. In higher dimensions no nice generating sets are known: this is one of the many reasons why working with Cremona groups in higher dimensions is very hard.

Chapter 5 discusses other group geometric properties of plane Cremona groups. While Chapter 2 presents a representation of the Cremona group in terms of isometries of an infinite hyperbolic space, this chapter deals with linear representations (there are none) and representations of subgroups of $\mathrm{SL}(n, \mathbb{Z})$, $n \geq 3$, inside the plane Cremona group.

Chapter 6 deals with results on finite subgroups of the plane Cremona groups. They have been of much interest for a very long time, and a short overview of the progress made in the last 80 years is given. The chapter focuses on the classification results of finite abelian and finite cyclic subgroups by Blanc and Dolgachev and Iskovskikh.

Chapter 7 is an extension of Chapter 6; it deals with infinite abelian subgroups of the plane Cremona group. It then moves on to the related topic of endomorphisms of Cremona groups, a subject already mentioned in Chapter 5.

Chapter 8 picks up the topic of Chapter 2 which is the action of the plane Cremona group on an infinite hyperbolic space by isometries. The action and its properties have been very fruitful and have played a vital role in many recent results on the plane Cremona group.

Chapter 9 has a more dynamical flavour. We first give three answers to the question "when is a birational self map of $\mathbb{P}^2_{\mathbb{C}}$ birationally conjugate to an automorphism?" We then recall some constructions of automorphisms of rational surfaces with positive entropy. And then we realize $\mathrm{SL}(2,\mathbb{Z})$ as a subgroup of automorphisms of a rational surface with the property that every element of infinite order has positive entropy.

Acknowledgments

I would like to thank all those working on Cremona groups and related topics. I am especially thankful for Serge Cantat and Dominique Cerveau who helped me discover this wonderful group.

I am grateful to the referees for very constructive recommendations.

Enfin merci à mon petit univers...

Julie Déserti

CHAPTER 1

Introduction

This chapter is devoted to recollections and first definitions.

In the first section, morphisms between varieties, blow-ups, Cremona groups and bubble space are introduced, the Zariski theorem, base-points, indeterminacy points are recalled, and examples of subgroups of the Cremona group are given, among them the group of automorphisms of $\mathbb{P}^n_\mathbb{C}$, the Jonquières group, the group of monomial maps.

The second section is devoted to divisors (prime divisors, Weil divisors, principal divisors, Picard group) and intersection theory.

The third section deals with a geometric definition of birational maps of the complex projective plane.

1.1. First definitions and examples

Denote by $\mathbb{P}^n_\mathbb{C}$ the complex projective space of dimension n. A *rational map*

$$\phi\colon V_1 \subset \mathbb{P}^n_\mathbb{C} \dashrightarrow V_2 \subset \mathbb{P}^k_\mathbb{C}$$

between two smooth projective complex varieties V_1 and V_2 is a regular map on a non-empty Zariski open subset of V_1 such that the image of the points where ϕ is well defined is contained in V_2. If ϕ is well defined on V_1 we say that ϕ is a *morphism* or a *regular map*, otherwise we denote by $\mathrm{Ind}(\phi)$ the set where ϕ is not defined, and call it the *indeterminacy set* of ϕ. A *birational map* between V_1 and V_2 is a rational map that admits an inverse which is rational. In other words it is an isomorphism between two non-empty Zariski open subsets of V_1 and V_2.

EXAMPLE 1. Let us give an example of a birational morphism. Let p be a point on a smooth algebraic surface S. We say that $\pi\colon Y \to S$ is a *blow-up* of p if

- ◇ Y is a smooth surface,
- ◇ $\pi_{|Y\smallsetminus\{\pi^{-1}(p)\}}\colon Y \smallsetminus \{\pi^{-1}(p)\} \to S \smallsetminus \{p\}$ is an isomorphism,
- ◇ and $\pi^{-1}(p) \simeq \mathbb{P}^1_\mathbb{C}$.

We call $\pi^{-1}(p)$ the *exceptional divisor*.

If $\pi\colon Y \to S$ and $\pi'\colon Y' \to S$ are two blow-ups of the same point p, then there exists an isomorphism $\varphi\colon Y \to Y'$ such that $\pi = \pi' \circ \varphi$. We can thus speak about *the* blow-up of $p \in S$.

Let us describe the blow-up of $(0:0:1)$ in $\mathbb{P}^2_\mathbb{C}$ endowed with the homogeneous coordinates $(z_0 : z_1 : z_2)$. Consider the affine chart $z_2 = 1$, *i.e.*, let us work in $\mathbb{C}^2$ with coordinates (z_0, z_1). Set

$$V = \big\{\big((z_0,z_1),(u:v)\big) \in \mathbb{C}^2 \times \mathbb{P}^1_\mathbb{C} \,|\, z_0v = z_1u\big\}.$$

Let $\pi\colon V\to\mathbb{C}^2$ be the morphism given by the first projection. Then

- $\diamond$ $\pi^{-1}(0,0)=\big\{\big((0,0),(u:v)\big)\,|\,(u:v)\in\mathbb{P}^1_\mathbb{C}\big\}$, so $\pi^{-1}(0,0)\simeq\mathbb{P}^1_\mathbb{C}$;
- $\diamond$ if $p=(z_0,z_1)$ is a point of $\mathbb{C}^2\smallsetminus\{(0,0)\}$, then

$$\pi^{-1}(p)=\big\{\big((z_0,z_1),(z_0:z_1)\big)\big\}\in V\smallsetminus\{\pi^{-1}(0,0)\},$$

and $\pi_{|V\smallsetminus\{\pi^{-1}(0,0)\}}$ is an isomorphism, the inverse being

$$(z_0,z_1)\mapsto\big((z_0,z_1),(z_0:z_1)\big).$$

In other words $V=\mathrm{Bl}_{(0,0)}\mathbb{P}^2_\mathbb{C}$ is the surface obtained by blowing up the complex projective plane at $(0:0:1)$, π is the blow-up of $(0:0:1)$, and $\pi^{-1}(0,0)$ is the exceptional divisor.

Let V be a complex algebraic variety, and let $\mathrm{Bir}(V)$ be the group of birational maps of V. The group $\mathrm{Bir}(\mathbb{P}^n_\mathbb{C})$ is called the *Cremona group*. If we fix homogeneous coordinates $(z_0:z_1:\ldots:z_n)$ of $\mathbb{P}^n_\mathbb{C}$, every element $\phi\in\mathrm{Bir}(\mathbb{P}^n_\mathbb{C})$ can be described by homogeneous polynomials of the same degree ϕ_0, ϕ_1, $\ldots$, $\phi_n\in\mathbb{C}[z_0,z_1,\ldots,z_n]$ without common factor of positive degree:

$$\phi\colon(z_0:z_1:\ldots:z_n)$$
$$\dashrightarrow\big(\phi_0(z_0,z_1,z_2,\ldots,z_n):\phi_1(z_0,z_1,z_2,\ldots,z_n):\ldots:\phi_n(z_0,z_1,z_2,\ldots,z_n)\big).$$

The *degree* of ϕ is the degree of the ϕ_i's. In the affine chart $z_0=1$, the map ϕ is given by $(\varphi_1,\varphi_2,\ldots,\varphi_n)$ where for any $1\leq i\leq n$,

$$\varphi_i=\frac{\phi_i(1,z_1,z_2,\ldots,z_n)}{\phi_0(1,z_1,z_2,\ldots,z_n)}\in\mathbb{C}(z_1,z_2,\ldots,z_n).$$

The subgroup of $\mathrm{Bir}(\mathbb{P}^n_\mathbb{C})$ consisting of elements ϕ such that all the φ_i are polynomials as well as the entries of ϕ^{-1} is exactly the *group* $\mathrm{Aut}(\mathbb{A}^n_\mathbb{C})$ *of polynomial automorphisms of the affine space* $\mathbb{A}^n_\mathbb{C}$.

Let S be a smooth projective surface. The *bubble space* $\mathcal{B}(S)$ is, roughly speaking, the set of all points that belong to S, or are infinitely near to S. Let us be more precise: consider all surfaces Y *above* S, *i.e.*, all birational morphisms $\pi\colon Y\to S$; we identify $p_1\in Y_1$ and $p_2\in Y_2$ if $\pi_1^{-1}\circ\pi_2$ is a local isomorphism in a neighborhood of p_2 that maps p_2 onto p_1. The bubble space $\mathcal{B}(S)$ is the union of all points of all surfaces above S modulo the equivalence relation generated by these identifications. A point $p\in\mathcal{B}(S)\cap S$ is a *proper point*. All points in $\mathcal{B}(S)$ that are not proper are called *infinitely near*.

Let S and S' be two smooth projective surfaces. Let $\phi\colon S\dashrightarrow S'$ be a birational map. By Zariski's theorem (*see for instance* [**7**]) we can write $\phi=\pi_2\circ\pi_1^{-1}$ where $\pi_1\colon Y\to S$ and $\pi_2\colon Y\to S'$ are finite sequences of blow-ups. We may assume that there is no (-1)-curve in Y contracted by both π_1 and π_2. We then say that $\pi_2\circ\pi_1^{-1}$ is a *minimal resolution* of ϕ. The *base-points* $\mathrm{Base}(\phi)$ of ϕ are the points blown up by π_1. The proper base-points of ϕ are precisely the *indeterminacy points* of ϕ.

A birational morphism $\pi\colon S\to S'$ induces a bijection $\pi_\bullet\colon\mathcal{B}(S)\to\mathcal{B}(S')\smallsetminus\mathrm{Base}(\pi^{-1})$. A birational map of smooth projective surfaces $\phi\colon S\dashrightarrow S'$ induces a bijection

$$\phi_\bullet\colon\mathcal{B}(S)\smallsetminus\mathrm{Base}(\phi)\to\mathcal{B}(S')\smallsetminus\mathrm{Base}(\phi^{-1})$$

by $\phi_\bullet=(\pi_2)_\bullet\circ(\pi_1)_\bullet^{-1}$ where $\pi_2\circ\pi_1^{-1}$ is a minimal resolution of ϕ.

Let us now give some subgroups of the Cremona group:

- First consider the automorphism group of $\mathbb{P}^n_\mathbb{C}$. It is the subgroup formed by *regular maps*, *i.e.*, maps well defined on $\mathbb{P}^n_\mathbb{C}$ and whose inverse is also well defined on $\mathbb{P}^n_\mathbb{C}$:

$$\mathrm{Aut}(\mathbb{P}^n_\mathbb{C}) = \big\{\phi \in \mathrm{Bir}(\mathbb{P}^n_\mathbb{C}) \,|\, \mathrm{Base}(\phi) = \mathrm{Base}(\phi^{-1}) = \emptyset\big\}.$$

  To any $M = \big(a_{i,j}\big)_{0\leq i,\, j\leq n} \in \mathrm{PGL}(n+1,\mathbb{C})$ corresponds an element of $\mathrm{Bir}(\mathbb{P}^n_\mathbb{C})$ of degree 1:

$$(z_0 : z_1 : \ldots : z_n) \mapsto \left(\sum_{j=0}^n a_{0,j}z_j : \sum_{j=0}^n a_{1,j}z_j : \ldots : \sum_{j=0}^n a_{n,j}z_j\right)$$

  and vice-versa. Such elements are biregular. Furthermore, Bezout's theorem implies that all biregular maps are linear. We thus have the following isomorphism:

$$\mathrm{Aut}(\mathbb{P}^n_\mathbb{C}) \simeq \mathrm{PGL}(n+1,\mathbb{C}).$$

- The n-dimensional subgroup of $\mathrm{Aut}(\mathbb{P}^n_\mathbb{C})$ consisting of diagonal automorphisms is denoted by D_n. Note that D_n is the torus of highest rank of $\mathrm{Bir}(\mathbb{P}^n_\mathbb{C})$[1].

- Start with the surface $\mathbb{P}^1_\mathbb{C} \times \mathbb{P}^1_\mathbb{C}$ considered as a smooth quadric in $\mathbb{P}^3_\mathbb{C}$; its automorphism group contains $\mathrm{PGL}(2,\mathbb{C}) \times \mathrm{PGL}(2,\mathbb{C})$. By the stereographic projection the quadric is birationally equivalent to the plane, so that $\mathrm{Bir}(\mathbb{P}^2_\mathbb{C})$ contains a copy of $\mathrm{PGL}(2,\mathbb{C}) \times \mathrm{PGL}(2,\mathbb{C})$ also.

  If G is a semi-simple algebraic group, H is a parabolic subgroup of G, and $V = {}^{\mathrm{G}}\!/_{\mathrm{H}}$ is a homogeneous variety of dimension n, then V is rational. Once a birational map $\pi\colon V \dashrightarrow \mathbb{P}^n_\mathbb{C}$ is given, $\pi \circ \mathrm{G} \circ \pi^{-1}$ determines an algebraic subgroup of $\mathrm{Bir}(\mathbb{P}^n_\mathbb{C})$.

- A *fibration* of a surface S is a rational map $\pi\colon S \dashrightarrow C$, where C is a curve, such that the general fibers are one-dimensional. Two fibrations $\pi_1\colon S \dashrightarrow C$ and $\pi_2\colon S \dashrightarrow C'$ are identified if there exists an open dense subset $\mathcal{U} \subset S$ that is contained in the domains of π_1 and π_2 such that $\pi_{1|\mathcal{U}}$ and $\pi_{2|\mathcal{U}}$ define the same set of fibers. We say that a group G *preserves* a fibration π if G permutes the fibers. A *rational fibration* of a rational surface S is a rational map $\pi\colon S \dashrightarrow \mathbb{P}^1_\mathbb{C}$ such that the general fiber is rational. The following statement due to Noether and Enriques says that, up to birational maps, there exists only one rational fibration of $\mathbb{P}^2_\mathbb{C}$:

  THEOREM 1.1 ([**7**]). *Let S be a surface. Let $\pi\colon S \dashrightarrow C$ be a rational fibration. Then there exists a birational map $\phi\colon C \times \mathbb{P}^1_\mathbb{C} \dashrightarrow S$ such that $\pi \circ \phi$ is the projection onto the first factor.*

  The *Jonquières subgroup* $\mathcal{J}$ of $\mathrm{Bir}(\mathbb{P}^2_\mathbb{C})$ is the subgroup of elements that preserve the pencil of lines through the point $(0:0:1) \in \mathbb{P}^2_\mathbb{C}$.

  Any subgroup of $\mathrm{Bir}(\mathbb{P}^2_\mathbb{C})$ that preserves a rational fibration is conjugate to a subgroup of $\mathcal{J}$ (Theorem 1.1).

[1] Indeed, according to [**19**] if G is an algebraic subgroup (*see* Chapter 3 for a definition) of $\mathrm{Bir}(\mathbb{P}^n_\mathbb{C})$ isomorphic to $(\mathbb{C}^*)^k$, then $k \leq n$, and if $k = n$, then G is conjugate to D_n.

With respect to affine coordinates $(z_0 : z_1 : 1)$ an element of $\mathcal{J}$ is of the form

$$(z_0, z_1) \dashrightarrow \left(\frac{\alpha z_0+\beta}{\gamma z_0+\delta}, \frac{A(z_0)z_1+B(z_0)}{C(z_0)z_1+D(z_0)}\right)$$

where $\begin{pmatrix} \alpha & \beta \\ \gamma & \delta \end{pmatrix}$ belongs to $\mathrm{PGL}(2,\mathbb{C})$ and $\begin{pmatrix} A & B \\ C & D \end{pmatrix}$ to $\mathrm{PGL}(2,\mathbb{C}(z_0))$. This induces an isomorphism

$$\mathcal{J} \simeq \mathrm{PGL}(2,\mathbb{C}) \rtimes \mathrm{PGL}(2,\mathbb{C}(z_0)).$$

- Let $M = (a_{i,j})_{1\leq i,\, j\leq n} \in M(n,\mathbb{Z})$ be a $n \times n$ matrix of integers. The matrix M determines a rational self map of $\mathbb{P}^n_\mathbb{C}$ given in the affine chart $z_0 = 1$ by

$$\phi_M\colon (z_1,\ldots,z_n) \mapsto \left(z_1^{a_{1,1}} z_2^{a_{1,2}} \ldots z_n^{a_{1,n}}, z_1^{a_{2,1}} z_2^{a_{2,2}} \ldots z_n^{a_{2,n}}, \ldots, z_1^{a_{n,1}} z_2^{a_{n,2}} \ldots z_n^{a_{n,n}}\right).$$

  The map ϕ_M is birational if and only if M belongs to $\mathrm{GL}(n,\mathbb{Z})$. This yields an injective homomorphism $\mathrm{GL}(n,\mathbb{Z}) \to \mathrm{Bir}(\mathbb{P}^n_\mathbb{C})$ whose image is called the *group of monomial maps* and is denoted $\mathrm{Mon}(n,\mathbb{C})$.
- The well-known result of Noether and Castelnuovo states that

  THEOREM 1.2 ([**1, 60**]). *The group* $\mathrm{Bir}(\mathbb{P}^2_\mathbb{C})$ *is generated by the involution*

$$\sigma_2\colon (z_0 : z_1 : z_2) \dashrightarrow (z_1z_2 : z_0z_2 : z_0z_1)$$

  and the group $\mathrm{Aut}(\mathbb{P}^2_\mathbb{C}) = \mathrm{PGL}(3,\mathbb{C})$.

  For $n \geq 3$ the Cremona group is not generated by $\mathrm{PGL}(n+1,\mathbb{C})$ and $\mathrm{Mon}(n,\mathbb{C})$ (*see* [**110, 155**]). In other words the subgroup

$$\langle \mathrm{PGL}(n+1,\mathbb{C}),\, \mathrm{Mon}(n,\mathbb{C})\rangle$$

  is a strict subgroup of $\mathrm{Bir}(\mathbb{P}^n_\mathbb{C})$. The finite index subgroup of $\langle \mathrm{PGL}(n+1,\mathbb{C}), \mathrm{Mon}(n,\mathbb{C})\rangle$ generated by $\mathrm{PGL}(n+1,\mathbb{C})$ and the involution

$$\sigma_n\colon (z_0 : z_1 : \ldots : z_n) \dashrightarrow \left(\prod_{\substack{i=0\\ i\neq 0}}^n z_i : \prod_{\substack{i=0\\ i\neq 1}}^n z_i : \ldots : \prod_{\substack{i=0\\ i\neq n}}^n z_i\right)$$

  has been studied in [**36, 81**]. The group $\mathrm{G}(n,\mathbb{C}) = \langle \sigma_n,\, \mathrm{PGL}(n+1,\mathbb{C})\rangle$ "looks like" $\mathrm{G}(2,\mathbb{C}) = \mathrm{Bir}(\mathbb{P}^2_\mathbb{C})$ in the following sense ([**81**]):
  - ◇ there is no non-trivial finite-dimensional linear representation of $\mathrm{G}(n,\mathbb{C})$ over any field;
  - ◇ the group $\mathrm{G}(n,\mathbb{C})$ is perfect, *i.e.*, $\big[\mathrm{G}(n,\mathbb{C}),\mathrm{G}(n,\mathbb{C})\big] = \mathrm{G}(n,\mathbb{C})$;
  - ◇ the group $\mathrm{G}(n,\mathbb{C})$ equipped with the Zariski topology is simple;
  - ◇ let φ be an automorphism of $\mathrm{Bir}(\mathbb{P}^n_\mathbb{C})$; there exist an automorphism κ of the field $\mathbb{C}$ and a birational self map ψ of $\mathbb{P}^n_\mathbb{C}$ such that

$$\varphi(\phi) = {}^\kappa(\psi\circ\phi\circ\psi^{-1}) \qquad \forall\, \phi\in \mathrm{G}(n,\mathbb{C}).$$

  We will deal with
  - ◇ the Noether and Castelnuovo theorem in §4.3.1 and §4.3.2;
  - ◇ the Hudson and Pan theorem in §4.3.3;
  - ◇ the fact that there is no non-trivial finite-dimensional linear representation of $\mathrm{G}(2,\mathbb{C})$ over any field in §5.1;

- ◇ the fact that $\mathrm{Bir}(\mathbb{P}^2_\mathbb{C}) = \mathrm{G}(2,\mathbb{C})$ is perfect in §5.2;
- ◇ the fact that $\mathrm{Bir}(\mathbb{P}^2_\mathbb{C}) = \mathrm{G}(2,\mathbb{C})$ equipped with the Zariski topology is simple in §3.4;
- ◇ the description of $\mathrm{Aut}(\mathrm{Bir}(\mathbb{P}^2_\mathbb{C})) = \mathrm{Aut}(\mathrm{G}(2,\mathbb{C}))$ in §7.1.

1.2. Divisors and intersection theory

Let V be an algebraic variety.

A *prime divisor* on V is an irreducible closed subset of V of codimension 1. For instance if V is a surface, then the prime divisors of V are the irreducible curves that lie on it; if V is the complex projective space, then the prime divisors are given by the zeros locus of irreducible homogeneous polynomials.

A *Weil divisor* on V is a formal finite sum of prime divisors with integer coefficients:

$$\sum_{i=1}^{m} a_i D_i \qquad m \in \mathbb{N},\ a_i \in \mathbb{Z},\ D_i \text{ prime divisor of } V.$$

Let us denote by $\mathrm{Div}(V)$ the set of all Weil divisors of V.

Let $f \in \mathbb{C}(V)^*$ be a rational function, and let D be a prime divisor. The *multiplicity* $\nu_f(D)$ of f at D is defined by

- ◇ $\nu_f(D) = k > 0$ if f vanishes on D at the order k;
- ◇ $\nu_f(D) = -k$ if f has a pole of order k on D;
- ◇ $\nu_f(D) = 0$ otherwise.

To any rational function $f \in \mathbb{C}(V)^*$ we associate a divisor $\mathrm{div}(f)$ defined by

$$\mathrm{div}(f) = \sum_{\substack{D \text{ prime} \\ \text{divisor}}} \nu_f(D)D.$$

Since $\nu_f(D)$ is zero for all but finitely many D the divisor $\mathrm{div}(f)$ belongs to $\mathrm{Div}(V)$. Divisors obtained like that are called *principal divisors*. The set of principal divisors form a subgroup of $\mathrm{Div}(V)$; indeed $\mathrm{div}(fg) = \mathrm{div}(f) + \mathrm{div}(g)$ for any f, $g \in \mathbb{C}(V)^*$.

Let us introduce an equivalence relation on $\mathrm{Div}(V)$. Two divisors D, D' are *linearly equivalent* if $D - D'$ is a principal divisor. The set of equivalence classes corresponds to the quotient of $\mathrm{Div}(V)$ by the subgroup of principal divisors. The *Picard group* of V is the group of isomorphism classes of line bundles on V; it is denoted $\mathrm{Pic}(V)$. When V is smooth the quotient of $\mathrm{Div}(V)$ by the subgroup of principal divisors is isomorphic to $\mathrm{Pic}(V)$.

EXAMPLE 2. Let us determine $\mathrm{Pic}(\mathbb{P}^n_\mathbb{C})$. Consider the morphism of groups

$$\theta\colon \mathrm{Div}(\mathbb{P}^n_\mathbb{C}) \to \mathbb{Z}$$

which associates to any divisor D of degree d the integer d. Note that $\ker\theta$ is the subgroup of principal divisors of $\mathbb{P}^n_\mathbb{C}$: let $D = \sum a_i D_i$ be an element of $\ker\theta$ where each D_i is a prime divisor given by an homogeneous polynomial $f_i \in \mathbb{C}[z_0, z_1, \ldots, z_n]$ of some degree d_i. Since $\sum a_i d_i = 0$, $f = \prod f_i^{a_i}$ belongs to $\mathbb{C}(\mathbb{P}^n_\mathbb{C})^*$. By construction $D = \mathrm{div}(f)$; hence D is a principal divisor. Conversely any principal divisor is equal to $\mathrm{div}(f)$ where $f = g/h$ for some homogeneous polynomials g, h of the same degree. Thus any principal divisor belongs to $\ker\theta$.

Since $\mathrm{Pic}(\mathbb{P}^n_\mathbb{C})$ is the quotient of $\mathrm{Div}(\mathbb{P}^n_\mathbb{C})$ by the subgroup of principal divisors, we get by restricting θ to the quotient an isomorphism between $\mathrm{Pic}(\mathbb{P}^n_\mathbb{C})$ and $\mathbb{Z}$. As a hyperplane is sent on 1 we obtain that $\mathrm{Pic}(\mathbb{P}^n_\mathbb{C}) = \mathbb{Z}H$ where H is the divisor of a hyperplane.

Let us now assume that $\dim V = 2$; set $V = S$. We can define the notion of intersection:

PROPOSITION 1.3 ([**109**]). *Let S be a smooth projective surface. There exists a unique bilinear symmetric form*

$$\mathrm{Div}(S) \times \mathrm{Div}(S) \to \mathbb{Z} \qquad (C, D) \mapsto C \cdot D$$

such that

- *if C and D are smooth curves with transverse intersections, then $C \cdot D = \#(C \cap D)$;*
- *if C and C' are linearly equivalent, then $C \cdot D = C' \cdot D$ for any D.*

In particular this yields an intersection form

$$\mathrm{Pic}(S) \times \mathrm{Pic}(S) \to \mathbb{Z} \qquad (C, D) \mapsto C \cdot D.$$

Let $\pi\colon \mathrm{Bl}_pS \to S$ be the blow-up of the point $p \in S$. The morphism π induces the map

$$\pi^*\colon \mathrm{Pic}(S) \to \mathrm{Pic}(\mathrm{Bl}_pS), \qquad C \mapsto \pi^{-1}(C).$$

If C is an irreducible curve on S, the *strict transform* $\widetilde{C}$ of C is $\widetilde{C} = \overline{\pi^{-1}(C \smallsetminus \{p\})}$.

If $C \subset S$ is a curve and if p is a point of S, let us define the *multiplicity* $m_p(C)$ *of C at p*. Recall that if V is a quasi-projective variety, and if q is a point of V, then $\mathcal{O}_{q,V}$ denotes the set of equivalence classes of pairs $(\mathcal{U}, \varphi)$ where φ belongs to $\mathbb{C}[\mathcal{U}]$, and $\mathcal{U} \subset V$ is an open subset such that $q \in \mathcal{U}$. Let $\mathfrak{m}$ be the maximal ideal of $\mathcal{O}_{p,S}$. If f is a local equation of C, then $m_p(C)$ is the integer k such that f belongs to $\mathfrak{m}^k \smallsetminus \mathfrak{m}^{k+1}$.

EXAMPLE 3. Assume that S is a rational surface. There exists a neighborhood $\mathcal{U}$ of p in S with $\mathcal{U} \subset \mathbb{C}^2$. We can assume that $p = (0, 0)$ in this affine neighborhood and that C is a curve described by the equation $\sum_{i=1}^n P_i(z_0, z_1) = 0$ where P_i is a homogeneous polynomial of degree i. The multiplicity $m_p(C)$ is the lowest i such that P_i is not equal to 0. The following properties hold:

- $m_p(C) \geq 0$,
- $m_p(C) = 0$ if and only if p does not belong to C,
- $m_p(C) = 1$ if and only if p is a smooth point of C.

Assume that C and D are distinct curves with no common component; we can define an integer $(C \cdot D)_p$ which counts the intersection of C and D at p:

- if either C or D does not pass through p, it is equal to 0;
- otherwise let f, resp. g, be some local equation of C, resp. D, in a neighborhood of p, and define $(C \cdot D)_p$ to be the dimension of $\mathcal{O}_{p,S}/_{(f,g)}$.

This number is related to $C \cdot D$ by the following statement:

PROPOSITION 1.4 ([**109**]). *If C and D are distinct curves without any common irreducible component on a smooth surface, then*

$$C \cdot D = \sum_{p \in C \cap D} (C \cdot D)_p.$$

In particular $C \cdot D \geq 0$.

Let C be a curve on S, and let p be a point of S. Take local coordinates z_0, z_1 at p such that $p=(0,0)$. Set $k=m_p(C)$. The curve C is thus given by

$$P_k(z_0,z_1)+P_{k+1}(z_0,z_1)+\ldots+P_r(z_0,z_1)=0$$

where the P_i's denote homogeneous polynomials of degree i. The blow-up of p can be viewed as $(u,v)\mapsto(uv,v)$, and the pull-back of C is given by

$$v^k\big(p_k(u,1)+vp_{k+1}(u,1)+\ldots+v^{r-k}p_r(u,1)\big)=0.$$

In other words the pull-back of C decomposes into k times the exceptional divisor $E=\pi^{-1}(0,0)=(v=0)$ and the strict transform. We can thus state:

LEMMA 1.5 ([**109**]). *Let S be a smooth surface. Let $\pi\colon \mathrm{Bl}_pS\to S$ be the blow-up of a point $p\in S$. If C is a curve on S, if $\widetilde{C}$ is its strict transform, and if $E=\pi^{-1}(p)$ is the exceptional divisor, then*

$$\pi^*(C)=\widetilde{C}+m_p(C)E.$$

We also have the following statement:

PROPOSITION 1.6 ([**109**]). *Let S be a smooth surface, let p be a point of S, and let $\pi\colon \mathrm{Bl}_pS\to S$ be the blow-up of p. Denote by $E\subset \mathrm{Bl}_pS$ the exceptional divisor $\pi^{-1}(p)\simeq\mathbb{P}^1_\mathbb{C}$. Then*

$$\mathrm{Pic}(\mathrm{Bl}_pS)=\pi^*\mathrm{Pic}(S)+\mathbb{Z}E.$$

The intersection form on Bl_pS is induced by the intersection form on S via the following formulas:

- ◇ *$\pi^*C\cdot\pi^*D=C\cdot D$ for any C, D in $\mathrm{Pic}(S)$;*
- ◇ *$\pi^*C\cdot E=0$ for any C in $\mathrm{Pic}(S)$;*
- ◇ *$E^2=E\cdot E=-1$;*
- ◇ *$\widetilde{C}^2=C^2-1$ for any smooth curve C passing through p and where $\widetilde{C}$ is the strict transform of C.*

If V is an algebraic variety, then the *nef cone* $\mathrm{Nef}(V)$ is the cone of divisors D such that $D\cdot C\geq 0$ for any curve C in V.

1.3. A geometric definition of birational maps

Let ϕ be the element of $\mathrm{Bir}(\mathbb{P}^2_\mathbb{C})$ given by

$$\phi\colon (z_0:z_1:z_2)\dashrightarrow\big(\phi_0(z_0,z_1,z_2):\phi_1(z_0,z_1,z_2):\phi_2(z_0,z_1,z_2)\big)$$

where the ϕ_i's are homogeneous polynomials of the same degree ν, and without common factor of positive degree. The *linear system* Λ_ϕ of ϕ is the strict pull-back of the system $\mathcal{O}_{\mathbb{P}^2_\mathbb{C}}(1)$ of lines of $\mathbb{P}^2_\mathbb{C}$ by φ.

REMARKS 1.7.

- ◇ If A is an automorphism of $\mathbb{P}^2_\mathbb{C}$, then $\Lambda_\phi=\Lambda_{A\circ\phi}$.
- ◇ The degree of the curves of Λ_ϕ is ν.

EXAMPLE 4. The linear system associated to σ_2 is the linear system of conics passing through $(1:0:0)$, $(0:1:0)$, and $(0:0:1)$.

REMARK 1.8. Let us define the linear system of a divisor and then mention the connection between the linear system of a divisor and the linear system of a birational map. Let D be a divisor on a surface S. Denote by $|D|$ the set of all effective divisors on S linearly equivalent to D. Every non-vanishing section of $\mathcal{O}_S(D)$ defines an element of $|D|$, namely its divisor of zeros; conversely every element of $|D|$ is the divisor of zeros of a non-vanishing section of $\mathcal{O}_S(D)$, defined up to scalar multiplication. Hence $|D|$ can be naturally identified with the projective space associated to the vector space $H^0(\mathcal{O}_S(D))$. A linear subspace P of $|D|$ is called a *linear system* on S; of course equivalently P can be defined by a vector subspace of $H^0(\mathcal{O}_S(D))$. The subspace P is *complete* if $P = |D|$. The *dimension* of P is its dimension as a projective space. A one-dimensional linear system is a *pencil*. A curve C is a *fixed component* of P if every divisor of P contains C. The *fixed part* of P is the biggest divisor that is contained in every element of P. A point p of S is a *base-point* of P if every divisor of P contains p. If the linear system has no fixed part, then it has only a finite number of fixed points; this number is bounded by D^2 for $D \in P$.

Let S be a surface. Then there is a bijection between

$$\big\{\text{rational maps } \phi\colon S \dashrightarrow \mathbb{P}^n_{\mathbb{C}} \text{ such that } \phi(S) \text{ is contained in no hyperplane}\big\}$$

and

$$\big\{\text{linear systems on } S \text{ without fixed part and of dimension } n\big\}.$$

This correspondence is constructed as follows: to the map ϕ we associate the linear system $\phi^*|H|$ where $|H|$ is the system of hyperplanes in $\mathbb{P}^n_{\mathbb{C}}$. Conversely let P be a linear system on S with no fixed part; denote by $\widehat{P}$ the projective dual space to P. Define a rational map $\phi\colon S \dashrightarrow \widehat{P}$ by sending $p \in S$ to the hyperplane in P consisting of the divisors passing through p: the map ϕ is defined at p if and only if p is not a base-point of P.

If p_1 is a point of indeterminacy of ϕ, then denote by $\pi_1\colon \mathrm{Bl}_{p_1}\mathbb{P}^2_{\mathbb{C}} \to \mathbb{P}^2_{\mathbb{C}}$ the blow-up of p_1 and by $\mathcal{E}_1$ the associated exceptional divisor. The map $\varphi_1 = \phi \circ \pi_1$ is a birational map from $\mathrm{Bl}_{p_1}\mathbb{P}^2_{\mathbb{C}}$ to $\mathbb{P}^2_{\mathbb{C}}$. If p_2 is a point of indeterminacy of φ_1, we blow-up p_2 via $\pi_2\colon \mathrm{Bl}_{p_1,p_2}\mathbb{P}^2_{\mathbb{C}} \to \mathbb{P}^2_{\mathbb{C}}$, and we set $\mathcal{E}_2 = \pi_2^{-1}(p_2)$. Again the map $\varphi_2 = \varphi_1 \circ \pi_1\colon \mathrm{Bl}_{p_1,p_2}\mathbb{P}^2_{\mathbb{C}} \dashrightarrow \mathbb{P}^2_{\mathbb{C}}$ is a birational map. We iterate this process until φ_r becomes a morphism. Set $E_i = (\pi_{i+1} \circ \ldots \circ \pi_r)^*\mathcal{E}_i$ and $\ell = (\pi_1 \circ \ldots \circ \pi_r)^*L$ where L is the divisor of a line. Applying r times Proposition 1.6 we get

$$\begin{cases} \mathrm{Pic}(\mathrm{Bl}_{p_1,p_2,\ldots,p_r}\mathbb{P}^2_{\mathbb{C}}) = \mathbb{Z}\ell \oplus \mathbb{Z}E_1 \oplus \mathbb{Z}E_2 \oplus \ldots \oplus \mathbb{Z}E_r, \\ \ell^2 = \ell \cdot \ell, \\ E_i^2 = E_i \cdot E_i = -1, \\ E_i \cdot E_j = 0 \quad \forall 1 \leq i \neq j \leq r, \\ E_i \cdot \ell = 0 \quad \forall 1 \leq i \leq r. \end{cases}$$

The curves of Λ_ϕ pass through the p_i's with multiplicity $m_{p_i}(\phi)$. Applying r times Lemma 1.5 the elements of Λ_{φ_r} are equivalent to

$$\nu L - \sum_{i=1}^{r} m_{p_i}(\phi)E_i$$

where L is the pull-back of a generic line in $\mathbb{P}^2_{\mathbb{C}}$. As a result the curves of Λ_{φ_r} have self intersection $\nu^2 - \sum_{i=1}^{r} m_{p_i}(\phi)^2$. Note that all the members of a linear system are linearly equivalent and that the dimension of Λ_{φ_r} is 2; the self intersection thus has to be non-negative by Proposition 1.4. As a consequence the number r exists; in other words ϕ has a finite number of base-points. By construction

$$\varphi_r\colon \mathrm{Bl}_{p_1,p_2,\ldots,p_r}\mathbb{P}^2_{\mathbb{C}} \to \mathbb{P}^2_{\mathbb{C}}$$

is a birational morphism which is the blow-up of the base-points of ϕ^{-1}. Consider a general line L of $\mathbb{P}^2_{\mathbb{C}}$ that does not pass through p_1, p_2, $\ldots$, p_r. Its pull-back $\varphi_r^{-1}(L)$ corresponds to a smooth curve on $\mathrm{Bl}_{p_1,p_2,\ldots,p_r}\mathbb{P}^2_{\mathbb{C}}$ which has self-intersection 1 and genus 0. Hence

$$\begin{cases} (\varphi_r^{-1}(L))^2 = 1, \\ \varphi_r^{-1}(L)\cdot K_{\mathrm{Bl}_{p_1,p_2,\ldots,p_r}\mathbb{P}^2_{\mathbb{C}}} = -3. \end{cases}$$

As the elements of Λ_{φ_r} are equivalent to $\nu L - \sum_{i=1}^r m_{p_i}(\phi)E_i$ and since

$$K_{\mathrm{Bl}_{p_1,p_2,\ldots,p_r}\mathbb{P}^2_{\mathbb{C}}} = -3L + \sum_{i=1}^{r} E_i$$

the following equalities hold:

$$\begin{cases} \displaystyle\sum_{i=1}^{r} m_{p_i}(\phi) = 3(\nu-1), \\ \displaystyle\sum_{i=1}^{r} m_{p_i}(\phi)^2 = \nu^2 - 1. \end{cases}$$

EXAMPLES 1.

- $\diamond$ If $\nu = 2$, then $r = 3$ and $m_{p_1}(\phi) = m_{p_2}(\phi) = m_{p_3}(\phi) = 1$.
- $\diamond$ If $\nu = 3$, then $r = 5$ and $m_{p_1}(\phi) = 2$, $m_{p_2}(\phi) = m_{p_3}(\phi) = m_{p_4}(\phi) = m_{p_5}(\phi) = 1$.

CHAPTER 2

An isometric action of the Cremona group on an infinite dimensional hyperbolic space

If S is a projective surface, the group $\mathrm{Bir}(S)$ of birational self maps of S acts faithfully by isometries on a hyperbolic space $\mathbb{H}^\infty(S)$ of infinite dimension. After recalling some notions of hyperbolic geometry in the first section of this chapter we describe this construction in the second section. Let us now give an outline of it before heading into details. Let S be a projective surface. If $\pi\colon Y \to S$ is a birational morphism, then one obtains an embedding $\pi^*\colon \mathrm{NS}(S) \to \mathrm{NS}(Y)$ of Néron-Severi groups. If $\pi_1\colon Y_1 \to S$ and $\pi_2\colon Y_2 \to S$ are two birational morphisms, then

- ◇ π_2 is above π_1 if $\pi_1^{-1} \circ \pi_2$ is a morphism,
- ◇ one can always find a third birational morphism $\pi_3\colon Y_3 \to S$ that is above π_1 and π_2.

Hence the inductive limit of all groups $\mathrm{NS}(Y_i)$ for all surfaces Y_i above S is well-defined; this limit $\mathcal{Z}(S)$ is the Picard-Manin space of S. The intersection forms on Y_i yield to a scalar product $\langle\, ,\rangle$ on $\mathcal{Z}(S)$.

Consider all surfaces Y above S, *i.e.*, all birational morphisms $\pi\colon Y \to S$. We identify $p_1 \in Y_1$ and $p_2 \in Y_2$ if $\pi_1^{-1}\circ\pi_2$ is a local isomorphism in a neighborhood of p_2 that maps p_2 onto p_1. The bubble space $\mathcal{B}(S)$ of S is the union of all points of all surfaces above S modulo the equivalence relation generated by these identifications. If p belongs to $\mathcal{B}(S)$, then we denote by $\mathbf{e}_p$ the divisor class of the exceptional divisor of the blow-up of p. The equalities $\mathbf{e}_p \cdot \mathbf{e}_p = -1$ and $\mathbf{e}_p \cdot \mathbf{e}_{p'} = 0$ hold by Proposition 1.6

The Néron-Severi group $\mathrm{NS}(S)$ is naturally embedded as a subgroup of the Picard-Manin space; this finite-dimensional lattice is orthogonal to $\mathbf{e}_p$ for any $p \in \mathcal{B}(S)$. More precisely

$$\mathcal{Z}(S) = \mathrm{NS}(S) \bigoplus_{p\in\mathcal{B}(S)} \mathbb{Z}\mathbf{e}_p.$$

As a result any element v of $\mathcal{Z}(S)$ can be written as a finite sum

$$v = w + \sum_{p\in\mathcal{B}(S)} m_p\mathbf{e}_p.$$

There is a completion process for which the completion $\mathrm{Z}(S)$ of $\mathcal{Z}(S)\otimes_{\mathbb{Z}}\mathbb{R}$ is

$$\mathrm{Z}(S) = \Big\{w + \sum_{p\in\mathcal{B}(S)} m_p\mathbf{e}_p \,|\, w \in \mathrm{NS}(\mathbb{R}, S),\ \sum_{p\in\mathcal{B}(S)} m_p^2 < \infty\Big\}.$$

The intersection form extends as a scalar product with signature $(1,\infty)$ on this space. The hyperbolic space $\mathbb{H}^\infty(S)$ of S is defined by

$$\mathbb{H}^\infty(S)=\big\{w\in \mathrm{Z}(S),\,|\,\langle w,\,w\rangle=1,\,\langle w,\,a\rangle>0 \text{ for all ample classes } a\in \mathrm{NS}(S)\big\}.$$

It is an infinite-dimensional analogue of the classical hyperbolic space $\mathbb{H}^n$. One can define a complete distance dist on $\mathbb{H}^\infty(S)$ by

$$\cosh(\mathrm{dist}(v,w))=\langle v,\,w\rangle.$$

Geodesics are intersections of $\mathbb{H}^\infty(S)$ with planes. The projection of $\mathbb{H}^\infty(S)$ to the projective space $\mathbb{P}(\mathrm{Z}(S))$ is one to one, and the boundary of its image is the projection of the cone of isotropic vectors of $\mathrm{Z}(S)$:

$$\partial\mathbb{H}^\infty(S)=\big\{\mathbb{R}_+v\,|\,v\in \mathrm{Z}(S),\,\langle v,\,v\rangle=0,\,\langle v,\,a\rangle>0 \text{ for all ample classes } a\in \mathrm{NS}(S)\big\}.$$

The important fact is that $\mathrm{Bir}(S)$ acts faithfully on $\mathrm{Z}(S)$ by continuous linear endomorphisms preserving the intersection form, the effective cone, the nef cone, $\mathcal{Z}(S)$ and also $\mathbb{H}^\infty(S)$.

If ϕ is an element of $\mathrm{Bir}(S)$, we denote by ϕ_* its action on $\mathrm{Z}(S)$: it is a linear isometry with respect to the intersection form; we also denote by ϕ_* the isometry of $\mathbb{H}^\infty(S)$ induced by this endomorphism of $\mathrm{Z}(S)$. Let f be an isometry of $\mathbb{H}^\infty(S)$; the translation length of f is

$$L(f)=\inf\big\{\mathrm{dist}(v,f(v))\,|\,v\in\mathbb{H}^\infty(S)\big\}.$$

If this infimum is a minimum, then

- ◇ either it is equal to 0, f has a fixed point in $\mathbb{H}^\infty(S)$, and f is elliptic;
- ◇ or it is positive, and f is loxodromic.

When the infimum is not realized, $L(f)$ is equal to 0, and f is parabolic.

This classification into three types holds for all isometries of $\mathbb{H}^\infty(S)$. For isometries ϕ_* induced by birational maps ϕ of S there is a dictionary between this classification and the geometric properties of ϕ. We give this dictionary in the third section.

2.1. Some hyperbolic geometry

Consider a real Hilbert space $\mathcal{H}$ of dimension n. Let $\mathbf{e}_0$ be a unit vector of $\mathcal{H}$, and let $\mathbf{e}_0^\perp$ be the orthogonal complement of the space $\mathbb{R}\mathbf{e}_0$. Denote by $(\mathbf{e}_i)_{i\in I}$ an orthonormal basis of $\mathbf{e}_0^\perp$. A scalar product with signature $(1,n-1)$ can be defined on $\mathcal{H}$ by setting

$$\langle u,\,v\rangle=a_0b_0-\sum_{i\in I}a_ib_i$$

for any two elements

$$u=a_0\mathbf{e}_0+\sum_{i\in I}a_i\mathbf{e}_i$$

and $v=b_0\mathbf{e}_0+\sum_{i\in I}b_i\mathbf{e}_i$ of $\mathcal{H}$. The set

$$\big\{v\in\mathcal{H}\,|\,\langle v,v\rangle=1\big\}$$

defines a hyperboloid with two connected components. Let $\mathbb{H}^{n-1}$ be the connected component of this hyperboloid that contains $\mathbf{e}_0$. A metric can be defined on $\mathbb{H}^{n-1}$ by

$$d(u,v) := \operatorname{arccosh}(\langle u,v\rangle).$$

REMARK 2.1. A useful model for $\mathbb{H}^2$ is the Poincaré model: $\mathbb{H}^2$ is identified to the upper half-plane $\{z \in \mathbb{C} \,|\, \mathrm{Im}(z) > 0\}$ with its Riemanniann metric given by $\mathrm{d}s^2 = \frac{x^2+y^2}{y^2}$. Its group of orientation preserving isometries coincides with $\mathrm{PSL}(2,\mathbb{R})$, acting by linear fractional transformations.

Let $(\mathcal{H}, \langle .,. \rangle)$ be a real Hilbert space of infinite dimension. Let $\mathbf{e}_0$ be a unit vector of $\mathcal{H}$, and let $\mathbf{e}_0^\perp$ be its orthogonal complement. Any element v of $\mathcal{H}$ can be written in a unique way as $v = v_{\mathbf{e}_0}\mathbf{e}_0 + v_{\mathbf{e}_0^\perp}$ where $v_{\mathbf{e}_0}$ belongs to $\mathbb{R}$ and $v_{\mathbf{e}_0^\perp}$ belongs to $\mathbf{e}_0^\perp$. Consider the symetric bilinear form $\mathcal{B}$ of $\mathcal{H}$ defined by

$$\mathcal{B}(x,y) = x_{\mathbf{e}_0}y_{\mathbf{e}_0} - \langle x_{\mathbf{e}_0^\perp}, y_{\mathbf{e}_0^\perp}\rangle;$$

it has signature $(1,\infty)$. Let $\mathbb{H}^\infty$ be the hyperboloid given by

$$\mathbb{H}^\infty = \{x \in \mathcal{H} \,|\, \mathcal{B}(x,x) = 1,\ \mathcal{B}(\mathbf{e}_0, x) > 0\}.$$

We consider on $\mathbb{H}^\infty$ the distance d defined by $\cosh d(x,y) = \mathcal{B}(x,y)$. The space $(\mathbb{H}^\infty, d)$ is a complete metric space of infinite dimension.

2.1.1. δ-hyperbolicity and CAT(-1) spaces.

Let (X,d) be a geodesic metric space. Let x, y, z be three points of X. We denote by $[p,q]$ the segment with endpoints p and q. A *geodesic triangle* with vertices x, y, z is the union of three geodesic segments $[x,y]$, $[y,z]$ and $[z,x]$. Let $\delta \geq 0$. If for any point $m \in [x,y]$ there is a point in $[y,z] \cup [z,x]$ at distance less than δ of m, and similarly for points on the other edges, then the triangle is said to be *δ-slim*. A *δ-hyperbolic space* is a geodesic metric space all of whose geodesic triangles are δ-slim. A δ-hyperbolic space is called a *Gromov hyperbolic space*.

EXAMPLES 2.

- ◇ Metric trees are 0-hyperbolic: all triangles are tripods.
- ◇ The hyperbolic plane is (-2)-hyperbolic. In fact the incircle of a geodesic triangle is the circle of largest diameter contained in the triangle, and any geodesic triangle lies in the interior of an ideal triangle, all of which are isometric with incircles of diameter $2\log 3$ (*see* [**65**]).
- ◇ The space $\mathbb{R}^2$ endowed with the euclidian metric is not δ-hyperbolic (for instance because of the existence of homotheties).

Let us now introduce CAT(-1) spaces[1]. Let (X, d_X) be a geodesic metric space. Consider a geodesic triangle T in X determined by the three points x, y, z and the data of three geodesics between two of these three points. A *comparison triangle* of T in the metric space $(X', d_{X'})$ is a triangle T' such that

$$\begin{cases} d_X(x,y) = d_{X'}(x',y'),\\ d_X(x,z) = d_{X'}(x',z'),\\ d_X(y,z) = d_{X'}(y',z'). \end{cases}$$

[1]The terminology corresponds to the initials of E. Cartan, A. Alexandrov and V. Toponogov.

Let p be a point of $[x, y] \subset T$. A point $p' \in [x', y'] \subset T'$ is a *comparison point* of p if $d_{X'}(x', p') = d_X(x, p)$.

The triangle T *satisfies the* CAT(-1) *inequality* if for any $(x, y) \in T^2$

$$d_X(x, y) \leq ||x' - y'||_{\mathbb{H}^2}$$

where T' is a comparison triangle of T in $\mathbb{H}^2$ and $x' \in T'$ (resp. $y' \in T'$) is a comparison point of x (resp. y).

The space X is CAT(-1) if all its triangles satisfy the CAT(-1) inequality.

REMARK 2.2. The CAT(-1) spaces are Gromov hyperbolic, but the converse is false.

Set $\mathcal{H}_{>0} = \{v \in \mathcal{H} \,|\, \langle v, v\rangle > 0\}$. The image of v by the map

$$\eta \colon \mathcal{H}_{>0} \to \mathbb{H}^\infty \qquad\qquad v \mapsto \frac{v}{\sqrt{\langle v, v\rangle}}$$

is called the normalization of v. Geometrically η associates to a point $v \in \mathcal{H}_{>0}$ the intersection of $\mathbb{H}^\infty$ with the line through v. Note that if the intersection of $\mathcal{H}$ with a vectorial subspace of dimension $n+1$ of $\mathcal{H}$ is not empty, then it is a copy of $\mathbb{H}^n$. In particular there exists a unique geodesic segment between two points of $\mathbb{H}^\infty$ obtained as the intersection of $\mathbb{H}^\infty$ with the plane that contains these two points. Hence any triangle of $\mathbb{H}^\infty$ is isometric to a triangle of $\mathbb{H}^2$. As a result $\mathbb{H}^\infty$ is CAT(-1) and δ-hyperbolic for the same constant δ as $\mathbb{H}^2$.

2.1.2. Boundary of $\mathbb{H}^\infty$. Let (X, d) be a geodesic metric space. Let T be a geodesic triangle of X given by x, y, $z \in X$ and geodesic segments between two of these three points. The triangle T *satisfies the* CAT(0) *inequality* if for any $(x, y) \in T^2$

$$d_X(x, y) \leq ||x' - y'||_{\mathbb{R}^2}$$

where $x' \in T'$ (resp. $y' \in T'$) is a comparison point of x (resp. y) and T' is a comparison triangle of T in $\mathbb{R}^2$.

The space X is CAT(0) if all its triangles satisfy the CAT(0) inequality.

REMARK 2.3. A CAT(-1) space is a CAT(0) space. In particular $\mathbb{H}^\infty$ is a CAT(0) space.

Since $\mathbb{H}^\infty$ is a CAT(0), complete metric space, there exists a notion of boundary at infinity that generalizes the notion of boundary of finite-dimensional Riemann varieties which are complete, simply connected, and with negative curvature. The *boundary of* $\mathbb{H}^\infty$ is defined by

$$\partial\mathbb{H}^\infty = \{v \in \mathcal{H} \,|\, \langle v, v\rangle = 0,\ \langle v, \mathbf{e}_0\rangle > 0\}.$$

A point of $\partial\mathbb{H}^\infty$ is called *point at infinity*.

2.1.3. Isometries. Denote by $\mathrm{O}_{1,n}(\mathbb{R})$ the group of linear transformations of $\mathcal{H}$ preserving the scalar product $\langle\,,\rangle$. The group of isometries $\mathrm{Isom}(\mathbb{H}^n)$ coincides with the index 2 subgroup $\mathrm{O}^+_{1,n}(\mathbb{R})$ of $\mathrm{O}(\mathcal{H})$ that preserves the chosen sheet $\mathbb{H}^n$ of the hyperboloid

$$\{u \in \mathcal{H} \,|\, \langle u, u\rangle = 1\}.$$

This group acts transitively on $\mathbb{H}^n$ and on its unit tangent bundle.

If h is an isometry of $\mathbb{H}^n$ and $v \in \mathcal{H}$ is an eigenvector of h with eigenvalue λ, then either $|\lambda| = 1$ or v is isotropic. Furthermore $\mathbb{H}^n$ is homeomorphic to a ball, so h has a least one eigenvector in $\mathbb{H}^n \cup \partial\mathbb{H}^n$. As a consequence according to [**49**] there are three types of isometries:

- ◇ h is *elliptic* if and only if h fixes a point $p \in \mathbb{H}^n$. Since $\langle\, ,\, \rangle$ is negative definite on $p^\perp$, h fixes pointwise $\mathbb{R}p$ and acts by rotation on $p^\perp$ with respect to $\langle\, ,\, \rangle$;
- ◇ h is *parabolic* if h is not elliptic and fixes a vector v in the isotropic cone. The line $\mathbb{R}v$ is uniquely determined by h. Let p be a point of $\mathbb{H}^n$; there exists an increasing sequence $(n_i) \in \mathbb{N}^\mathbb{N}$ such that $(h^{n_i}(p))_{i\in\mathbb{N}}$ converges to the boundary point determined by v.
- ◇ h is *loxodromic* if and only if h has an eigenvector v_h^+ with eigenvalue $\lambda > 1$. Note that v_h^+ is unique up to scalar multiplication. There is another unique isotropic eigenline $\mathbb{R}v_h^-$ corresponding to the eigenvalue $\frac{1}{\lambda}$. On the orthogonal complement of $\mathbb{R}v_h^- \oplus \mathbb{R}v_h^+$ the isometry h acts as a rotation with respect to $\langle\, ,\, \rangle$. The boundary points determined by v_h^- and v_h^+ are the two fixed points of h in $\mathbb{H}^n \cup \partial\mathbb{H}^n$; the first one is an attracting fixed point $\alpha(h)$, the second one is a repelling fixed point $\omega(h)$.

To an isometry h of $\mathbb{H}^n$ one can associate the *translation length* of h:

$$L(h) = \inf\big\{d(h(p),p)\,|p \in \mathbb{H}^n\big\}.$$

The isometry h is elliptic if and only if $L(h) = 0$, and the infimum is achieved, *i.e.*, h has a fixed point in $\mathbb{H}^n$. The isometry h is parabolic if and only if $L(h) = 0$, and the infinimum is not achieved. The isometry h is loxodromic if and only if $L(h) > 0$. In that case

- ◇ $\exp(L(h))$ is the largest eigenvalue of h,
- ◇ and $d(p, h^n(p))$ grows like $nL(h)$ as n goes to infinity for any point $p \in \mathbb{H}^n$.

2.2. The isometric action of Bir(S) on an infinite-dimensional hyperbolic space

2.2.1. The Picard-Manin space. Let S be a smooth, irreducible, projective, complex surface. As we see in Chapter 1 the Picard group $\mathrm{Pic}(S)$ is the quotient of the abelian group of divisors by the subgroup of principal divisors ([**109**]). The intersection between curves extends to a quadratic form, the so-called intersection form:

$$\mathrm{Pic}(S) \times \mathrm{Pic}(S) \to \mathbb{Z}, \qquad\qquad (C, D) \mapsto C \cdot D.$$

The quotient of $\mathrm{Pic}(S)$ by the subgroup of divisors E such that $E \cdot D = 0$ for all divisor classes D is the *Néron-Severi group* $\mathrm{NS}(S)$. In case of rational surfaces we have $\mathrm{NS}(S) = \mathrm{Pic}(S)$. The Néron-Severi group is a free abelian group, and its rank, the *Picard number* is finite. The pull-back of a birational morphism $\pi\colon Y \to S$ yields an injection from $\mathrm{Pic}(S)$ into $\mathrm{Pic}(Y)$; we thus get an injection from $\mathrm{NS}(S)$ into $\mathrm{NS}(Y)$. The morphism $\pi\colon Y \to S$ can be written as a finite sequence of blow-ups. Let $\mathbf{e}_1$, $\mathbf{e}_2$, $\dots$, $\mathbf{e}_k \subset Y$ be the class of the irreducible components of the exceptional divisor of π, that is, the classes contracted by π. We have the following decomposition:

$$\mathrm{NS}(Y) = \mathrm{NS}(S) \oplus \mathbb{Z}\mathbf{e}_1 \oplus \mathbb{Z}\mathbf{e}_2 \oplus \ldots \oplus \mathbb{Z}\mathbf{e}_k \tag{2.1}$$

which is orthogonal with respect to the intersection form.

Consider $\pi_1\colon Y \to S$ and $\pi_2\colon Y' \to S$, two birational morphisms of smooth projective surfaces. We say that π_1 is *above* π_2 if $\pi_2^{-1}\circ\pi_1$ is a morphism. For any two birational morphisms $\pi_1\colon Y \to S$ and $\pi_2\colon Y' \to S$ there exists a birational morphism $\pi_3\colon Y'' \to S$ that lies above π_1 and π_2.

Let us consider the set of all birational morphisms of smooth projective surfaces $\pi\colon Y \to S$. The corresponding embeddings of the Néron-Severi groups $\mathrm{NS}(S) \to \mathrm{NS}(Y)$ form a directed family; the direct limit

$$\mathcal{Z}(S) := \lim_{\pi\colon Y\to S} \mathrm{NS}(Y)$$

thus exists. It is called the *Picard Manin space* of S. The intersection forms on the groups $\mathrm{NS}(Y)$ induce a quadratic form on $\mathcal{Z}(S)$ of signature $(1,\infty)$.

Let p be a point of the bubble space of S. Denote by $\mathbf{e}_p$ the divisor class of the exceptional divisor of the blow-up of p in the corresponding Néron-Severi group. One deduces from (2.1) the following decomposition:

$$\mathcal{Z}(S) = \mathrm{NS}(S) \oplus \bigoplus_{p\in\mathcal{B}(S)} \mathbb{Z}\mathbf{e}_p.$$

Furthermore according to Proposition 1.6 the following properties hold

$$\begin{cases} \mathbf{e}_p\cdot\mathbf{e}_p = -1, \\ \mathbf{e}_p\cdot\mathbf{e}_q = 0 \text{ for all } p\neq q. \end{cases}$$

2.2.2. The hyperbolic space $\mathbb{H}^\infty(S)$. Let S be a smooth projective surface, and let $\mathcal{Z}(S)$ be its Picard-Manin space. We define $\mathrm{Z}(S)$ to be the completion of the real vector space $\mathcal{Z}(S)\otimes\mathbb{R}$

$$\mathrm{Z}(S) = \Big\{v + \sum_{p\in\mathcal{B}(S)} m_p\mathbf{e}_p \,|\, v\in\mathrm{NS}(S)\otimes\mathbb{R},\ m_p\in\mathbb{R},\ \sum_{p\in\mathcal{B}(S)} m_p^2 < \infty\Big\}.$$

The intersection form extends continuously to a quadratic form on $\mathrm{Z}(S)$ with signature $(1,\infty)$. Let $\mathrm{Isom}(\mathrm{Z}(S))$ be the group of isometries of $\mathrm{Z}(S)$ with respect to the intersection form. The set of vectors $v\in\mathrm{Z}(S)$ such that $\langle v,v\rangle = 1$ is a hyperboloid. The subset

$$\mathbb{H}^\infty(S) = \big\{v\in\mathrm{Z}(S)\,|\,\langle v,v\rangle = 1,\ \langle v,\mathbf{e}_0\rangle > 0\big\}$$

is the sheet of that hyperboloid containing ample classes of $\mathrm{NS}(S,\mathbb{R})$. Let $\mathrm{Isom}(\mathbb{H}^\infty(S))$ be the subgroup of $\mathrm{Isom}(\mathcal{Z}(S))$ that preserves $\mathbb{H}^\infty(S)$. The space $\mathbb{H}^\infty(S)$ equipped with the distance defined by

$$\cosh(d(v,v')) = \langle v,v'\rangle$$

is isometric to a hyperbolic space $\mathbb{H}^\infty$. Let $\partial\mathbb{H}^\infty(S)$ be the boundary of $\mathbb{H}^\infty(S)$. To simplify we will often write $\mathbb{H}^\infty$ (resp. $\partial\mathbb{H}^\infty$) instead of $\mathbb{H}^\infty(S)$ (resp. $\partial\mathbb{H}^\infty(S)$).

2.2.3. An isometric action of $\mathrm{Bir}(S)$. Let us now describe the action of $\mathrm{Bir}(S)$ on $\mathbb{H}^\infty$ (*see* [**52, 138**]). Let $\phi\colon Y \to S$ be a birational morphism of smooth projective surfaces. Denote by $p_1, p_2, \ldots, p_n \in \mathcal{B}(S)$ the points blown up by ϕ. Denote by $\mathbf{e}_{p_i}$ the irreducible component of the exceptional divisor contracted to p_i. One has

$$\mathrm{NS}(Y) = \mathrm{NS}(S)\oplus\mathbb{Z}\mathbf{e}_{p_1}\oplus\mathbb{Z}\mathbf{e}_{p_2}\oplus\ldots\oplus\mathbb{Z}\mathbf{e}_{p_n}.$$

The morphism ϕ induces the isomorphism $\phi_*\colon \mathcal{Z}(Y) \to \mathcal{Z}(S)$ defined by

$$\begin{cases} \phi_*(\mathbf{e}_p) = \mathbf{e}_{\phi_\bullet(p)} & \forall\, p \in \mathcal{B}(Y) \smallsetminus \mathrm{Base}(\phi), \\ \phi_*(\mathbf{e}_{p_i}) = \mathbf{e}_{p_i} & \forall\, 1 \leq i \leq n, \\ \phi_*(D) = D & \forall\, D \in \mathrm{NS}(S) \subset \mathrm{NS}(Y). \end{cases}$$

Let $\phi\colon Y \dashrightarrow S$ be a birational map of smooth projective surfaces. Let $\pi_2 \circ \pi_1^{-1}$ be a minimal resolution of ϕ. The map ϕ induces an isomorphism $\phi_*\colon \mathcal{Z}(Y) \to \mathcal{Z}(S)$ defined by

$$\phi_* = (\pi_2)_* \circ (\pi_1)_*^{-1}.$$

Let S be a smooth projective surface. Any element ϕ of $\mathrm{Bir}(S)$ induces an isomorphism $\phi_*\colon \mathcal{Z}(S) \to \mathcal{Z}(S)$, and ϕ_* yields an automorphism of $\mathcal{Z}(S) \otimes \mathbb{R}$ which extends to an automorphism of the completion $\mathrm{Z}(S)$ and preserves the intersection form.

Let ϕ be a birational self map of $\mathbb{P}^2_{\mathbb{C}}$. Assume that ϕ has degree d. Then the base-point $\mathbf{e}_0$, *i.e.*, the class of a line in $\mathbb{P}^2_{\mathbb{C}}$, is mapped by ϕ_* to the finite sum

$$d\mathbf{e}_0 - \sum_i m_i \mathbf{e}_{p_i}$$

where each m_i is a positive integer and $\mathbf{e}_{p_i}$ are the classes of the exceptional divisors corresponding to the base-points of ϕ^{-1}. For instance if $\phi = \sigma_2$ is the standard Cremona involution, then

$$(\sigma_2)_* \mathbf{e}_0 = 2\mathbf{e}_0 - \mathbf{e}_{p_1} - \mathbf{e}_{p_2} - \mathbf{e}_{p_3}$$

where $p_1 = (1:0:0)$, $p_2 = (0:1:0)$ and $p_3 = (0:0:1)$.

REMARK 2.4. An invariant structure is given by the canonical form. The canonical class of $\mathbb{P}^2_{\mathbb{C}}$ blown up in n points $p_1, p_2, \ldots, p_n$ is equal to $-3\mathbf{e}_0 - \sum_{j=1}^n \mathbf{e}_{p_j}$. By taking intersection products one obtains a linear form ω_∞ defined by

$$\omega_\infty\colon \mathcal{Z}(\mathbb{P}^2_{\mathbb{C}}) \to \mathbb{Z}, \qquad m_0\mathbf{e}_0 - \sum_{j=1}^n m_j \mathbf{e}_{p_j} \mapsto -3m_0 + \sum_{j=1}^n m_j.$$

Since the isometric action of $\mathrm{Bir}(\mathbb{P}^2_{\mathbb{C}})$ on $\mathcal{Z}(\mathbb{P}^2_{\mathbb{C}})$ preserves the linear form ω_∞ we get the following equalities already obtained in §1.3: if $\phi_*\mathbf{e}_0 = d\mathbf{e}_0 - \sum_{j=1}^n m_j \mathbf{e}_{p_j}$, then

$$\begin{cases} d^2 = 1 + \displaystyle\sum_{j=1}^n m_j^2, \\ 3d - 3 = \displaystyle\sum_{j=1}^n m_j. \end{cases}$$

EXAMPLE 5. Let us understand the isometry $(\sigma_2)_*$. Denote by p_1, p_2 and p_3 the base-points of σ_2, and set $S = \mathrm{Bl}_{p_1,p_2,p_3}\mathbb{P}^2_{\mathbb{C}}$. The involution σ_2 lifts to an automorphism $\widetilde{\sigma_2}$ on S. The Néron-Severi group $\mathrm{NS}(S)$ of S is the lattice of rank 4 generated by the class $\mathbf{e}_0$, coming from the class of a line in $\mathbb{P}^2_{\mathbb{C}}$, and the classes $\mathbf{e}_i = \mathbf{e}_{p_i}$ given by the three exceptional divisors. The action of $\widetilde{\sigma_2}$ on $\mathrm{NS}(S)$ is given by

$$\begin{cases} (\widetilde{\sigma_2})_*\mathbf{e}_0 = 2\mathbf{e}_0 - \mathbf{e}_1 - \mathbf{e}_2 - \mathbf{e}_3, \\ (\widetilde{\sigma_2})_*\mathbf{e}_1 = \mathbf{e}_0 - \mathbf{e}_2 - \mathbf{e}_3, \\ (\widetilde{\sigma_2})_*\mathbf{e}_2 = \mathbf{e}_0 - \mathbf{e}_1 - \mathbf{e}_3, \\ (\widetilde{\sigma_2})_*\mathbf{e}_3 = \mathbf{e}_0 - \mathbf{e}_1 - \mathbf{e}_2. \end{cases}$$

Then $(\widetilde{\sigma_2})_*$ coincides on $\mathrm{NS}(S)$ with the reflection with respect to $\mathbf{e}_0-\mathbf{e}_1-\mathbf{e}_2-\mathbf{e}_3$:

$$(\widetilde{\sigma_2})_*(p) = p + \langle p, \mathbf{e}_0 - \mathbf{e}_1 - \mathbf{e}_2 - \mathbf{e}_3\rangle \qquad \forall\, p \in \mathrm{NS}(S)$$

Let us blow-up all points of S; we thus obtain a basis of $\mathrm{Z}(\mathbb{P}^2_{\mathbb{C}})$:

$$\mathrm{Z}(\mathbb{P}^2_{\mathbb{C}}) = \mathrm{NS}(S) \bigoplus_{p\in\mathcal{B}(S)} \mathbb{Z}\mathbf{e}_p.$$

The isometry $(\sigma_2)_*$ of $\mathcal{Z}(\mathbb{P}^2_{\mathbb{C}})$ acts on $\mathrm{NS}(S)$ as the reflection $(\widetilde{\sigma_2})_*$ and permutes each vector $\mathbf{e}_p$ with $\mathbf{e}_{\sigma_2(p)}$.

2.3. Types and degree growth

Consider an ample class $\mathbf{h} \in \mathrm{NS}(S, \mathbb{R})$ with self-intersection 1. The *degree* of $\phi \in \mathrm{Bir}(S)$ with respect to the polarization $\mathbf{h}$ is defined by

$$\deg_{\mathbf{h}} \phi = \langle \phi_*(\mathbf{h}), \mathbf{h}\rangle = \cosh(d(\mathbf{h}, \phi_*\mathbf{h})).$$

Note that if $S = \mathbb{P}^2_{\mathbb{C}}$ and $\mathbf{h} = \mathbf{e}_0$ is the class of a line, then $\deg_{\mathbf{h}} \phi$ is the degree of ϕ as defined in Chapter 1.

A birational map ϕ of a projective surface S is

- ◇ *virtually isotopic to the identity* if there is a positive iterate ϕ^n of ϕ and a birational map $\psi\colon Z \dashrightarrow S$ such that $\psi^{-1} \circ \phi^n \circ \psi$ is an element of $\mathrm{Aut}(Z)^0$;
- ◇ a *Jonquières twist* if ϕ preserves a one-parameter family of rational curves on S, but ϕ is not virtually isotopic to the identity;
- ◇ a *Halphen twist* if ϕ preserves a one-parameter family of genus one curves on S, but ϕ is not virtually isotopic to the identity.

Furthermore the Jonquières twists (resp. Halphen twists) preserve a unique fibration ([**85**]).

REMARK 2.5. If ϕ is a Jonquières (resp. Halphen) twist, then, after conjugacy by a birational map $\psi\colon Z \dashrightarrow S$, ϕ permutes the fibers of a rational (resp. genus one) fibration $\pi\colon Z \to B$. If z is the divisor class of the generic fiber of the fibration, then z is an isotropic vector in $\mathcal{Z}(S)$ fixed by ϕ_*. In particular ϕ_* cannot be loxodromic.

Let $\mathcal{C}$ and $\mathcal{C}'$ be two smooth cubic curves in the complex projective plane. By the Bezout theorem, $\mathcal{C}$ and $\mathcal{C}'$ intersect in nine points denoted p_1, p_2, $\ldots$, p_9. There is a pencil of cubic curves passing through these nine points. Let us blow-up p_1, p_2, $\ldots$, p_9. We get a rational surface S with a fibration $\pi\colon S \to \mathbb{P}^1_{\mathbb{C}}$ whose fibers are genus 1 curves. More generally let us consider a pencil of curves of degree $3m$ for $m \in \mathbb{Z}_+$, blow-up its base-points and denote by S the surface we get. Such a pencil of genus 1 curves is called a *Halphen pencil*, and such a surface is called a *Halphen surface of index m*.

DEFINITION. A surface S is a *Halphen* one if $|-mK_S|$ satisfies the three following properties

- ◇ it is one-dimensional,
- ◇ it has no fixed component,
- ◇ it is base-point free.

According to [**56**] up to birational conjugacy

- ◇ every pencil of genus 1 curves of $\mathbb{P}^2_{\mathbb{C}}$ is a Halphen pencil,

◇ Halphen surfaces are the only examples of rational elliptic surfaces.

LEMMA 2.6 ([**172**]). *Let S be a Halphen surface. Let ϕ be an element of* $\mathrm{Bir}(S)$ *that preserves the Halphen pencil. Then ϕ belongs to* $\mathrm{Aut}(S)$.

Up to conjugacy by birational maps every pencil of genus 1 curves of $\mathbb{P}^2_\mathbb{C}$ is a Halphen pencil and Halphen surfaces are the only examples of rational elliptic surfaces ([**56**]) so Lemma 2.6 implies:

COROLLARY 2.7. *A subgroup* G *of* $\mathrm{Bir}(\mathbb{P}^2_\mathbb{C})$ *that preserves a pencil of genus* 1 *curves is conjugate to a subgroup of the automorphism group of some Halphen surface.*

PROOF OF LEMMA 2.6. The Halphen pencil is defined by a multiple of the class of the anticanonical divisor $-K_S$. As a result any birational map of a Halphen surface that preserves the Halphen fibration preserves the class of the canonical divisor K_S. Assume by contradiction that ϕ is not an automorphism. Take a minimal resolution of ϕ:

$$\begin{array}{ccc} & Z & \\ {}^{\pi}\swarrow & & \searrow^{\eta} \\ S & \overset{\phi}{\dashrightarrow} & S \end{array}$$

Denote by E_i and F_i the total pull-backs of the exceptional curves. On the one hand

$$K_Z = \eta^*(K_S) + \sum E_i,$$

and on the other hand

$$K_Z = \pi^*(K_S) + \sum F_i.$$

The map ϕ preserves K_S, so $\eta^*(K_S) = \pi^*(K_S)$, and hence $\sum E_i = \sum F_i$. By assumption ϕ is not an automorphism, *i.e.*, $\sum E_i$ contains at least one (-1)-curve E_k. Hence both

$$E_k \cdot \left(\sum E_i\right) = -1$$

and

$$E_k \cdot \left(\sum F_i\right) = -1$$

hold. This implies that E_k is contained in the support of $\sum F_i$: contradiction with the minimality of the resolution. □

REMARK 2.8. The automorphism groups of Halphen surfaces are studied in [**101**] and in [**56**].

On the contrary Jonquières twists are not conjugate to automorphisms of projective surfaces ([**33**, **85**]).

Let S be a projective complex surface with a polarization H. Let $\phi\colon S \dashrightarrow S$ be a birational map. The *dynamical degree* of ϕ is defined by

$$\lambda(\phi) = \lim_{n\to+\infty} \deg_H(\phi^n)^{1/n}.$$

DEFINITIONS. An element ϕ of $\mathrm{Bir}(\mathbb{P}^2_\mathbb{C})$ is called *elliptic* (resp. *parabolic*, resp. *loxodromic*) if the corresponding isometry ϕ_* is elliptic (resp. parabolic, resp. loxodromic).

The map ϕ is loxodromic if and only if $\lambda(\phi) > 1$. As a consequence when $\phi \in \mathrm{Bir}(\mathbb{P}^2_\mathbb{C})$, $\lambda(\phi) > 1$, the isometry ϕ_* preserves a unique geodesic line $\mathrm{Ax}(\phi) \subset \mathbb{H}^\infty$ called the *axis* of ϕ. This line is the intersection of $\mathbb{H}^\infty$ with a plane $P_\phi \subset \mathrm{Z}(\mathbb{P}^2_\mathbb{C})$ which intersects the isotropic cone of $\mathrm{Z}(\mathbb{P}^2_\mathbb{C})$ in two lines $\mathbb{R}v^+_{\phi_*}$ and $\mathbb{R}v^-_{\phi_*}$ such that

$$\phi_*(p) = \lambda(\phi)^{\pm 1} p$$

for all $p \in \mathbb{R}v^{\pm}_{\phi_*}$ (the lines $\mathbb{R}v^+_{\phi_*}$ and $\mathbb{R}v^-_{\phi_*}$ correspond to $\omega(\phi)$ and $\alpha(\phi)$ with the notations of §2.1.3).

Take $\alpha \in \mathbb{R}v^-_{\phi_*}$ and $\omega \in \mathbb{R}v^+_{\phi_*}$ normalized so that $\langle \alpha, \omega\rangle = 1$. The point $p = \frac{\alpha+\omega}{\sqrt{2}}$ lies on $\mathrm{Ax}(\phi)$. Since $\phi_*(p) = \frac{\lambda(\phi)^{-1}\alpha + \lambda(\phi)\omega}{\sqrt{2}}$ one obtains

$$\begin{aligned} \exp(L(\phi_*)) + \frac{1}{\exp(L(\phi_*))} &= 2\cosh(d(p, \phi_*(p))) \\ &= 2\langle p, \phi_*(p)\rangle \\ &= \lambda(\phi) + \frac{1}{\lambda(\phi)}. \end{aligned}$$

The translation length is thus equal to $\log \lambda(\phi)$. Consequently $\lambda(\phi)$ does not depend on the polarization and is invariant under conjugacy.

There is a correspondence between the dynamical behavior of a birational map ϕ of S, in particular its degree, and the type of the induced isometry on $\mathbb{H}^\infty$:

THEOREM 2.9 ([**50, 85, 101**]). *Let S be a smooth projective complex surface with a fixed polarization H. Let $\phi\colon S \dashrightarrow S$ be a birational map. Then one of the following holds:*

- *ϕ is elliptic, $(\deg_H \phi^n)_n$ is bounded, and ϕ is virtually isotopic to the identity;*
- *ϕ is parabolic and*

  either $\deg_H \phi^n \sim cn$ for some positive constant c and ϕ is a Jonquières twist;

  or $\deg_H \phi^n \sim cn^2$ for some positive constant c and ϕ is a Halphen twist;
- *ϕ is loxodromic and $\deg_H \phi^n = c\lambda(\phi)^n + O(1)$ for some positive constant c.*

EXAMPLES 3.

- Any birational map of finite order is elliptic. Any automorphism of $\mathbb{P}^2_\mathbb{C}$ is elliptic. Any element of the group

  $$\big\{(z_0, z_1) \mapsto (\alpha z_0 + P(z_1), \beta z_1 + \gamma) \,|\, \alpha,\, \beta \in \mathbb{C}^*,\, \gamma \in \mathbb{C},\, P \in \mathbb{C}[z_1]\big\}$$

  is elliptic.
- Any element of $\mathcal{J}$ of the form

  $$(z_0, z_1) \dashrightarrow \left(z_0, \frac{a(z_0)z_1 + b(z_0)}{c(z_0)z_1 + d(z_0)}\right)$$

  with $\frac{(\mathrm{tr}\, M)^2}{\det M} \in \mathbb{C}(z_0) \smallsetminus \mathbb{C}$ where

  $$M = \begin{pmatrix} a(z_0) & b(z_0) \\ c(z_0) & d(z_0) \end{pmatrix}$$

  is a Jonquières twist ([**61**]).

- Consider the family of birational self maps of $\mathbb{P}^2_\mathbb{C}$) given in the affine chart $z_2 = 1$ by

$$\phi_\varepsilon\colon (z_0, z_1) \dashrightarrow \left(z_1 + 1 - \varepsilon, z_0\frac{z_1 - \varepsilon}{z_1 + 1}\right).$$

  If
  - $\varepsilon = -1$, then ϕ_ε is elliptic;
  - $\varepsilon \in \{0, 1\}$, then ϕ_ε is a Jonquières twist;
  - $\varepsilon \in \left\{\frac{1}{2}, \frac{1}{3}\right\}$, then ϕ_ε is a Halphen twist;
  - $\varepsilon \in \bigcup_{k\geq 4} \frac{1}{k}$, then ϕ_ε is loxodromic.

  This family has been introduced in [**85**].
- If $\phi\colon (z_0, z_1) \dashrightarrow (z_1, z_0 + z_1^2)$, then $\deg(\phi^n) = (\deg\phi)^n = 2^n$. If $\psi\colon (z_0, z_1) \dashrightarrow (z_0^2 z_1, z_0 z_1)$, then $\deg\psi^n \sim \left(\frac{3+\sqrt{5}}{2}\right)^n$; in particular $\deg(\psi^n) \neq (\deg\psi)^n$.
- Let us finish with a more geometric example. Consider the elliptic curve $E = {}^{\mathbb{C}}\!/\!_{\mathbb{Z}[\mathbf{i}]}$. The linear action of the group $\mathrm{GL}(2, \mathbb{Z}[\mathbf{i}])$ on the complex plane preserves the lattice $\mathbb{Z}[\mathbf{i}]\times\mathbb{Z}[\mathbf{i}]$. This yields to an action of $\mathrm{GL}(2, \mathbb{Z}[\mathbf{i}])$ by regular automorphisms on the abelian surface $S = E \times E$. Since this action commutes with $(z_0, z_1) \mapsto (\mathbf{i}z_0, \mathbf{i}z_1)$ one gets a morphism from $\mathrm{PGL}(2, \mathbb{Z}[\mathbf{i}])$ to $\mathrm{Aut}\left({}^{S}\!/\!_{(z_0, z_1)} \mapsto (\mathbf{i}z_0, \mathbf{i}z_1)\right)$. As ${}^{S}\!/\!_{(z_0, z_1)} \mapsto (\mathbf{i}z_0, \mathbf{i}z_1)$ is rational one obtains an embedding of $\mathrm{PGL}(2, \mathbb{Z}[\mathbf{i}])$ into $\mathrm{Bir}(\mathbb{P}^2_\mathbb{C})$.

Any element virtually isotopic to the identity is *regularizable*, that is, birationally conjugate to an automorphism. What can we say about two birational maps virtually isotopic to the identity? We will see that if they commute they are simultaneously regularizable. Before proving it let us introduce a new notion.

DEFINITIONS. An element $\phi \in \mathrm{Bir}(\mathbb{P}^2_\mathbb{C})$ is *algebraically stable* if $\deg\phi^n = (\deg\phi)^n$ for all $n \geq 0$.

More generally if S is a compact complex surface, then $\phi \in \mathrm{Bir}(S)$ is *algebraically stable* if $(\phi^*)^n = (\phi^n)^*$ for all $n \geq 0$.

A geometric characterization of algebraically stable maps is the following: $\phi \in \mathrm{Bir}(S)$ is algebraically stable if and only if there is no curve $C \subset S$ such that $\phi^k(C) \in \mathrm{Ind}(\phi)$ for some integer k. Let us give an idea of the fact that this geometric characterization is equivalent to the Definition when $S = \mathbb{P}^2_\mathbb{C}$. If $\phi^k\big(\mathcal{C} \smallsetminus \mathrm{Ind}(\phi)\big) \subset \mathrm{Ind}(\phi)$, then all the components of $\phi \circ \phi^k$ have a common factor that defines the equation of $\mathcal{C}$. Then $\deg(\phi \circ \phi^k) < (\deg\phi)(\deg\phi^k)$. The converse holds.

Diller and Favre proved the following result:

PROPOSITION 2.10 ([**85**]). *Let S be a compact complex surface, and let ϕ be a birational self map of S. There exists a composition of finitely many point blow-ups that lifts ϕ to an algebraically stable map.*

Before giving the proof, let us give its idea. Assume that ϕ is not algebraically stable. In other words there exist a curve $C \subset S$ and an integer k such that C is blown down onto p_1 and $p_k = \phi^{k-1}(p_1)$ belongs to $\mathrm{Ind}(\phi)$. The idea of Diller and Favre to get an algebraically stable map is the following: after blowing up the points $p_i = \phi^i(p_1)$, $i = 1, \ldots, k$, the orbit of C consists of curves. Doing this for any element of $\mathrm{Exc}(\phi)$ whose iterate belongs to $\mathrm{Ind}(\phi)$ one gets the statement (note that the cardinal of $\mathrm{Exc}(\phi)$ is finite, so the process ends).

PROOF. Let us write ϕ as follows: $\phi = \phi_n \circ \phi_{n-1} \circ \ldots \circ \phi_1$ where

- ◇ $\phi_i \colon S_{i-1} \to S_i$;
- ◇ $S_0 = S_n = S$;
- ◇ and
  - (i) either ϕ_i blows up a point $p_i = \mathrm{Ind}(\phi_i) \in S_i$, and we denote by $V_{i+1} = \mathrm{Exc}(\phi_i^{-1}) \subset S_{i+1}$ the exceptional divisor of ϕ_i^{-1};
  - (ii) or ϕ_i blows down the exceptional divisor $E_i \subset S_i$; in this case we set $q_{i+1} := \phi_i(E_i) \in S_{i+1}$.

For any $j \in \mathbb{N}$ set $S_j := S_{j \bmod n}$ and $\phi_j := \phi_{j \bmod n}$.

Assume that ϕ is not algebraically stable. Then there exist integers $1 \leq i \leq N$ such that ϕ_i blows down E_i and

$$\phi_{N-1} \circ \phi_{N-2} \circ \ldots \circ \phi_i(E_i) = p_N \in \mathrm{Ind}(\phi_N).$$

Choosing a pair (i, N) of minimal length we can assume that for all $i < j \leq N$

$$m_j := \phi_j \circ \phi_{j-1} \circ \ldots \circ \phi_i(E_i) = \phi_j \circ \phi_{j-1} \circ \ldots \circ \phi_{i+1}(q_{i+1})$$

does not belong to $\mathrm{Ind}(\phi_i) \cup \mathrm{Exc}(\phi_i)$.

First blow up S_N at $m_N = p_N$. Then

- ◇ ϕ_N lifts to a biholomorphism $\widehat{\phi}_N$ of $\mathrm{Bl}_{p_N} S_N$;
- ◇ $\widehat{\phi}_{N-1}$ either blows up the two distinct points m_{N-1} and p_{N-1} or blows up m_{N-1} and blows down $E_{N-1} \not\ni m_{N-1}$;
- ◇ $\sum \mathrm{Card}\big(\phi_j(\mathrm{Exc}(\phi_j))\big) = \sum \mathrm{Card}\big(\widehat{\phi}_j(\mathrm{Exc}(\widehat{\phi}_j))\big)$.

Note that modifying S_N means modifying S_{N+n}, S_{N-n}, $\ldots$; nevertheless blowing up a point m_j does not interfere with the behavior of the map ϕ_j around m_{N+n}, m_{N-n}, $\ldots$ (indeed if $j_1 = j_2 \bmod n$ but $j_1 \neq j_2$, then the points m_{j_1}, m_{j_2} of $S_1 = S_2$ are distinct), and these points can be blow-up independently.

Similarly blow-up m_{N-1}, m_{N-2}, $\ldots$, m_{i+2}. At each step $\sum \mathrm{Card}(\phi_j(\mathrm{Exc}(\phi_j)))$ remains constant. Let us finish by blowing up $m_{i+1} = \phi_i(E_i)$; the situation is then different: ϕ_i becomes a biholomorphism $\widehat{\phi}_i$. The number of components of $\mathrm{Exc}(\phi_i)$ thus reduces from 1 to 0. As a consequence

$$\sum \mathrm{Card}\big(\widehat{\phi}_j(\mathrm{Exc}(\widehat{\phi}_j))\big) = \sum \mathrm{Card}\big(\phi_j(\mathrm{Exc}(\phi_j))\big) - 1. \tag{2.2}$$

Repeating the above argument finitely many times either we produce an algebraically stable map $\widehat{\phi} = \widehat{\phi}_N \circ \widehat{\phi}_{N-1} \circ \ldots \circ \widehat{\phi}_1$, or thanks to (2.2), we eliminate all exceptional components of the factors of ϕ. In both cases we get an algebraically stable map. □

LEMMA 2.11 ([**76**]). *Let ϕ, ψ be two birational self maps of a compact complex surface S. Assume that ϕ and ψ are both virtually isotopic to the identity. Assume that ϕ and ψ commute.*

There exist a surface Y and a birational map $\zeta \colon Y \dashrightarrow S$ such that

- ◇ $\zeta^{-1} \circ \phi^\ell \circ \zeta \in \mathrm{Aut}(Y)^0$ *for some integer* ℓ,
- ◇ $\zeta^{-1} \circ \psi \circ \zeta$ *is algebraically stable.*

PROOF. Since ϕ is virtually isotopic to the identity we can assume that up to birational conjugacy and finite index ϕ is an automorphism of S. Let $N(\psi)$ be the minimal number of blow-ups needed to make ψ algebraically stable (such a $N(\psi)$ exists according to Proposition 2.10). If $N(\psi) = 0$, then $\zeta = \mathrm{id}$ suits. Assume that Lemma 2.11 holds when $N(\psi) \leq j$. Consider a pair (ϕ, ψ) of birational self maps of S such that

- ◇ ϕ and ψ are both virtually isotopic to the identity,
- ◇ ϕ and ψ commute,
- ◇ $N(\psi) = j + 1$.

Since ψ is not algebraically stable there exists a curve C blown down by ψ and such that $\psi^q(C)$ is a point of indeterminacy p of ψ for some integer q. The maps ψ and ϕ commute, so an iterate ϕ^k of ϕ fixes the irreducible components of $\mathrm{Ind}(\psi)$. Let us blow-up p via π. On the one hand $\pi^{-1} \circ \phi^k \circ \pi$ is an automorphism because p is fixed by ϕ^k and on the other hand $N(\pi^{-1} \circ \psi \circ \pi) = j$. One can thus conclude by induction that there exist a surface Y and a birational map $\zeta \colon Y \dashrightarrow S$ such that $\zeta^{-1} \circ \phi^\ell \circ \zeta \in \mathrm{Aut}(Y)^0$ for some integer ℓ and $\zeta^{-1} \circ \psi \circ \zeta$ is algebraically stable. □

PROPOSITION 2.12 ([**76**]). *Let ϕ, ψ be two birational self maps of a surface S. Assume that ϕ and ψ are both virtually isotopic to the identity. Assume that ϕ and ψ commute.*

Then there exist a surface Z and a birational map $\pi \colon Z \dashrightarrow S$ such that

- ◇ *$\pi^{-1} \circ \phi \circ \pi$ and $\pi^{-1} \circ \psi \circ \pi$ belong to $\mathrm{Aut}(Z)$;*
- ◇ *$\pi^{-1} \circ \phi^k \circ \pi$ and $\pi^{-1} \circ \psi^k \circ \pi$ belong to $\mathrm{Aut}(Z)^0$ for some integer k.*

PROOF. By assumption there exist a surface $\widetilde{S}$, a birational map $\eta \colon \widetilde{S} \dashrightarrow S$, and an integer n such that $\eta^{-1} \circ \phi \circ \eta$ belongs to $\mathrm{Aut}(S)$ and $\eta^{-1} \circ \phi^n \circ \eta$ belongs to $\mathrm{Aut}(S)^0$. Let us now work on $\widetilde{S}$; to simplify denote by ϕ the automorphism $\eta^{-1} \circ \phi^n \circ \eta$ and by ψ the birational map $\eta^{-1} \circ \psi \circ \eta$.

According to Lemma 2.11 there exist a surface Y, a birational map $v \colon Y \dashrightarrow S$, and an integer ℓ such that $\zeta^{-1} \circ \widetilde{\phi}^\ell \circ \zeta$ belongs to $\mathrm{Aut}(Y)^0$ and $\zeta^{-1} \circ \widetilde{\psi} \circ \zeta$ is algebraically stable.

Set $\overline{\phi} = \zeta^{-1} \circ \widetilde{\phi}^i \circ \zeta$ and $\overline{\psi} = \zeta^{-1} \circ \widetilde{\psi} \circ \zeta$. To get an automorphism from $\overline{\psi}$ let us blow-down curves in $\mathrm{Exc}(\overline{\psi}^{-1})$. But curves blown down by $\overline{\psi}^{-1}$ are of self-intersection < 0 and $\overline{\phi}$ fixes such curves since $\overline{\phi}$ is isotopic to the identity. We conclude by using the fact that $\mathrm{Card}\big(\mathrm{Exc}(\overline{\psi}^{-1})\big)$ is finite. □

2.4. On the hyperbolicity of graphs associated to the Cremona group

To reinforce the analogy between the mapping class group and the plane Cremona group Lonjou looked for a graph analogous to the curve graph and such that the Cremona group acts on it trivially in [**133**].

A candidate is the graph introduced by Wright (Chapter 4 §4.2.2 and [**180**]).

As we have recalled in Chapter 4 §4.2.2 the complex C is a simplicial complex of dimension 2 and 1-connected on which $\mathrm{Bir}(\mathbb{P}^2_\mathbb{C})$ acts. Since Lonjou is interested in the Gromov hyperbolicity property, she is only interested in the 1-skeleton of C. She proved that the diameter of this non-locally finite graph is infinite ([**133**, Corollary

2.7]). She then focuses on the following question "Is this graph Gromov hyperbolic ?"[2] The answer is no:

THEOREM 2.13 ([**133**]). *The Wright graph is not Gromov hyperbolic.*

The first point of the proof is to note that the Wright graph is quasi-isometric to a graph related to the system of generators of $\mathrm{Bir}(\mathbb{P}^2_\mathbb{C})$ given by $\mathrm{PGL}(3,\mathbb{C})$ and the Jonquières maps. It is an analogue of the Cayley graph in the case of a finitely generated group. The vertices of this graph are the elements of $\mathrm{Bir}(\mathbb{P}^2_\mathbb{C})$ modulo pre-composition by an automorphism of $\mathbb{P}^2_\mathbb{C}$. An edge connects two vertices if there exists a Jonquières map that permutes the two vertices. The distance between two vertices ϕ, ψ in $\mathrm{Bir}(\mathbb{P}^2_\mathbb{C})$ is the minimal number of Jonquières maps needed to decompose $\psi^{-1}\circ\phi$ (in [**35**] Blanc and Furter called this integer the translation length of $\psi^{-1}\circ\phi$. They gave an algorithm to compute this length. They also got that the diameter of the Wright graph is infinite).

The second point is to prove that this graph contains a subgraph quasi-isometric to $\mathbb{Z}^2$ (*see* [**133**, Theorem 2.12]). She took two Halphen twists that commute. They generate a subgroup isomorphic to $\mathbb{Z}^2$. Using some results of [**35**] she established that the action of this subgroup on one of the vertices of the graph induces the desired graph[3].

Then Lonjou constructed two graphs associated to a Voronï tessellation of the Cremona group introduced in [**132**]; she proved that

- ◇ one of these graphs is quasi-isometric to the Wright graph;
- ◇ the second one is Gromov hyperbolic.

[2]Minosyan and Osin note that if the answer to this question is yes, the results of [**68**] allow us to give a new proof of the non-simplicity of $\mathrm{Bir}(\mathbb{P}^2_\mathbb{C})$ (*see* [**143, 144**]).

[3]The Cayley graph of the modular group of a compact surface of genus $g\geq 2$ is not Gromov hyperbolic; indeed, this group has subgroups isomorphic to $\mathbb{Z}^2$ (for instance generated by two Dehn twists along two disjoint closed curves).

CHAPTER 3

Algebraic subgroups of the Cremona group

The first section of this chapter deals with the algebraic structure of the n-dimensional Cremona group, the fact that it is not an algebraic group of infinite dimension if $n \geq 2$, the obstruction to this, which is of a topological nature. By contrast, the existence of a Euclidean topology on the Cremona group which extends that of its classical subgroups and makes it a topological group is recalled. More precisely in [**47**] Shafarevich asked

> "Can one introduce a universal structure of an infinite-dimensional group in the group of all automorphisms (resp. all birational automorphisms) of arbitrary algebraic variety?"

We will see that the answer to this question is no ([**34**]). For any algebraic variety V defined over $\mathbb{C}$ there is a natural notion of families of elements of $\mathrm{Bir}(\mathbb{P}^n_\mathbb{C})$ parameterized by V. These are maps $V(\mathbb{C}) \to \mathrm{Bir}(\mathbb{P}^n_\mathbb{C})$ compatible with the structures of algebraic varieties. Note that $\mathrm{Bir}(\mathbb{P}^1_\mathbb{C}) \simeq \mathrm{PGL}(2,\mathbb{C})$ and families $V \dashrightarrow \mathrm{Bir}(\mathbb{P}^1_\mathbb{C})$ correspond to morphisms of algebraic varieties. If $n \geq 2$ the set $\mathrm{Bir}_d(\mathbb{P}^n_\mathbb{C})$ of all birational maps of $\mathbb{P}^n_\mathbb{C}$ of degree d has the structure of an algebraic variety defined over $\mathbb{C}$ such that the families $V \to \mathrm{Bir}_d(\mathbb{P}^n_\mathbb{C})$ correspond to morphisms of algebraic varieties ([**34**]). So $\mathrm{Bir}(\mathbb{P}^n_\mathbb{C})$ decomposes into a disjoint infinite union of algebraic varieties, having unbounded dimension. Blanc and Furter established the following statement:

THEOREM 3.1 ([**34**]). *Let $n \geq 2$. There is no structure of algebraic variety of infinite dimension on $\mathrm{Bir}(\mathbb{P}^n_\mathbb{C})$ such that families $V \to \mathrm{Bir}(\mathbb{P}^n_\mathbb{C})$ would correspond to morphisms of algebraic varieties.*

The lack of structure comes from the degeneration of maps of degree d into maps of smaller degree. A family of birational self maps of $\mathbb{P}^2_\mathbb{C}$ of degree d which depends on a parameter t may degenerate for certain values of t onto a non-reduced expression of the type $P\,\mathrm{id} = P(z_0 : z_1 : z_2)$ where P denotes a homogeneous polynomial of degree $d-1$. Consider, for instance, the family

$$\begin{aligned}\phi_{a,b,c} \quad : \quad & (z_0 : z_1 : z_2) \dashrightarrow \\ & \big(z_0(az_2^2 + cz_0z_2 + bz_0^2) : z_1(az_2^2 + (b+c)z_0z_2 + (a+b)z_0^2) : \\ & \qquad\qquad\qquad\qquad z_2(az_2^2 + cz_0z_2 + bz_0^2)\big)\end{aligned}$$

parameterized by the nodal plane cubic $a^3 + b^3 = abc$. The family $(\phi_{a,b,c})$ is globally defined by formulas of degree 3, but each element $\phi_{a,b,c}$ has degree ≤ 2 and there is no global parameterization by homogeneous formulas of degree 2. In fact the obstruction to a positive answer to Shafarevich question comes only from the topology:

THEOREM 3.2 ([**34**]). *There is no* $\mathbb{C}$*-algebraic variety of infinite dimension that is homeomorphic to* $\mathrm{Bir}(\mathbb{P}^n_{\mathbb{C}})$.

In 2010 in the question session of the workshop "Subgroups of the Cremona group" in Edinburgh, Serre asked the following question:

"Is it possible to introduce such topology on $\mathrm{Bir}(\mathbb{P}^2_{\mathbb{C}})$ that is compatible with $\mathrm{PGL}(3,\mathbb{C})$ and $\mathrm{PGL}(2,\mathbb{C})\times\mathrm{PGL}(2,\mathbb{C})$?"

We will see that Blanc and Furter gave a positive answer to this question:

THEOREM 3.3 ([**34**]). *Let* $n\geq 1$ *be an integer. There is a natural topology on* $\mathrm{Bir}(\mathbb{P}^n_{\mathbb{C}})$, *called the Euclidean topology, such that:*

- ◇ $\mathrm{Bir}(\mathbb{P}^n_{\mathbb{C}})$, *endowed with the Euclidean topology, is a Hausdorff topological group,*
- ◇ *the restriction of the Euclidean topology to algebraic subgroups, in particular to* $\mathrm{PGL}(n+1,\mathbb{C})$ *and* $\mathrm{PGL}(2,\mathbb{C})^n$, *is the classical Euclidean topology.*

In the literature an algebraic subgroup G of $\mathrm{Bir}(V)$ corresponds to taking an algebraic group G and a morphism $\mathrm{G}\to\mathrm{Bir}(V)$ that is a group morphism and whose schematic kernel is trivial. We will see that in the case of $V=\mathbb{P}^n_{\mathbb{C}}$ one can give a more intrinsic definition (Corollary 3.11) which corresponds to taking closed subgroups of $\mathrm{Bir}(\mathbb{P}^n_{\mathbb{C}})$ of bounded degree and that these two definitions agree (Lemma 3.12).

An element $\phi\in\mathrm{Bir}(\mathbb{P}^n_{\mathbb{C}})$ is algebraic if it is contained in an algebraic subgroup G of $\mathrm{Bir}(\mathbb{P}^n_{\mathbb{C}})$. It is equivalent to saying that the sequence $(\deg\phi^n)_{n\in\mathbb{N}}$ is bounded. According to [**34**] the group G is thus an affine algebraic group. As a consequence ϕ decomposes as $\phi=\phi_s\circ\phi_u$ where ϕ_s is a semi-simple element of G and ϕ_u a unipotent element of G. This decomposition does not depend on G (*see* [**156**]). In particular there is a natural notion of semi-simple and unipotent elements of $\mathrm{Bir}(\mathbb{P}^n_{\mathbb{C}})$. As we will see, G could even be chosen to be the abelian algebraic subgroup $\overline{\{\phi^i\,|\,i\in\mathbb{Z}\}}$ of $\mathrm{Bir}(\mathbb{P}^n_{\mathbb{C}})$. In all linear algebraic groups the set of unipotent elements is closed; Popov asked if it is the case in the context of the Cremona group. A natural and related question is the following one: is the set $\mathrm{Bir}(\mathbb{P}^n_{\mathbb{C}})_{\mathrm{alg}}$ of algebraic elements of $\mathrm{Bir}(\mathbb{P}^n_{\mathbb{C}})$ closed? The second section deals with the answers to these questions (Theorem 3.31).

In the third section the classification of maximal algebraic subgroups of the plane Cremona group is given.

In the fourth section we give a sketch of the proof of the fact that $\mathrm{Bir}(\mathbb{P}^n_{\mathbb{C}})$ is topologically simple when endowed with the Zariski topology, *i.e.*, it does not contain any non-trivial closed normal strict subgroup. The main ingredients of the proof are some clever deformation arguments.

The fifth section is devoted to a modern proof of the regularization theorem of Weil which says that for every rational action ρ of an algebraic group G on a variety X there exist a variety Y with a regular action μ of G and a dominant rational map $\phi\colon X\dashrightarrow Y$ with the following properties: for any $(g,p)\in\mathrm{G}\times X$ such that

- ◇ ρ is defined in (g,p);
- ◇ ϕ is defined in p and $\rho(g,p)$;
- ◇ μ is defined in $(g,\phi(p))$;

we have $\phi(\rho(g,p))=\mu(g,\phi(p))$.

3.1. Topologies and algebraic subgroups of $\mathrm{Bir}(\mathbb{P}^n_{\mathbb{C}})$

3.1.1. Zariski topology. Take an irreducible algebraic variety V. A *family of birational self maps of $\mathbb{P}^n_{\mathbb{C}}$ parameterized by V* is a birational self map

$$\varphi\colon V\times\mathbb{P}^n_{\mathbb{C}} \dashrightarrow V\times\mathbb{P}^n_{\mathbb{C}}$$

such that

- ◇ φ determines an isomorphism between two open subsets $\mathcal{U}$ and $\mathcal{V}$ of $V\times\mathbb{P}^n_{\mathbb{C}}$ such that the first projection pr_1 maps both $\mathcal{U}$ and $\mathcal{V}$ surjectively onto V,
- ◇ $\varphi(v,x)=\big(v,\mathrm{pr}_2(\varphi(v,x))\big)$ where pr_2 denotes the second projection; hence each $\varphi_v=\mathrm{pr}_2(\varphi(v,\cdot))$ is a birational self map of $\mathbb{P}^n_{\mathbb{C}}$.

The map $v\mapsto\varphi_v$ is called a *morphism* from the parameter space V to $\mathrm{Bir}(\mathbb{P}^n_{\mathbb{C}})$.

A subset $S\subset\mathrm{Bir}(\mathbb{P}^n_{\mathbb{C}})$ is *closed* if for any algebraic variety V and any morphism $V\to\mathrm{Bir}(\mathbb{P}^n_{\mathbb{C}})$ its preimage is closed.

This yields a topology on $\mathrm{Bir}(\mathbb{P}^n_{\mathbb{C}})$ called the *Zariski topology*.

REMARK 3.4. For any $\phi\in\mathrm{Bir}(\mathbb{P}^n_{\mathbb{C}})$ the maps from $\mathrm{Bir}(\mathbb{P}^n_{\mathbb{C}})$ into itself given by

$$\psi\mapsto\psi\circ\varphi,\qquad\qquad \psi\mapsto\varphi\circ\psi,\qquad\qquad \psi\mapsto\psi^{-1}$$

are homeomorphisms of $\mathrm{Bir}(\mathbb{P}^n_{\mathbb{C}})$ with respect to the Zariski topology.

Indeed let V be an irreducible algebraic variety. If f, $g\colon V\times\mathbb{P}^n_{\mathbb{C}}\to V\times\mathbb{P}^n_{\mathbb{C}}$ are two V-birational maps inducing morphisms $V\to\mathrm{Bir}(\mathbb{P}^n_{\mathbb{C}})$, then $f\circ g$ and f^{-1} are again V-birational maps that induce morphisms $V\to\mathrm{Bir}(\mathbb{P}^n_{\mathbb{C}})$.

Let $\mathrm{Bir}_{\leq d}(\mathbb{P}^n_{\mathbb{C}})$ (resp. $\mathrm{Bir}_d(\mathbb{P}^n_{\mathbb{C}})$) be the set of elements of $\mathrm{Bir}(\mathbb{P}^n_{\mathbb{C}})$ of degree $\leq d$ (resp. of degree d); we have the following increasing sequence:

$$\mathrm{Aut}(\mathbb{P}^n_{\mathbb{C}})=\mathrm{Bir}_{\leq 1}(\mathbb{P}^n_{\mathbb{C}})\subseteq\mathrm{Bir}_{\leq 2}(\mathbb{P}^n_{\mathbb{C}})\subseteq\mathrm{Bir}_{\leq 3}(\mathbb{P}^n_{\mathbb{C}})\subseteq\ldots$$

whose union gives the Cremona group. We will see that $\mathrm{Bir}_{\leq d}(\mathbb{P}^n_{\mathbb{C}})$ is closed in $\mathrm{Bir}(\mathbb{P}^n_{\mathbb{C}})$ and the topology of $\mathrm{Bir}(\mathbb{P}^n_{\mathbb{C}})$ is the inductive topology induced by the above sequence. As a result it suffices to describe the topology of $\mathrm{Bir}_{\leq d}(\mathbb{P}^n_{\mathbb{C}})$ to understand the topology of $\mathrm{Bir}(\mathbb{P}^n_{\mathbb{C}})$.

Take a positive integer d. Let W_d be the set of equivalence classes of non-zero $(n+1)$-uples $(\phi_0,\phi_1,\ldots,\phi_n)$ of homogeneous polynomials $\phi_i\in\mathbb{C}[z_0,z_1,\ldots,z_n]$ of degree d where $(\phi_0,\phi_1,\ldots,\phi_n)$ is equivalent to $(\lambda\phi_0,\lambda\phi_1,\ldots,\lambda\phi_n)$ for any $\lambda\in\mathbb{C}^*$. We denote by $(\phi_0:\phi_1:\ldots:\phi_n)$ the equivalence class of $(\phi_0,\phi_1,\ldots,\phi_n)$. Let $H_d\subseteq W_d$ be the set of elements $\phi=(\phi_0:\phi_1:\ldots:\phi_n)\in W_d$ such that the rational map $\psi_\phi\colon\mathbb{P}^n_{\mathbb{C}}\dashrightarrow\mathbb{P}^n_{\mathbb{C}}$ given by

$$(z_0:z_1:\ldots:z_n)\dashrightarrow\big(\phi_0(z_0,z_1,\ldots,z_n):\phi_1(z_0,z_1,\ldots,z_n):\ldots:\phi_n(z_0,z_1,\ldots,z_n)\big)$$

is birational. The map

$$H_d\to\mathrm{Bir}(\mathbb{P}^n_{\mathbb{C}})\qquad\qquad \phi\mapsto\psi_\phi$$

is denoted by π_d.

LEMMA 3.5 ([**34**]). *The following properties hold:*

- ◇ *The set W_d is isomorphic to $\mathbb{P}^k_{\mathbb{C}}$ where $k=(n+1)\binom{d+n}{d}-1$.*
- ◇ *The set H_d is locally closed in W_d; thus it inherits from W_d the structure of an algebraic variety.*
- ◇ *The map $\pi_d\colon H_d\to\mathrm{Bir}(\mathbb{P}^n_{\mathbb{C}})$ is a morphism, and $\pi_d(H_d)$ is the set $\mathrm{Bir}_{\leq d}(\mathbb{P}^n_{\mathbb{C}})$.*

- *For all ϕ in $\mathrm{Bir}_{\leq d}(\mathbb{P}^n_\mathbb{C})$ the set $\pi_d^{-1}(\phi)$ is closed in W_d, so in H_d as well.*
- *If $S \subset H_\ell$ ($\ell \geq 1$) is closed, then $\pi_d^{-1}(\pi_\ell(S))$ is closed in H_d.*

Hence W_d and H_d are naturally algebraic varieties, $\mathrm{Bir}_d(\mathbb{P}^n_\mathbb{C})$ also, but not $\mathrm{Bir}_{\leq d}(\mathbb{P}^n_\mathbb{C})$.

PROOF OF LEMMA 3.5.

- The set of homogeneous polynomials of degree d in $(n+1)$ variables is a $\mathbb{C}$-vector space of dimension $\binom{d+n}{d}$; this implies the first assertion.
- Denote by $Y \subseteq W_{d^{n-1}} \times W_d$ the set defined by

$$\big\{(\varphi,\phi) \in W_{d^{n-1}} \times W_d \,|\, \varphi \circ \phi = P\,\mathrm{id} \text{ for some } P \in \mathbb{C}[z_0, z_1, \ldots, z_n]_{d^n}\big\}.$$

If P is non-zero, then the rational maps ψ_ϕ and ψ_φ are birational and inverses of each other.

If P is zero, then ψ_ϕ contracts the entire set $\mathbb{P}^n_\mathbb{C}$ onto a strict subvariety included in the set $\big\{\varphi_1 = \varphi_2 = \ldots = \varphi_n = 0\big\}$.

In particular for any pair (φ,ϕ) of Y the rational map ψ_ϕ is birational if and only if its Jacobian is non-zero.

As a consequence any element $\phi \in H_d$ corresponds to at least one pair (φ,ϕ) in Y (indeed according to [**4**] the inverse of a birational self map of $\mathbb{P}^n_\mathbb{C}$ of degree d has degree $\leq d^{n-1}$).

The description of Y shows that it is closed in $W_{d^{n-1}} \times W_d$. The image $\mathrm{pr}_2(Y)$ of Y by the second projection pr_2 is closed in W_d since $W_{d^{n-1}}$ is a complete variety and pr_2 a Zariski closed morphism. One can write H_d as $\mathcal{U} \cap \mathrm{pr}_2(Y)$ where $\mathcal{U} \subseteq W_d$ is the open set of elements having a non-zero Jacobian. As a result H_d is locally closed in W_d and closed in $\mathcal{U}$.
- Consider the H_d-rational map ϕ defined by

$$f\colon H_d \times \mathbb{P}^n_\mathbb{C} \dashrightarrow H_d \times \mathbb{P}^n_\mathbb{C} \qquad\qquad (\varphi, z) \dashrightarrow (\varphi, \varphi(z)).$$

Set $J = \det\left(\left(\frac{\partial \varphi_i}{\partial x_j}\right)_{0 \leq i,\, j \leq n}\right)$. Let $\mathcal{V} \subset H_d \times \mathbb{P}^n_\mathbb{C}$ be the open set where J is not zero.

CLAIM 3.6 ([**34**]). *The restriction $f_{|\mathcal{V}}$ of f to $\mathcal{V}$ is an open immersion.*

Hence π_d is a morphism and it follows from the construction of H_d that the image of π_d is $\mathrm{Bir}_{\leq d}(\mathbb{P}^n_\mathbb{C})$.
- Let ϕ be an element in $\mathrm{Bir}(\mathbb{P}^n_\mathbb{C})_{\leq d}$. It corresponds to a birational self map ψ_ϕ of $\mathbb{P}^n_\mathbb{C}$ given by

$$\psi_\phi\colon (z_0 : z_1 : \ldots : z_n) \dashrightarrow (\phi_0(z_0, z_1, \ldots, z_n) : \phi_1(z_0, z_1, \ldots, z_n) : \ldots : \phi_n(z_0, z_1, \ldots, z_n))$$

for some homogeneous polynomials of degree $k \leq d$ having no common divisor. Then

$$(\pi_d)^{-1}(\psi_\phi) = \big\{(\varphi_0 : \varphi_1 : \ldots : \varphi_n) \in W_d \,|\, \varphi_i \phi_j = \varphi_j \phi_i \quad \forall 1 \leq i < j \leq n\big\} \subset H_d.$$

This set is thus closed in W_d, and so in H_d.
- If ℓ is a positive integer and F a closed subset of H_ℓ, then we denote by Y_F the subset of $Y \times \overline{F}$ (where $Y \subset W_{d^{n-1}} \times W_d$ is as above and $\overline{F}$ is the closure of F in W_ℓ) given by

$$Y_F = \big\{((\zeta, \phi), \varphi) \,|\, \phi \text{ and } \varphi \text{ yield the same map } \mathbb{P}^n_\mathbb{C} \dashrightarrow \mathbb{P}^n_\mathbb{C}\big\}.$$

In other words

$$Y_F = \{((\zeta,\phi),\varphi) \,|\, \phi_i\varphi_j = \phi_j\varphi_i \quad \forall i,\, j\}.$$

Hence Y_F is closed in $Y\times\overline{F}$ and also in $W_{d^{n-1}}\times W_d\times W_\ell$. The subset $\mathrm{pr}_2(Y_F)$ of W_d is closed in W_d, and so in $\mathrm{pr}_2(Y)$; as a result $\mathrm{pr}_2(Y_F)\cap\mathcal{U}$ is closed in $\mathrm{pr}_2(Y)\cap\mathcal{U}$. We conclude using the fact that $\mathrm{pr}_2(Y_F)\cap\mathcal{U} = (\pi_d)^{-1}(\pi_\ell(F))$ and $\mathrm{pr}_2(Y)\cap\mathcal{U} = H_d$. □

LEMMA 3.7 ([**34**]). *Let V be an irreducible algebraic variety, and let $\upsilon\colon V\to \mathrm{Bir}(\mathbb{P}^n_{\mathbb{C}})$ be a morphism. There exists an open affine covering $(\mathcal{V}_i)_{i\in I}$ of V such that for each i there exist an integer d_i and a morphism $\upsilon_i\colon \mathcal{V}_i\to H_{d_i}$ such that $\upsilon_{|\mathcal{V}_i} = \pi_{d_i}\circ\upsilon_i$.*

PROOF. Consider a morphism $\tau\colon V\to\mathrm{Bir}(\mathbb{P}^n_{\mathbb{C}})$ given by a V-birational map

$$\phi\colon V\times\mathbb{P}^n_{\mathbb{C}} \dashrightarrow V\times\mathbb{P}^n_{\mathbb{C}}$$

which restricts to an open immersion on an open set $\mathcal{U}$. Take a point p_0 in V. Let $\mathcal{V}_0\subset V$ be an open affine set containing p_0. Take an element $w_0 = (p_0, y)$ of $\mathcal{U}$. Let us fix homogeneous coordinates $(z_0 : z_1 : \ldots : z_n)$ on $\mathbb{P}^n_{\mathbb{C}}$ such that

- ◇ $y = (1 : 0 : 0 : \ldots : 0)$,
- ◇ $\phi(w_0)$ does not belong to the plane $z_0 = 0$.

Let us denote by $\mathbb{A}^n_{\mathbb{C}}\subset\mathbb{P}^n_{\mathbb{C}}$ the affine set where $z_0 = 1$;

$$x_1 = \frac{z_1}{z_0} \qquad x_2 = \frac{z_2}{z_0} \qquad \ldots \qquad x_n = \frac{z_n}{z_0}$$

are natural affine coordinates of $\mathbb{A}^n_{\mathbb{C}}$. The map ϕ restricts to a rational map of $\mathcal{V}_0\times\mathbb{P}^n_{\mathbb{C}}$ defined at w_0. Its composition with the projection on the ith coordinate is a rational function on $\mathcal{V}_0\times\mathbb{A}^n_{\mathbb{C}}$ defined at w_0. Hence $\phi_{|\mathcal{V}_0\times\mathbb{A}^n_{\mathbb{C}}}$ can be written in a neighborhood of w_0 as

$$(v, x_1, x_2, \ldots, x_n) \mapsto \left(\frac{R_1}{Q_1}, \frac{R_2}{Q_2}, \ldots, \frac{R_n}{Q_n}\right)$$

for some R_i, Q_i in $\mathbb{C}[V][x_1, x_2, \ldots, x_n]$ such that none of the Q_i vanish at w_0. As a result ϕ is given in a neighborhood of w_0 by

$$(v, (z_0 : z_1 : \ldots : z_n)) \mapsto (P_0 : P_1 : \ldots : P_n)$$

where the $P_i\in\mathbb{C}[\mathcal{V}_0][z_0, z_1, \ldots, z_n]$ are homogeneous polynomials of the same degree d_0 such that not all vanish at w_0. Denote by $\mathcal{U}_0$ the set of points of $(V\times\mathbb{P}^n_{\mathbb{C}})\cap\mathcal{U}$ where at least one of the P_i does not vanish; $\mathcal{U}_0$ is an open subset of $V\times\mathbb{P}^n_{\mathbb{C}}$. Its projection $\mathrm{pr}_1(\mathcal{U}_0)$ on V is an open subset of $\mathcal{V}_0$ containing p_0. There thus exists an affine open subset $\widetilde{A_0}\subseteq\mathrm{pr}_1(\mathcal{U}_0)$ containing p_0. The n-uple $(P_0, P_1, \ldots, P_n)$ yields to a morphism $\upsilon_0\colon\widetilde{A_0}\to H_d$. By construction $\upsilon_{|\widetilde{A_0}} = \pi_d\circ\upsilon_0$. If we repeat this process for any point of V we get an affine covering. □

Lemma 3.7 implies the following one:

COROLLARY 3.8 ([**34**]).

- ◇ *A set $S\subseteq\mathrm{Bir}(\mathbb{P}^n_{\mathbb{C}})$ is closed if and only if $\pi_d^{-1}(S)$ is closed in H_d for any $d\geq 1$.*
- ◇ *For any d,the set $\mathrm{Bir}_{\leq d}(\mathbb{P}^n_{\mathbb{C}})$ is closed in $\mathrm{Bir}(\mathbb{P}^n_{\mathbb{C}})$.*
- ◇ *For any d, the map $\pi_d\colon H_d\to\mathrm{Bir}_{\leq d}(\mathbb{P}^n_{\mathbb{C}})$ is surjective, continuous and closed. In particular it is a topological quotient map.*

PROOF. Let us prove the first assertion. Assume that S is closed in $\mathrm{Bir}(\mathbb{P}^n_{\mathbb{C}})$. Recall that a subset of $\mathrm{Bir}(\mathbb{P}^n_{\mathbb{C}})$ is closed in $\mathrm{Bir}(\mathbb{P}^n_{\mathbb{C}})$ if and only if its preimage by any morphism is closed. Since any $\pi_d\colon H_d \to \mathrm{Bir}(\mathbb{P}^n_{\mathbb{C}})$ is a morphism, $\pi_d^{-1}(S)$ is thus closed in H_d.

Conversely, suppose that $\pi_d^{-1}(S)$ is closed in H_d for any d. Let V be an irreducible algebraic variety, and let $\upsilon\colon V \to \mathrm{Bir}(\mathbb{P}^n_{\mathbb{C}})$ be a morphism. According to Lemma 3.7 there exists an open affine covering $(\mathcal{V}_i)_{i\in I}$ of V such that for any i there exist

- $\diamond$ an integer d_i,
- $\diamond$ a morphism $\upsilon_i\colon \mathcal{V}_i \to H_{d_i}$,

with $\upsilon_{|\mathcal{V}_i} = \pi_{d_i} \circ \upsilon_i$. As $\pi_{d_i}^{-1}(S)$ is closed and $\upsilon^{-1}(S) \cap \mathcal{V}_i = \upsilon_i^{-1}(\pi_{d_i}^{-1}(F))$ one gets that $\upsilon^{-1}(S) \cap \mathcal{V}_i$ is closed in $\mathcal{V}_i$ for any i. As a result, $\upsilon^{-1}(S)$ is closed.

We will now prove the second assertion. According to the first assertion it suffices to prove that

$$\pi_\ell^{-1}\big(\mathrm{Bir}_{\leq d}(\mathbb{P}^n_{\mathbb{C}})\big) = \pi_\ell^{-1}(\pi_d(H_d))$$

is closed in H_ℓ for any ℓ. This follows from Lemma 3.5.

Finally let us prove the third assertion. The surjectivity follows from the construction of H_d and π_d (*see* [**34**]). Since π_d is a morphism, π_d is continuous. Let $S \subseteq H_d$ be a closed subset. According to Lemma 3.5 the set $\pi_\ell^{-1}(\pi_d(S))$ is closed in H_ℓ for any ℓ. The first assertion allows us to conclude. □

The first and third assertions of Corollary 3.8 imply:

PROPOSITION 3.9 ([**34**]). *The Zariski topology of* $\mathrm{Bir}(\mathbb{P}^n_{\mathbb{C}})$ *is the inductive limit topology given by the Zariski topologies of* $\mathrm{Bir}_{\leq d}(\mathbb{P}^n_{\mathbb{C}})$, $d \in \mathbb{N}$, *which are the quotient topology of*

$$\pi_d\colon H_d \to \mathrm{Bir}_{\leq d}(\mathbb{P}^n_{\mathbb{C}})$$

where H_d *is endowed with its Zariski topology.*

3.1.2. Algebraic subgroups. An *algebraic subgroup* of $\mathrm{Bir}(\mathbb{P}^n_{\mathbb{C}})$ is a subgroup $\mathrm{G} \subset \mathrm{Bir}(\mathbb{P}^n_{\mathbb{C}})$ which is the image of an algebraic group H by a homomorphism υ such that $\upsilon\colon \mathrm{H} \to \mathrm{Bir}(\mathbb{P}^n_{\mathbb{C}})$ is a morphism; by Lemma 3.7 any algebraic group is contained in some $\mathrm{Bir}_{\leq d}(\mathbb{P}^n_{\mathbb{C}})$, *i.e.*, any algebraic group has *bounded degree*. Corollary 3.11 allows us to give a more intrinsic definition of algebraic groups which corresponds to taking closed subgroups of $\mathrm{Bir}(\mathbb{P}^n_{\mathbb{C}})$ of bounded degree. Lemma 3.12 shows that these two definitions agree.

PROPOSITION 3.10 ([**34**]). *Let* G *be a subgroup of* $\mathrm{Bir}(\mathbb{P}^n_{\mathbb{C}})$. *Assume that*

- $\diamond$ G *is closed for the Zariski topology;*
- $\diamond$ G *is connected for the Zariski topology;*
- $\diamond$ $\mathrm{G} \subset \mathrm{Bir}_{\leq d}(\mathbb{P}^n_{\mathbb{C}})$ *for some integer* d.

If d *is choosen minimal, then the set* $(\pi_d)^{-1}\big(\mathrm{G} \cap \mathrm{Bir}_d(\mathbb{P}^n_{\mathbb{C}})\big)$ *is non-empty. Let us denote by* K *the closure of* $(\pi_d)^{-1}\big(\mathrm{G} \cap \mathrm{Bir}_d(\mathbb{P}^n_{\mathbb{C}})\big)$ *in* H_d. *Then*

- $\diamond$ π_d *induces a homeomorphism* $\mathrm{K} \to \mathrm{G}$;
- $\diamond$ *if* V *is an irreducible algebraic variety, the morphisms* $V \to \mathrm{Bir}(\mathbb{P}^n_{\mathbb{C}})$ *having image in* G *correspond, via* π_d, *to the morphisms of algebraic varieties* $V \to \mathrm{K}$;

- *the liftings to* K *of the maps*

$$\mathrm{G}\times\mathrm{G}\to\mathrm{G},\ (\varphi,\psi) \qquad\qquad \mathrm{G}\to\mathrm{G},\ \varphi\mapsto\varphi^{-1}$$

give rise to morphisms of algebraic varieties $\mathrm{K}\times\mathrm{K}\to\mathrm{K}$ *and* $\mathrm{K}\to\mathrm{K}$.

This gives G *a unique structure of the algebraic group.*

COROLLARY 3.11 ([**34**]). *Let* G *be a subgroup of* $\mathrm{Bir}(\mathbb{P}^n_{\mathbb{C}})$. *Assume that* G *is*

- *closed for the Zariski topology,*
- *of bounded degree.*

Then there exist an algebraic group K *together with a morphism* $\mathrm{K}\to\mathrm{Bir}(\mathbb{P}^n_{\mathbb{C}})$ *inducing a homeomorphism* $\pi\colon\mathrm{K}\to\mathrm{G}$ *such that:*

- π *is a group homomorphism,*
- *and for any irreducible algebraic variety* V *the morphisms* $V\to\mathrm{Bir}(\mathbb{P}^n_{\mathbb{C}})$ *having their image in* G *correspond, via* π, *to the morphisms of algebraic varieties* $V\to\mathrm{K}$.

PROOF. Let us first prove that G has a finite number of irreducible components. The group G is closed in $\mathrm{Bir}_{\leq d}(\mathbb{P}^n_{\mathbb{C}})$; hence its preimage $(\pi_d)^{-1}(\mathrm{G})$ is also closed in H_d. It thus has a finite number of irreducible components C_1, C_2, $\ldots$, C_r. The sets $\pi_d(C_1)$, $\pi_d(C_2)$, $\ldots$, $\pi_d(C_r)$ are closed and irreducible and cover G (the third assertion of Corollary 3.8). If we keep the maximal ones with respect to inclusion we get the irreducible components of G.

As for algebraic groups ([**111**, §7.3]) one can show that:

- exactly one irreducible component of G passes through id;
- this irreducible component is a closed normal subgroup of finite index in G whose cosets are the connected as well as irreducible components of G.

This allows us to reduce to the connected case; Proposition 3.10 allows us to conclude. □

LEMMA 3.12 ([**34**]). *Let* A *be an algebraic group and let* $\rho\colon\mathrm{A}\to\mathrm{Bir}(\mathbb{P}^n_{\mathbb{C}})$ *be a morphism that is also a group homomorphism.*

Then the image G *of* A *is a closed subgroup of* $\mathrm{Bir}(\mathbb{P}^n_{\mathbb{C}})$ *of bounded degree.*

If $\pi\colon\mathrm{K}\to\mathrm{G}$ *is the homeomorphism constructed in Corollary* 3.11, *there exists a unique morphism of algebraic groups* $\widetilde{\rho}\colon\mathrm{A}\to\mathrm{K}$ *such that* $\rho=\pi\circ\widetilde{\rho}$.

PROOF. Lemma 3.7 asserts that $\mathrm{G}=\rho(\mathrm{A})$ has bounded degree. The closure $\overline{\mathrm{G}}$ of G is a subgroup of $\mathrm{Bir}(\mathbb{P}^n_{\mathbb{C}})$; indeed inversion being a homeomorphism $\overline{\mathrm{G}}^{-1}=\overline{\mathrm{G}^{-1}}=\overline{\mathrm{G}}$. Similarly translation by $g\in\mathrm{G}$ is a homeomorphism; thus $g\overline{\mathrm{K}}=\overline{g\mathrm{K}}=\overline{\mathrm{K}}$, that is, $\mathrm{G}\overline{\mathrm{G}}\subset\overline{\mathrm{G}}$. In turn, if $g\in\overline{\mathrm{G}}$, then $\mathrm{G}g\subset\overline{\mathrm{G}}$, so $\overline{\mathrm{G}}g=\overline{\mathrm{G}g}\subset\overline{\mathrm{G}}$. As a result, $\overline{\mathrm{G}}$ is a subgroup of $\mathrm{Bir}(\mathbb{P}^n_{\mathbb{C}})$.

According to Corollary 3.11 there exist a canonical homeomorphism $\mathrm{K}\to\overline{\mathrm{G}}$ where K is an algebraic group, and a lift $\widetilde{\rho}\colon\mathrm{A}\to\mathrm{H}$ of the morphism $\rho\colon\mathrm{A}\to\mathrm{Bir}(\mathbb{P}^n_{\mathbb{C}})$ whose image is contained in $\overline{\mathrm{G}}$. As ρ is a group homomorphism, $\widetilde{\rho}$ is a morphism of algebraic groups; hence its image is closed, so $\mathrm{im}\,\rho=\mathrm{K}$. Therefore, $\overline{\mathrm{G}}=\mathrm{G}$. □

PROPOSITION 3.13 ([**34**]). *Any algebraic subgroup of* $\mathrm{Bir}(\mathbb{P}^n_{\mathbb{C}})$ *is affine.*

SKETCH OF THE PROOF. Let G be an algebraic subgroup of $\mathrm{Bir}(\mathbb{P}^n_{\mathbb{C}})$. One can show that G is linear, and this reduces to the connected case. By the regularization theorem of Weil (*see* §3.5) the group G acts by automorphisms on some (smooth) rational variety V. Assume that $\alpha_V\colon V\to A(V)$ is the Albanese morphism. According to the Nishi-Matsumura theorem, the induced action of G on $A(V)$ factors through a morphism $A(\mathrm{G})\to A(V)$ with finite kernel (*see* for instance [**46**]). But V is rational; hence $A(V)$ is trivial and so does $A(\mathrm{G})$. The structure theorem of Chevalley asserts that G is affine (*see for instance* [**158**]). □

Let us finish with some examples:

- ◇ The Cremona group in one variable $\mathrm{Bir}(\mathbb{P}^1_\mathbb{C})$ coincides with the group of linear projective transformations $\mathrm{PGL}(2,\mathbb{C})$; it is an algebraic group of dimension 3.
- ◇ In dimension 2 the Cremona group contains the two following algebraic subgroups:
  - the group $\mathrm{PGL}(3,\mathbb{C})$ of automorphisms of $\mathbb{P}^2_\mathbb{C}$;
  - the group $\mathrm{PGL}(2,\mathbb{C})\times\mathrm{PGL}(2,\mathbb{C})$ obtained as follows: the surface $\mathbb{P}^1_\mathbb{C}\times\mathbb{P}^1_\mathbb{C}$ can be considered as a smooth quadric in $\mathbb{P}^3_\mathbb{C}$ whose automorphism group contains $\mathrm{PGL}(2,\mathbb{C})\times\mathrm{PGL}(2,\mathbb{C})$; by stereographic projection the quadric is birationally equivalent to $\mathbb{P}^2_\mathbb{C}$. Hence $\mathrm{Bir}(\mathbb{P}^2_\mathbb{C})$ also contains a copy of $\mathrm{PGL}(2,\mathbb{C})\times\mathrm{PGL}(2,\mathbb{C})$.
- ◇ More generally, $\mathrm{Aut}(\mathbb{P}^n_\mathbb{C})=\mathrm{PGL}(n+1,\mathbb{C})$ is an algebraic subgroup of $\mathrm{Bir}(\mathbb{P}^n_\mathbb{C})$ and

$$\underbrace{\mathrm{PGL}(2,\mathbb{C})\times\mathrm{PGL}(2,\mathbb{C})\times\ldots\times\mathrm{PGL}(2,\mathbb{C})}_{n\text{ times}}$$

  is an algebraic subgroup of

$$\mathrm{Aut}(\underbrace{\mathbb{P}^1_\mathbb{C}\times\mathbb{P}^1_\mathbb{C}\times\ldots\mathbb{P}^1_\mathbb{C}}_{n\text{ times}})\subset\mathrm{Bir}(\mathbb{P}^n_\mathbb{C}).$$

- ◇ If G is a semi-simple algebraic group, H is a parabolic subgroup of G and $V={}^{\mathrm{G}}\!/_{\mathrm{H}}$, then the homogeneous variety V of dimension n is rational; $\pi\circ\mathrm{G}\circ\pi^{-1}$ determines an algebraic subgroup of $\mathrm{Bir}(\mathbb{P}^n_\mathbb{C})$ for any birational map $\pi\colon V\dashrightarrow\mathbb{P}^n_\mathbb{C}$.

3.1.3. Euclidean topology. We can put the Euclidean topology on a complex algebraic group; this gives any algebraic group the structure of a topological group. Recall that the Euclidean topology is finer than the Zariski one.

Let $n\geq 1$ be an integer. The group $\mathrm{Bir}(\mathbb{P}^1_\mathbb{C})=\mathrm{Aut}(\mathbb{P}^2_\mathbb{C})=\mathrm{PGL}(2,\mathbb{C})$ is obviously a topological group. Assume now that $n\geq 2$; we will

- ◇ first, define the Euclidean topology on $\mathrm{Bir}_{\leq d}(\mathbb{P}^n_\mathbb{C})$ and show that the natural inclusion $\mathrm{Bir}_{\leq d}(\mathbb{P}^n_\mathbb{C})\hookrightarrow\mathrm{Bir}_{\leq d+1}(\mathbb{P}^n_\mathbb{C})$ is a closed embedding;
- ◇ second, define the Euclidean topology on $\mathrm{Bir}(\mathbb{P}^n_\mathbb{C})$ as the inductive limit topology induced by those of $\mathrm{Bir}_{\leq d}(\mathbb{P}^n_\mathbb{C})$, that is a subset $F\subset\mathrm{Bir}(\mathbb{P}^n_\mathbb{C})$ is closed if and only if $F\cap\mathrm{Bir}_{\leq d}(\mathbb{P}^n_\mathbb{C})$ is closed in $\mathrm{Bir}_{\leq d}(\mathbb{P}^n_\mathbb{C})$ for each d. Finally, we will prove that $\mathrm{Bir}(\mathbb{P}^n_\mathbb{C})$ endowed with the Euclidean topology is a topological group;
- ◇ third, give some remarks and properties.

3.1.4. The Euclidean topology on $\mathrm{Bir}_{\leq d}(\mathbb{P}^n_\mathbb{C})$. Let us recall that W_d is a projective space and H_d is locally closed in W_d for the Zariski topology (Lemma 3.5).

Let us put the Euclidean topology on W_d: the distance between $(p_0 : p_1 : \ldots : p_n)$ and $(q_0 : q_1 : \ldots : q_n)$ is (*see* [**176**])

$$\frac{\sum_{i<j} |p_i q_j - p_j q_i|^2}{\left(\sum_i |p_i|^2\right)\left(\sum_i |q_i|^2\right)}.$$

We then put the induced topology on H_d. The behavior of the Zariski topology on $\mathrm{Bir}(\mathbb{P}^n_{\mathbb{C}})$ leads to:

DEFINITION. The *Euclidean topology* on $\mathrm{Bir}_{\leq d}(\mathbb{P}^n_{\mathbb{C}})$ is the quotient topology induced by the surjective map $\pi_d \colon H_d \to \mathrm{Bir}_{\leq d}(\mathbb{P}^n_{\mathbb{C}})$ where we put the Euclidean topology on H_d.

Recall that if $f \colon X \to Y$ is a quotient map between topological spaces, A is a subspace of X, A is open, and $A = f^{-1}(f(A))$, then the induced map $A \to f(A)$ is a quotient map ([**44**, Chapter I, §3.6]). Set

$$H_{d,d} = (\pi_d)^{-1}(\mathrm{Bir}_d(\mathbb{P}^n_{\mathbb{C}})).$$

As $(\pi_d)^{-1}(\mathrm{Bir}_{\leq d-1}(\mathbb{P}^n_{\mathbb{C}}))$ is closed in H_d, $H_{d,d}$ is open in H_d for the Zariski topology and hence also for the Euclidean topology; π_d restricts to a homeomorphism $H_{d,d} \to \mathrm{Bir}_d(\mathbb{P}^n_{\mathbb{C}})$ for any $d \geq 1$.

LEMMA 3.14 ([**34**]). *Let $d \geq 1$ be an integer. The spaces W_d and H_d are locally compact metric spaces endowed with the Euclidean topology.*

In particular, the sets W_d, H_d, and $\mathrm{Bir}_{\leq d}(\mathbb{P}^n_{\mathbb{C}})$ are sequential spaces: a subset F is closed if the limit of every convergent sequence with values in F belongs to F.

PROOF. The construction of the topology implies that W_d and H_d are metric spaces. As W_d is compact and H_d is locally closed in W_d (Lemma 3.5), the set H_d is locally compact. But metric spaces are sequential spaces and quotients of sequential spaces are sequential ([**95**]). □

We now would like to prove that the topological map $\pi_d \colon H_d \to \mathrm{Bir}_{\leq d}(\mathbb{P}^n_{\mathbb{C}})$ is proper and the topological space $\mathrm{Bir}_{\leq d}(\mathbb{P}^n_{\mathbb{C}})$ is locally compact. Recall that a map $f \colon X \to Y$ between two topological spaces is *proper* if it is continuous and universally closed: for each topological space Z the map $f \times \mathrm{id}_Z \colon X \times Z \to Y \times Z$ is closed ([**44**]). A topological space is *locally compact* if it is Hausdorff and if each of its points has a compact neighborhood. If $f \colon X \to Y$ is a quotient map between topological spaces such that X is locally compact, then f is proper if and only if it is closed and the preimages of points are compact. This implies furthermore that Y is locally compact. According to Lemma 3.5 for any ϕ in $\mathrm{Bir}_{\leq d}(\mathbb{P}^n_{\mathbb{C}})$ the set $(\pi_d)^{-1}(\phi)$ is closed in the compact space W_d, so $(\pi_d)^{-1}(Y)$ is compact. The topological space H_d being locally compact (Lemma 3.14), to prove that π_d is proper it suffices to prove that π_d is closed.

CLAIM 3.15. *The map $\pi_d \colon H_d \to \mathrm{Bir}_{\leq d}(\mathbb{P}^n_{\mathbb{C}})$ is proper.*

PROOF. Let $F \subset H_d$ be a closed subset. To prove that $\pi_d(F)$ is closed in $\mathrm{Bir}_{\leq d}(\mathbb{P}^n_{\mathbb{C}})$ amounts to proving that the saturated set $\widehat{F} = (\pi_d)^{-1}(\pi_d(F))$ is closed in H_d. Consider a sequence $(\varphi_i)_{i \in \mathbb{N}}$ of elements in $\widehat{F}$ which converges to $\varphi \in H_d$. Let us show that φ belongs to $\widehat{F}$. Since π_d is by construction continuous, the sequence $\big(\pi_d(\varphi_i)\big)_{i\in\mathbb{N}}$ converges to $\pi_d(\varphi)$ in $\mathrm{Bir}_{\leq d}(\mathbb{P}^n_\mathbb{C})$. Taking a subsequence of $\big(\pi_d(\varphi_i)\big)_{i\in\mathbb{N}}$ if needed, we may suppose that the degree of all $\pi_d(\varphi_i)$ is constant equal to some $m \leq d$.

- ⋄ Assume $m = d$, then $(\pi_d)^{-1}(\pi_d(\varphi_i)) = \{\varphi_i\}$ for each i. As a result each φ_i belongs to F, so φ belongs to $F \subset \widehat{F}$ as wanted.
- ⋄ Suppose $m < d$. Set $k = d - m \geq 1$. For any i there exists a non-zero homogeneous polynomial $a_i \in \mathbb{C}[z_0, z_1, \ldots, z_n]$ of degree k such that

$$\varphi_i = \big(a_i f_{i,0} : a_i f_{i,1} : \ldots : a_i f_{i,n}\big)$$

and $(f_{i,0} : f_{i,1} : \ldots : f_{i,n}) \in W_m$ corresponds to a birational map of degree $m < d$. Each a_i is defined up to a constant and $\mathbb{P}(\mathbb{C}[z_0, z_1, \ldots, z_n])$ is compact, so, taking a subsequence if needed, we can suppose that $(a_i)_{i\in\mathbb{N}}$ converges to a non-zero homogeneous polynomial $a \in \mathbb{C}[z_0, z_1, \ldots, z_n]$ of degree k.

Taking a subsequence if needed we can assume that $\{(f_{i,0} : f_{i,1} : \ldots : f_{i,n})\}_{i\in\mathbb{N}}$ converges to an element $(f_0 : f_1 : \ldots : f_n)$ of the projective space W_m. Since $(\varphi_i)_{i\in\mathbb{N}}$ converges to φ we get that

$$\varphi = (af_0 : af_1 : \ldots : af_n) \in H_d.$$

As φ_i belongs to $\widehat{F} = (\pi_d)^{-1}(\pi_d(F))$ for any i, there exists φ_i' in F such that $\pi_d(\varphi_i') = \pi_d(\varphi_i)$. Consequently

$$\varphi_i' = \big(b_i f_{i,0} : b_i f_{i,1} : \ldots : b_i f_{i,n}\big)$$

for some non-zero homogeneous polynomial $b_i \in \mathbb{C}[z_0, z_1, \ldots, z_n]$ of degree k. As before we can assume that $(b_i)_{i\in\mathbb{N}}$ converges to a non-zero homogeneous polynomial $b \in \mathbb{C}[z_0, z_1, \ldots, z_n]$ of degree k. The sequence $(\varphi_i')_{i\in\mathbb{N}}$ converges to $(bf_0 : bf_1 : \ldots : bf_n)$ and F is closed, thus $(bf_0 : bf_1 : \ldots : bf_n)$ belongs to F. This implies that $\varphi = (af_0 : af_1 : \ldots : af_n)$ belongs to $\widehat{F}$. □

We can thus state:

LEMMA 3.16 ([**34**]). *Let $d \geq 1$ be an integer. Then*

- ⋄ *the topological map $\pi_d \colon H_d \to \mathrm{Bir}_{\leq d}(\mathbb{P}^n_\mathbb{C})$ is proper (and closed);*
- ⋄ *the topological space $\mathrm{Bir}_{\leq d}(\mathbb{P}^n_\mathbb{C})$ is locally compact (and Hausdorff).*

LEMMA 3.17 ([**34**]). *Let $d \geq 0$ be an integer. The natural injection*

$$\iota_d \colon \mathrm{Bir}_{\leq d}(\mathbb{P}^n_\mathbb{C}) \to \mathrm{Bir}_{\leq d+1}(\mathbb{P}^n_\mathbb{C})$$

is a closed embedding, that is, a homeomorphism onto its image which is closed in $\mathrm{Bir}_{\leq d+1}(\mathbb{P}^n_\mathbb{C})$.

PROOF. Consider the map

$$\widehat{\iota_d} \colon H_d \to H_{d+1}, \qquad (f_0 : f_1 : \ldots : f_n) \mapsto (z_0f_0 : z_0f_1 : \ldots : z_0f_n).$$

It is a morphism of algebraic varieties that is a closed immersion. As a result it is continuous and closed with respect to the Euclidean topology. The diagram

$$\begin{array}{ccc} H_d & \xrightarrow{\widehat{\iota_d}} & H_{d+1} \\ \downarrow{\scriptstyle \pi_d} & & \downarrow{\scriptstyle \pi_{d+1}} \\ \mathrm{Bir}_{\leq d}(\mathbb{P}^n_{\mathbb{C}}) & \xrightarrow[\iota_d]{} & \mathrm{Bir}_{\leq d+1}(\mathbb{P}^n_{\mathbb{C}}) \end{array}$$

commutes.

The continuity of $\widehat{\iota_d}$ implies the continuity of ι_d: let $\mathcal{U}$ be an open subset of $\mathrm{Bir}_{\leq d+1}(\mathbb{P}^n_{\mathbb{C}})$; the equality $(\pi_d)^{-1}((\iota_d)^{-1}(\mathcal{U})) = (\pi_{d+1} \circ \widehat{\iota_d})^{-1}(\mathcal{U})$ shows that $(\pi_d)^{-1}((\iota_d)^{-1}(\mathcal{U}))$ is open in H_d, that is, $(\iota_d)^{-1}(\mathcal{U})$ is open in $\mathrm{Bir}_{\leq d}(\mathbb{P}^n_{\mathbb{C}})$. □

3.1.5. The Euclidean topology on $\mathrm{Bir}(\mathbb{P}^n_{\mathbb{C}})$**.** Thanks to Lemma 3.17 one can put on $\mathrm{Bir}(\mathbb{P}^n_{\mathbb{C}})$ the inductive limit topology given by the $\mathrm{Bir}_{\leq d}(\mathbb{P}^n_{\mathbb{C}})$: a subset of $\mathrm{Bir}(\mathbb{P}^n_{\mathbb{C}})$ is closed (resp. open) if and only if its intersection with any $\mathrm{Bir}_{\leq d}(\mathbb{P}^n_{\mathbb{C}})$ is closed (resp. open). In particular the injections $\mathrm{Bir}_{\leq d}(\mathbb{P}^n_{\mathbb{C}}) \hookrightarrow \mathrm{Bir}(\mathbb{P}^n_{\mathbb{C}})$ are closed embeddings. This topology is called the *Euclidean topology* of $\mathrm{Bir}(\mathbb{P}^n_{\mathbb{C}})$. Let us now prove that $\mathrm{Bir}(\mathbb{P}^n_{\mathbb{C}})$ is a topological group endowed with the Euclidean topology.

LEMMA 3.18 ([**34**]). *Let $d \geq 1$ be an integer. The map*

$$\mathcal{I}_d \colon \mathrm{Bir}_{\leq d}(\mathbb{P}^n_{\mathbb{C}}) \to \mathrm{Bir}_{\leq d^{n-1}}(\mathbb{P}^n_{\mathbb{C}}), \qquad \phi \mapsto \phi^{-1}$$

is continuous.

PROOF. As in Lemma 3.5 we consider the set $Y \subset W_{d^{n-1}} \times W_d$ defined by

$$Y = \big\{(\varphi, \phi) \in W_{d^{n-1}} \times W_d \,|\, \varphi \circ \phi = P\,\mathrm{id} \text{ for some } P \in \mathbb{C}[z_0, z_1, \ldots, z_n]_d\big\}.$$

Let $\mathcal{U} \subset W_d$ (resp. $\mathcal{U}' \subset W_{d^{n-1}}$) be the set of elements having a non-zero Jacobian. The set Y is closed in $W_{d^{n-1}} \times W_d$ (see the proof of Lemma 3.5) and the set $\mathcal{U}$ is open in W_d. As a consequence

$$L = Y \cap (W_{d^{n-1}} \times \mathcal{U}) = Y \cap (\mathcal{U}' \times \mathcal{U})$$

is locally closed in the algebraic variety $W_{d^{n-1}} \times W_d$.

The projection on the first factor is a morphism $\eta_1 \colon L \to H_{d^{n-1}}$ which is not surjective in general. The projection on the second factor induces a surjective morphism $\eta_2 \colon L \to H_d$. By construction, the diagram

$$\begin{array}{ccccc} H_d & \xleftarrow{\eta_2} & L \subset W_{d^{n-1}} \times W_d & \xrightarrow{\eta_1} & H_{d^{n-1}} \\ \downarrow{\scriptstyle \pi_d} & & & & \downarrow{\scriptstyle \pi_{d-1}} \\ \mathrm{Bir}_{\leq d}(\mathbb{P}^n_{\mathbb{C}}) & & \xrightarrow[\mathcal{I}_d]{} & & \mathrm{Bir}_{\leq d^{n-1}}(\mathbb{P}^n_{\mathbb{C}}) \end{array}$$

commutes.

Let us prove that η_2 is a closed map for the Euclidean topology. The set $W_{d^{n-1}}$ is compact, so the second projection $W_{d^{n-1}} \times W_d \to W_d$ is a closed map. Its restriction $\eta_2' \colon Y \to W_d$ to the closed subset Y of $W_{d^{n-1}} \times W_d$ is a closed map. Since $L = (\eta_2')^{-1}(H_d)$, we get that η_2 is a closed map[1].

---

[1]Let us recall that if $\varphi \colon A \to B$ is a continuous closed map between topological spaces and C is any subset of B, then φ induces a continuous closed map $\varphi^{-1}(C) \to C$.

As the diagram is commutative for any $F \subset \mathrm{Bir}_{\leq d^{n-1}}(\mathbb{P}^n_{\mathbb{C}})$ we have

$$\eta_2\big((\pi_{d^{n-1}} \circ \eta_1)^{-1}(F)\big) = (\mathcal{I}_d \circ \pi_d)^{-1}(F);$$

furthermore this set corresponds to elements $(\phi_0 : \phi_1 : \ldots : \phi_n) \in W_d$ such that the rational map ψ_ϕ is the inverse of an element of F. Assume that F is closed in $\mathrm{Bir}_{\leq d^{n-1}}(\mathbb{P}^n_{\mathbb{C}})$. The maps η_1 and $\pi_{d^{n-1}}$ are continuous for the Euclidean topology; hence $(\pi_{d^{n-1}} \circ \eta_1)^{-1}(F)$ is closed in L. Lemma 3.16 asserts that

$$\pi_d^{-1}(\mathcal{I}_d^{-1}(F)) = \eta_2\big((\pi_{d^{n-1}} \circ \eta_1)^{-1}(F)\big)$$

is closed in H_d and $\mathcal{I}_d^{-1}(F)$ is closed in $\mathrm{Bir}_{\leq d}(\mathbb{P}^n_{\mathbb{C}})$. □

Let us introduce the map $\mathcal{I}$ defined by

$$\mathcal{I}\colon \mathrm{Bir}(\mathbb{P}^n_{\mathbb{C}}) \to \mathrm{Bir}(\mathbb{P}^n_{\mathbb{C}}), \qquad \phi \mapsto \phi^{-1}.$$

The degree of the inverse of a birational self map of $\mathbb{P}^n_{\mathbb{C}}$ of degree d has degree at most d^{n-1}. Consequently for any $d \geq 1$ the map $\mathcal{I}$ restricts to an injective map

$$\mathcal{I}_d\colon \mathrm{Bir}_{\leq d}(\mathbb{P}^n_{\mathbb{C}}) \to \mathrm{Bir}_{\leq d^{n-1}}(\mathbb{P}^n_{\mathbb{C}}).$$

According to Lemma 3.18 the map $\mathcal{I}_d$ is continuous. The definition of the topology of $\mathrm{Bir}(\mathbb{P}^n_{\mathbb{C}})$ implies that $\mathcal{I}$ is continuous. Since $\mathcal{I} = \mathcal{I}^{-1}$ one has:

COROLLARY 3.19 ([**34**]). *The map*

$$\mathcal{I}\colon \mathrm{Bir}(\mathbb{P}^n_{\mathbb{C}}) \to \mathrm{Bir}(\mathbb{P}^n_{\mathbb{C}}), \qquad \phi \mapsto \phi^{-1}$$

is a homeomorphism.

Let us now look at the composition of two birational maps.

LEMMA 3.20 ([**34**]). *For any d, $k \geq 1$, the map*

$$\chi_{d,k}\colon \mathrm{Bir}_{\leq d}(\mathbb{P}^n_{\mathbb{C}}) \times \mathrm{Bir}_{\leq k}(\mathbb{P}^n_{\mathbb{C}}) \to \mathrm{Bir}_{\leq dk}(\mathbb{P}^n_{\mathbb{C}}), \qquad (\phi, \psi) \mapsto \phi \circ \psi$$

is continuous.

PROOF. Let us consider the map $\widehat{\chi_{d,k}}\colon H_d \times H_k \to H_{dk}$ given by

$$\big((\phi_0 : \phi_1 : \ldots : \phi_n), (\psi_0 : \psi_1 : \ldots : \psi_n)\big) \mapsto \big(\phi_n(\psi_0, \psi_1, \ldots, \psi_n)\big) : \ldots : \phi_n(\psi_0, \psi_1, \ldots, \psi_n)\big).$$

The diagram

$$\begin{array}{ccc} H_d \times H_k & \xrightarrow{\ \widehat{\chi_{d,k}}\ } & H_{dk} \\ \Big\downarrow{\scriptstyle \pi_d \times \pi_k} & & \Big\downarrow{\scriptstyle \pi_{dk}} \\ \mathrm{Bir}_{\leq d}(\mathbb{P}^n_{\mathbb{C}}) \times \mathrm{Bir}_{\leq k}(\mathbb{P}^n_{\mathbb{C}}) & \xrightarrow[\ \mathcal{I}_d\ ]{} & \mathrm{Bir}_{\leq dk}(\mathbb{P}^n_{\mathbb{C}}) \end{array}$$

commutes.

The map $\widehat{\chi_{d,k}}$ is a morphism of algebraic varieties, so is continuous for the Euclidean topology. Therefore, if F is a closed subset of $\mathrm{Bir}_{\leq dk}(\mathbb{P}^n_{\mathbb{C}})$, then $(\pi_{dk} \circ \widehat{\chi_{d,k}})^{-1}(F)$ is closed in $H_d \times H_k$. But the diagram is commutative, so

$$(\pi_d \circ \widehat{\chi_{d,k}})(F) = (\pi_d \times \pi_k)^{-1}\big((\chi_{d,k})^{-1}(F)\big).$$

The product of two proper maps is proper ([**44**, Chapter 1,§10.1]); as a consequence $\pi_d \times \pi_k$ is proper and hence closed. This implies that $\pi_d \times \pi_k$ is a quotient map. Hence $(\chi_{d,k})^{-1}(F)$ is closed and $\chi_{d,k}$ is continuous. □

According to Lemma 3.20 the map

$$\chi_{d,k}\colon \mathrm{Bir}_{\leq d}(\mathbb{P}^n_\mathbb{C}) \times \mathrm{Bir}_{\leq k}(\mathbb{P}^n_\mathbb{C}) \to \mathrm{Bir}_{\leq dk}(\mathbb{P}^n_\mathbb{C})$$

is continuous for each d, $k \geq 1$. As a consequence by definition of the topology of $\mathrm{Bir}(\mathbb{P}^n_\mathbb{C})$ we get:

COROLLARY 3.21 ([**34**]). *The map*

$$\mathrm{Bir}(\mathbb{P}^n_\mathbb{C}) \times \mathrm{Bir}(\mathbb{P}^n_\mathbb{C}) \to \mathrm{Bir}(\mathbb{P}^n_\mathbb{C}), \qquad (\phi, \psi) \mapsto \phi \circ \psi$$

is continuous.

Corollaries 3.19 and 3.21 complete the proof of:

THEOREM 3.22 ([**34**]). *The n-dimensional Cremona group endowed with the Euclidean topology is a topological group.*

Let us give a statement about the restriction of the topology on algebraic subgroups:

PROPOSITION 3.23 ([**34**]). *Let* G *be a Zariski closed subgroup of* $\mathrm{Bir}(\mathbb{P}^n_\mathbb{C})$ *of bounded degree, let* K *be its associated algebraic group (Corollary* 3.11*). We put on* G *the restriction of the Euclidean topology of* $\mathrm{Bir}(\mathbb{P}^n_\mathbb{C})$*, and we get the Euclidean topology on* K *via the bijection* $\pi\colon \mathrm{K} \to \mathrm{G}$ *which becomes a homeomorphism.*

3.1.6. Properties of the Euclidean topology of $\mathrm{Bir}(\mathbb{P}^n_\mathbb{C})$.

LEMMA 3.24. *Any convergent sequence of* $\mathrm{Bir}(\mathbb{P}^n_\mathbb{C})$ *has bounded degree.*

PROOF. If the sequence $(\varphi_i)_{i\in\mathbb{N}}$ of $\mathrm{Bir}(\mathbb{P}^n_\mathbb{C})$ converges to φ, then $\{\varphi_i \,|\, i \in \mathbb{N}\} \cup \{\varphi\}$ is compact, so contained in $\mathrm{Bir}_{\leq d}(\mathbb{P}^n_\mathbb{C})$ for some d. □

LEMMA 3.25. *The topological group* $\mathrm{Bir}(\mathbb{P}^n_\mathbb{C})$ *is Hausdorff.*

PROOF. According to [**44**, III, §2.5, Prop. 13] a topological group is Hausdorff if and only if the trivial one-element subgroup is closed. Any point of $\mathrm{Bir}(\mathbb{P}^n_\mathbb{C})$ is closed in some $\mathrm{Bir}_{\leq d}(\mathbb{P}^n_\mathbb{C})$ (Lemma 3.5), so it is closed in $\mathrm{Bir}(\mathbb{P}^n_\mathbb{C})$. As a result, $\mathrm{Bir}(\mathbb{P}^n_\mathbb{C})$ is Hausdorff. □

LEMMA 3.26. *Any compact subset of* $\mathrm{Bir}(\mathbb{P}^n_\mathbb{C})$ *is contained in* $\mathrm{Bir}_{\leq d}(\mathbb{P}^n_\mathbb{C})$ *for some d.*

PROOF. Assume by contradiction that $\mathrm{Bir}(\mathbb{P}^n_\mathbb{C})$ contains a compact subset K such that $(\varphi_i)_{i\in\mathbb{N}}$ is a sequence of elements of K with $\deg \varphi_{i+1} > \deg \varphi_i$ for each i. Let us consider $K' = \{\varphi_i \,|\, i \in \mathbb{N}\}$. On the one hand it is a closed subset of the compact set K; hence it is compact. On the other hand the intersection of any subset of K' with $\mathrm{Bir}_{\leq d}(\mathbb{P}^n_\mathbb{C})$ is closed, so K′ is an infinite set endowed with the discrete topology; in particular it cannot be compact: contradiction. □

LEMMA 3.27. *For $n \geq 2$ the topological space* $\mathrm{Bir}(\mathbb{P}^n_\mathbb{C})$ *is not locally compact.*

PROOF. Let $\mathcal{U} \subset \mathrm{Bir}(\mathbb{P}^n_\mathbb{C})$ be an open neighborhood of the identity. Since any compact subset of $\mathrm{Bir}(\mathbb{P}^n_\mathbb{C})$ is contained in $\mathrm{Bir}_{\leq d}(\mathbb{P}^n_\mathbb{C})$ (Lemma 3.26) for some d, to prove that $\mathcal{U}$ is not contained in any compact subset of $\mathrm{Bir}(\mathbb{P}^n_\mathbb{C})$, it suffices to show that $\mathcal{U}$ contains elements of arbitrarily large degree. For any integers m, $k \geq 1$, let us consider the birational map given in the affine chart $z_0 = 1$ by

$$f_{m,k}\colon (z_1, z_2, \ldots, z_n) \dashrightarrow \left(z_1 + \frac{1}{k} z_2^m, z_2, \ldots, z_n\right).$$

Fixing m we note that the sequence $(f_{m,k})_{k\geq 1}$ converges to the identity; in particular $f_{m,k}$ belongs to $\mathcal{U}$ when k is large enough. □

LEMMA 3.28. *For $n \geq 2$ the topological space* $\mathrm{Bir}(\mathbb{P}^n_\mathbb{C})$ *is not metrisable.*

PROOF. Consider the inclusion

$$\begin{aligned} \mathbb{C}[z_2] &\hookrightarrow \mathrm{Aut}(\mathbb{C}^n) \subset \mathrm{Bir}(\mathbb{P}^n_\mathbb{C}), \\ P &\hookrightarrow \big((z_1, z_2, \ldots, z_n) \dashrightarrow (z_1 + P(z_2), z_2, z_3, \ldots, z_n)\big). \end{aligned}$$

Observe that $\mathbb{C}[z_2]$ is closed in $\mathrm{Bir}(\mathbb{P}^n_\mathbb{C})$ and that for any d the induced topology on $\mathbb{C}[z_2]_{\leq d}$ is the topology as a vector space (or as an algebraic group). The induced topology on $\mathbb{C}[z_2]$ is thus the inductive limit topology given by

$$\mathbb{C}[z_2]_{\leq 1} \subset \mathbb{C}[z_2]_{\leq 2} \subset \ldots.$$

For any sequence $\ell = (\ell_n)_{n\in\mathbb{N}}$ of positive integers the set

$$\left\{ \sum_{i=0}^{d} a_i X_i \,|\, |a_i| < \frac{1}{\ell_i} \right\}$$

is open in $\mathbb{C}[z_2]$. This implies that $\mathbb{C}[z_2]$ is not countable and hence not metrisable. The same holds for $\mathrm{Bir}(\mathbb{P}^n_\mathbb{C})$. □

LEMMA 3.29. *The topological group* $\mathrm{Bir}(\mathbb{P}^n_\mathbb{C})$ *is compactly generated if and only if $n \leq 2$.*

PROOF. The group $\mathrm{Bir}(\mathbb{P}^1_\mathbb{C}) = \mathrm{PGL}(2,\mathbb{C})$ is a linear algebraic group; consequently it is compactly generated.

By the classical Noether and Castelnuovo Theorem the group $\mathrm{Bir}(\mathbb{P}^2_\mathbb{C})$ is generated by $\mathrm{Aut}(\mathbb{P}^2_\mathbb{C}) = \mathrm{PGL}(3,\mathbb{C})$ and the standard involution σ_2. The linear algebraic group $\mathrm{Aut}(\mathbb{P}^2_\mathbb{C}) = \mathrm{PGL}(3,\mathbb{C})$ being compactly generated, $\mathrm{Bir}(\mathbb{P}^2_\mathbb{C})$ is compactly generated.

Assume $n \geq 3$. The group $\mathrm{Bir}(\mathbb{P}^n_\mathbb{C})$ is not generated by $\mathrm{Bir}_{\leq d}(\mathbb{P}^n_\mathbb{C})$ for any integer d because the birational type of the hypersurfaces that are contracted by some element of $\mathrm{Bir}_{\leq d}(\mathbb{P}^n_\mathbb{C})$ is bounded (*see* [**155**] for more details or Chapter 4, §4.3.3). The fact that $\mathrm{Bir}(\mathbb{P}^n_\mathbb{C})$ is not compactly generated follows from Lemma 3.26. □

REMARK 3.30. Theorem 3.1 holds for any field, Theorem 3.2 holds for any algebraically closed field, and Theorem 3.3 holds for a (locally compact) local field.

3.2. Algebraic elements of the Cremona group

The goal of this section is the study of algebraic elements; in particular we will show that the set of all these elements is a countable union of closed subsets but it is not closed.

In this section the considered topology is the Zariski topology.

An element $\phi \in \mathrm{Bir}(\mathbb{P}^n_\mathbb{C})$ is *algebraic* if it is contained in an algebraic subgroup G of $\mathrm{Bir}(\mathbb{P}^n_\mathbb{C})$.

Let us denote by $\mathrm{Bir}(\mathbb{P}^n_\mathbb{C})_{\mathrm{alg}}$ the set of algebraic elements of $\mathrm{Bir}(\mathbb{P}^n_\mathbb{C})$.

THEOREM 3.31 ([**30**]). *Let $n \geq 2$.*

- *There are a closed subset* $U \subset \mathrm{Bir}(\mathbb{P}^n_{\mathbb{C}})$ *canonically homeomorphic to* $\mathbb{A}^1_{\mathbb{C}}$ *and a family of birational maps* $U \to \mathrm{Bir}(\mathbb{P}^n_{\mathbb{C}})$ *such that algebraic elements of* U *are unipotent and correspond to elements of the subgroup of* $(\mathbb{C},+)$ *generated by* 1;
- *there is a closed subset* $S \subset \mathrm{Bir}(\mathbb{P}^n_{\mathbb{C}})$ *such that algebraic elements of* S *are semi-simple and correspond to elements of*

$$\{(a,\xi) \in \mathbb{A}^1_{\mathbb{C}} \times (\mathbb{A}^1_{\mathbb{C}} \smallsetminus \{0\}) \,|\, a = \xi^k\}$$

for some $k \in \mathbb{Z}$.

In particular $\mathrm{Bir}(\mathbb{P}^n_{\mathbb{C}})_{alg}$ *and the set of unipotent elements of* $\mathrm{Bir}(\mathbb{P}^n_{\mathbb{C}})$ *are not closed in* $\mathrm{Bir}(\mathbb{P}^n_{\mathbb{C}})$.

Furthermore we will see that $\mathrm{Bir}(\mathbb{P}^n_{\mathbb{C}})_{\mathrm{alg}}$ is a countable union of closed sets of $\mathrm{Bir}(\mathbb{P}^n_{\mathbb{C}})$.

LEMMA 3.32 ([**30**]). *Let* ϕ *be an element of* $\mathrm{Bir}(\mathbb{P}^n_{\mathbb{C}})$. *The closure of* $\{\phi^k \,|\, k \in \mathbb{Z}\}$ *in* $\mathrm{Bir}(\mathbb{P}^n_{\mathbb{C}})$ *is a closed abelian subgroup of* $\mathrm{Bir}(\mathbb{P}^n_{\mathbb{C}})$.

PROOF. Let us denote by Ω the closure of $\{\phi^k \,|\, k \in \mathbb{Z}\}$ in $\mathrm{Bir}(\mathbb{P}^n_{\mathbb{C}})$. For any $j \in \mathbb{Z}$ the set $\phi^j(\Omega)$ is a closed subset of $\mathrm{Bir}(\mathbb{P}^n_{\mathbb{C}})$. It contains $\{\phi^k \,|\, k \in \mathbb{Z}\}$; thus it contains Ω. As a result $\phi^k(\Omega) = \Omega$ for any $k \in \mathbb{Z}$. Set

$$M = \{\psi \in \mathrm{Bir}(\mathbb{P}^n_{\mathbb{C}}) \,|\, \psi(\Omega) \subset \Omega\} = \bigcap_{\omega \in \Omega} \Omega\omega^{-1}.$$

As M is closed and contains $\{\phi^k \,|\, k \in \mathbb{Z}\}$, the set M contains Ω. Therefore, M is closed under composition. Similarly the set $\{\psi^{-1} \,|\, \psi \in \Omega\}$ is closed in $\mathrm{Bir}(\mathbb{P}^n_{\mathbb{C}})$ and contains $\{\phi^k \,|\, k \in \mathbb{Z}\}$. The set Ω is then a subgroup of $\mathrm{Bir}(\mathbb{P}^n_{\mathbb{C}})$.

Let us now prove that Ω is abelian. The centralizer

$$\mathrm{Cent}(\varphi) = \{\psi \in \mathrm{Bir}(\mathbb{P}^n_{\mathbb{C}}) \,|\, \psi \circ \varphi = \varphi \circ \psi\}$$

of an element φ of $\mathrm{Bir}(\mathbb{P}^n_{\mathbb{C}})$ is the preimage of the identity by the continuous map

$$\mathrm{Bir}(\mathbb{P}^n_{\mathbb{C}}) \to \mathrm{Bir}(\mathbb{P}^n_{\mathbb{C}}) \qquad \psi \mapsto \psi \circ \varphi \circ \psi^{-1} \circ \varphi^{-1}.$$

Since a point of $\mathrm{Bir}(\mathbb{P}^n_{\mathbb{C}})$ is closed (Lemma 3.5), $\mathrm{Cent}(\phi)$ is closed.

The closed subgroup $\mathrm{Cent}(\phi)$ of $\mathrm{Bir}(\mathbb{P}^n_{\mathbb{C}})$ contains $\{\phi^j \,|\, j \in \mathbb{Z}\}$; hence it contains Ω. Consequently each element of Ω commutes with ϕ. The set

$$\{\psi \in \mathrm{Bir}(\mathbb{P}^n_{\mathbb{C}}) \,|\, \psi \circ \omega = \omega \circ \psi \quad \forall \omega \in \Omega\} = \bigcap_{\omega \in \Omega} \mathrm{Cent}(\omega)$$

is closed and contains $\{\phi^j \,|\, j \in \mathbb{Z}\}$, so it contains Ω. Therefore, Ω is abelian. □

PROPOSITION 3.33 ([**30**]). *Let* ϕ *be an element of* $\mathrm{Bir}(\mathbb{P}^n_{\mathbb{C}})$.

- *If the sequence* $(\deg \phi^k)_{k \in \mathbb{N}}$ *is unbounded, then* ϕ *is not contained in any algebraic subgroup of* $\mathrm{Bir}(\mathbb{P}^n_{\mathbb{C}})$.
- *If the sequence* $(\deg \phi^k)_{k \in \mathbb{N}}$ *is bounded, then* $\{\phi^j \,|\, j \in \mathbb{Z}\}$ *is an abelian algebraic subgroup of* $\mathrm{Bir}(\mathbb{P}^n_{\mathbb{C}})$.

A direct consequence is the following result:

COROLLARY 3.34 ([**30**]). *Let* ϕ *be a birational self map of* $\mathbb{P}^n_{\mathbb{C}}$. *The following assertions are equivalent:*

- *the map* ϕ *is algebraic;*

◇ *the sequence* $(\deg \phi^k)_{k\in\mathbb{N}}$ *is bounded, i.e.,* ϕ *is elliptic.*

PROOF OF PROPOSITION 3.33. The first assertion follows from Lemma 3.12.

Let us now focus on the second assertion. Assume that the sequence $(\deg \phi^k)_{k\in\mathbb{N}}$ is bounded. According to [**4**] one has for any k,

$$\deg \phi^{-k} \leq (\deg \phi^k)^{n-1}.$$

As a consequence, the set $\{\phi^j \mid j \in \mathbb{Z}\}$ is contained in $\mathrm{Bir}(\mathbb{P}^n_{\mathbb{C}})_{\leq d}$ for some d, and so does the closure Ω of $\{\phi^j \mid j \in \mathbb{Z}\}$. Lemma 3.32 allows us to conclude. □

PROPOSITION 3.35. *For any* k, $d \in \mathbb{N}$, *set*

$$\mathrm{Bir}(\mathbb{P}^n_{\mathbb{C}})_{k,d} = \{\phi \in \mathrm{Bir}(\mathbb{P}^n_{\mathbb{C}}) \mid \deg \phi^k \leq d\}$$

and

$$\mathrm{Bir}(\mathbb{P}^n_{\mathbb{C}})_{\infty,d} = \{\phi \in \mathrm{Bir}(\mathbb{P}^n_{\mathbb{C}}) \mid \deg \phi^k \leq d \,\forall\, k \in \mathbb{N}\}.$$

Then

◇ *the set* $\mathrm{Bir}(\mathbb{P}^n_{\mathbb{C}})_{k,d}$ *is closed in* $\mathrm{Bir}(\mathbb{P}^n_{\mathbb{C}})$*;*

◇ *the set* $\mathrm{Bir}(\mathbb{P}^n_{\mathbb{C}})_{\infty,d} = \bigcap_{i\in\mathbb{N}} \mathrm{Bir}(\mathbb{P}^n_{\mathbb{C}})_{i,d}$ *is closed in* $\mathrm{Bir}(\mathbb{P}^n_{\mathbb{C}})$*;*

◇ *the set* $\mathrm{Bir}(\mathbb{P}^n_{\mathbb{C}})_{alg}$ *of all algebraic elements of* $\mathrm{Bir}(\mathbb{P}^n_{\mathbb{C}})$ *coincides with the union of all* $\mathrm{Bir}(\mathbb{P}^n_{\mathbb{C}})_{\infty,d}$, $d \geq 1$.

PROOF. The set $\mathrm{Bir}(\mathbb{P}^n_{\mathbb{C}})_{\leq d}$ is closed in $\mathrm{Bir}(\mathbb{P}^n_{\mathbb{C}})$ for any d (Corollary 3.8), and the map

$$\mathrm{Bir}(\mathbb{P}^n_{\mathbb{C}}) \to \mathrm{Bir}(\mathbb{P}^n_{\mathbb{C}}), \qquad\qquad \varphi \mapsto \varphi^k$$

is continuous (Remark 3.4); the set $\mathrm{Bir}(\mathbb{P}^n_{\mathbb{C}})_{k,d}$ is thus closed in $\mathrm{Bir}(\mathbb{P}^n_{\mathbb{C}})$.

The first assertion clearly implies the second one.

The third assertion follows from Corollary 3.34. □

Let us now deal with the first assertion of Theorem 3.31. Assume $n \geq 2$. Consider the morphism $\rho\colon \mathbb{A}^1_{\mathbb{C}} \to \mathrm{Bir}(\mathbb{P}^n_{\mathbb{C}})$ given by

$$a \mapsto \big((z_0 : z_1 : \ldots : z_n) \dashrightarrow (z_0z_1 : z_1(z_1+z_0) : z_2(z_1+az_0) : z_3z_1 : z_4z_1 : \ldots : z_nz_1)\big).$$

It is clearly injective. Let $\widetilde{\rho}\colon \mathbb{P}^1_{\mathbb{C}} \to W_2$ be the closed embedding given by

$$(\alpha : \beta) \to (\alpha z_0z_1 : \alpha z_1(z_1+z_0) : z_2(z_1+az_0) : \alpha z_3z_1 : \alpha z_4z_1 : \ldots : \alpha z_nz_1).$$

Note that $\widehat{\rho}((0 : 1))$ does not belong to H_2. However for any $t \in \mathbb{A}^1_{\mathbb{C}}$ one has $\mathrm{pr}_2(\widehat{\rho}((1 : t)) = \rho(t)$. The restriction is to $\mathbb{A}^1_{\mathbb{C}}$; thus it yields a closed embedding $\mathbb{A}^1_{\mathbb{C}} \to H_2$. According to Corollary 3.8 the restriction of π_2 to $\widehat{\rho}(\mathbb{P}^1_{\mathbb{C}} \smallsetminus \{(0:1)\})$ is an homeomorphism.

PROPOSITION 3.36.

◇ *For* $t \in \mathbb{C}$ *the following conditions are equivalent:*
  - $\rho(t)$ *is algebraic,*
  - $\rho(t)$ *is unipotent,*
  - $\rho(t)$ *is conjugate to* $\rho(0)\colon (z_1, z_2, \ldots, z_n) \to (z_1+1, z_2, \ldots, z_n)$,
  - t *belongs to the subgroup of* $(\mathbb{C},+)$ *generated by* 1.

◇ *The pull-back by* ρ *of the set of algebraic elements is not closed.*

PROOF.

◇ A direct computation yields to

$$\rho(a)^k\colon (z_1,z_2,\ldots,z_n) \mapsto \left(z_1+k, z_2\,\frac{(z_1+a)(z_1+a+1)\ldots(z_1+a+k-1)}{z_1(z_1+1)\ldots(z_1+m-1)}, z_3,z_4,\ldots,z_n\right)$$

In particular the second coordinate of $\rho(a)^k(z_1,z_2,\ldots,z_n)$ is

$$z_2\,\frac{\displaystyle\prod_{i=0}^{k-1}(z_1+a+i)}{\displaystyle\prod_{i=0}^{k-1}(z_1+i)}.$$

If a does not belong to the subgroup of $(\mathbb{C},+)$ generated by 1, then the degree growth of $\rho(a)^k$ is linear which implies that $\rho(a)$ is not algebraic.

If a belongs to the subgroup of $(\mathbb{C},+)$ generated by 1, then

$$\deg\rho(a)^k \leq |k|+1 \qquad \forall\, k\in\mathbb{N}.$$

As a consequence $\rho(a)$ is algebraic. Furthermore $\rho(a)$ is conjugate to

$$\rho(0)\colon (z_1,z_2,\ldots,z_n)\mapsto(z_1+1,z_2,\ldots,z_n)$$

via

$$(z_1,z_2,\ldots,z_n) \dashrightarrow \left(z_1, \frac{z_2}{z_1(z_1+1)\ldots(z_1+a-1)}, z_3,z_4,\ldots,z_n\right)$$

if $a>0$ or via

$$(z_1,z_2,\ldots,z_n) \dashrightarrow \Big(z_1, z_2z_1(z_1-1)\ldots(z_1+a), z_3,z_4,\ldots,z_n\Big)$$

if $a<0$[2]. In particular $\rho(a)$ is unipotent.

◇ The second assertion follows from the first one and the fact that the subgroup of $(\mathbb{C},+)$ generated by 1 is not closed. □

Finally let us prove the second assertion of Theorem 3.31. Assume $n\geq 2$. Consider the morphism

$$\rho\colon \mathbb{A}^1_\mathbb{C}\times(\mathbb{A}^1_\mathbb{C}\smallsetminus\{0\})\to\mathrm{Bir}(\mathbb{P}^n_\mathbb{C})$$

given by

$$(a,\xi)\mapsto\Big((z_0:z_1:\ldots:z_n)\dashrightarrow$$
$$(z_0(z_1+z_0):\xi z_1(z_1+z_0):z_2(z_1+az_0):z_3(z_1+z_0):\ldots:z_n(z_1+z_0))\Big).$$

It is injective. Let $\widehat{\rho}\colon\mathbb{P}^2_\mathbb{C}\to W_2$ be the closed embedding given by

$$(\alpha:\beta:\gamma)\dashrightarrow\Big(\alpha z_0(z_1+z_0):\gamma z_1(z_1+z_0):z_2(\alpha z_1+\beta z_0):\alpha z_3(z_1+z_0):\ldots:\alpha z_n(z_1+z_0)\Big).$$

[2]Let us recall that a belongs to the subgroup of $(\mathbb{C},+)$ generated by 1.

Note that

$$(z_0 : z_1 : \ldots : z_n)$$
$$\dashrightarrow \Big(\alpha z_0(z_1+z_0) : \gamma z_1(z_1+z_0) : z_2(\alpha z_1+\beta z_0) : \alpha z_3(z_1+z_0) : \ldots : \alpha z_n(z_1+z_0)\Big)$$

is a birational map if and only if $\alpha\gamma \neq 0$. This yields a closed embedding

$$\mathbb{A}^1_{\mathbb{C}} \times (\mathbb{A}^1_{\mathbb{C}} \smallsetminus \{0\}) \to H_2, \qquad (a,\xi) \mapsto \widehat{\rho}((1 : a : \xi)).$$

Furthermore $\mathrm{pr}_2(\widehat{\rho}(1 : a : \xi)) = \rho(a,\xi)$. Proposition 3.33 says that the restriction of π_2 to the image is a homeomorphism.

PROPOSITION 3.37.

◇ *For $(a,\xi) \in \mathbb{A}^1_{\mathbb{C}} \times (\mathbb{A}^1_{\mathbb{C}} \smallsetminus \{0\})$ the following conditions are equivalent:*
- *$\rho(a,\xi)$ is algebraic,*
- *$\rho(a,\xi)$ is semi-simple,*
- *$\rho(a,\xi)$ is conjugate to $\rho(1,\xi)\colon (z_1, z_2, \ldots, z_n) \mapsto (\xi z_1, z_2, z_3, \ldots, z_n)$,*
- *there exists $k \in \mathbb{Z}$ such that $a = \xi^k$.*

◇ *The pull-back by ρ of the set of algebraic elements is not closed.*

PROOF.

◇ Note that

$$\rho(a,\xi)^k\colon (z_1, z_2, \ldots, z_n) \dashrightarrow \left(\xi^k z_1, z_2\,\frac{(z_1+a)(\xi z_1+a)\ldots(\xi^k z_1+a)}{(z_1+1)(\xi z_1+1)\ldots(\xi^{k-1}z_1+1)}, z_3, z_4, \ldots, z_n\right).$$

In particular the second coordinate of $\rho(a,\xi)^k$ is

$$z_2\,\frac{\displaystyle\prod_{i=0}^{k-1}(\xi^i z_1 + a)}{\displaystyle\prod_{i=0}^{k-1}(\xi^i z_1 + 1)}.$$

If a does not belong to $\langle\xi\rangle \subset (\mathbb{C},\cdot)$, then the degree growth of $\rho(a,\xi)^k$ is linear; hence $\rho(a,\xi)$ is not algebraic.

If a belongs to $\langle\xi\rangle \subset (\mathbb{C},\cdot)$, then $a = \xi^k$ for some $k \in \mathbb{Z}$ and for any $j \in \mathbb{N}$,

$$\deg \rho(a,\xi)^j \leq |k| + 1,$$

so $\rho(a,\xi)$ is algebraic. Note that $\rho(a,\xi)$ is conjugate to $\rho(1,\xi)$ via

$$(z_1, z_2, \ldots, z_n) \dashrightarrow \left(z_1, \frac{z_2}{z_1(z_1+1)\ldots(z_1+a-1)}, z_3, z_4, \ldots, z_n\right)$$

if $k > 0$ and via

$$(z_1, z_2, \ldots, z_n) \dashrightarrow \Big(z_1, z_2 z_1(z_1-1)\ldots(z_1+a), z_3, z_4, \ldots, z_n\Big)$$

if $k < 0$.

◇ The second assertion follows from the first one and the fact that

$$\Big\{(a,\xi) \in \mathbb{A}^1_{\mathbb{C}} \times (\mathbb{A}^1_{\mathbb{C}} \smallsetminus \{0\}) \,|\, a = \xi^k \text{ for some } k \in \mathbb{Z}\Big\}$$

is not closed. □

REMARK 3.38. Note that all the results of this section hold for $\mathrm{Bir}(\mathbb{P}^n_{\Bbbk})$ whose $\Bbbk$ is an algebraically closed field of characteristic 0.

3.3. Classification of maximal algebraic subgroups of $\mathrm{Bir}(\mathbb{P}^2_{\mathbb{C}})$

In [**26**] the author gives a complete classification of maximal algebraic subgroups of the plane Cremona group and provides algebraic varieties that parametrize the conjugacy classes. The algebraic subgroups of $\mathrm{Bir}(\mathbb{P}^n_{\mathbb{C}})$ have been studied for a long time. Enriques established in [**91**] the complete classification of maximal connected algebraic subgroups of $\mathrm{Bir}(\mathbb{P}^2_{\mathbb{C}})$: every such subgroup is the conjugate of the identity component of the automorphism group of a minimal rational surface. A modern proof was given in [**169**]. The case of $\mathrm{Bir}(\mathbb{P}^3_{\mathbb{C}})$ was treated by Enriques and Fano and, more recently, by Umemura ([**167**–**169**]). Demazure has studied the smooth connected subgroups of $\mathrm{Bir}(\mathbb{P}^n_{\mathbb{C}})$ that contain a split torus of dimension n (*see* [**74**]). Only a few results are known for non-connected subgroups even in dimension 2. Nevertheless there are a lot of statements in the case of finite subgroups which are algebraic ones ([**6**, **8**, **9**, **23**, **24**, **69**, **89**, **115**, **178**]) and we deal with them in Chapter 6. But these results do not show which finite groups are maximal algebraic subgroups. As mentioned in [**89**] there are some remaining open questions like the description of the algebraic varieties that parameterize conjugacy classes of finite subgroups G of $\mathrm{Bir}(\mathbb{P}^2_{\mathbb{C}})$. Blanc gives an answer to this question for

- ◇ abelian finite subgroups G of $\mathrm{Bir}(\mathbb{P}^2_{\mathbb{C}})$ whose elements do not fix a curve of positive genus ([**25**]);
- ◇ finite cyclic subgroups of $\mathrm{Bir}(\mathbb{P}^2_{\mathbb{C}})$ (*see* [**28**]);
- ◇ maximal algebraic subgroups of $\mathrm{Bir}(\mathbb{P}^2_{\mathbb{C}})$ (*see* [**26**]).

Before specifying Blanc results let us recall some notions. If S is a projective smooth rational surface and G is a subgroup of $\mathrm{Aut}(S)$, we say that (G,S) is a *pair*. A birational map $\varphi\colon X \dashrightarrow Y$ is G-*equivariant* if the inclusion $\varphi\circ\mathrm{G}\circ\varphi^{-1}\subset\mathrm{Aut}(Y)$ holds. The pair (G,S) is *minimal* if every birational G-equivariant morphism $\varphi\colon S\dashrightarrow S'$ where S' is a projective, smooth surface, is an isomorphism. A morphism $\pi\colon S\to\mathbb{P}^1_{\mathbb{C}}$ is a *conic bundle* if all generic fibers of π are isomorphic to $\mathbb{P}^1_{\mathbb{C}}$ and if there exists a finite number of singular fibers which are the transverse union of two curves isomorphic to $\mathbb{P}^1_{\mathbb{C}}$.

3.3.1. Del Pezzo surfaces and their automorphism groups.

A *del Pezzo surface* is a smooth projective surface S such that the anti-canonical divisor $-K_S$ is ample. Let us recall the classification of del Pezzo surfaces. The number $d=K_S^2$ is called the *degree* of S. By Noether's formula $1\leq d\leq 9$. For $d\geq 3$, the anticanonical linear system $|-K_S|$ maps S onto a non-singular surface of degree d in $\mathbb{P}^d_{\mathbb{C}}$. If $d=9$, then $S\simeq\mathbb{P}^2_{\mathbb{C}}$. If $d=8$, then $S\simeq\mathbb{P}^1_{\mathbb{C}}\times\mathbb{P}^1_{\mathbb{C}}$ or $S\simeq\mathbb{F}_1$. For $d\leq 7$ a del Pezzo surface S is isomorphic to the blow-up of $n=9-d$ points in $\mathbb{P}^2_{\mathbb{C}}$ in general position, that is

- ◇ no three of them are colinear,
- ◇ no six are on the same conic,
- ◇ if $n=8$, then the points are not on a plane cubic which has one of them as its singular point.

There exist ([**88**, Chapter 8])

- ◇ a unique isomorphism class of del Pezzo surfaces of degree 5 (resp. 6, resp. 7, resp. 9),
- ◇ two isomorphism classes of del Pezzo surfaces of degree 8,

◇ and infinitely many isomorphism classes of del Pezzo surfaces of degree 1, (resp. 2, resp. 3, resp. 4).

We will see that automorphism groups of del Pezzo surfaces are algebraic subgroups of $\mathrm{Bir}(\mathbb{P}^2_\mathbb{C})$ and that they are finite if and only if the degree of the corresponding surface is ≤ 5. If S is a del Pezzo surface of degree 5, then $\mathrm{Aut}(S) = \mathfrak{S}_5$. Automorphism groups of del Pezzo surfaces of degree ≤ 4 are described in [**89**, §10]. In particular the authors got the following:

THEOREM 3.39 ([**89**]). *If the automorphism group of a del Pezzo surface is finite, then it has order at most* 648.

LEMMA 3.40 ([**172**]). *If the automorphism group of a del Pezzo surface is finite, then it can be embedded into* $\mathrm{GL}(8,\mathbb{C})$.

PROOF. Let S be a del Pezzo surface such that $\mathrm{Aut}(S)$ is finite. Then $\deg S \leq 5$ and S is isomorphic to $\mathrm{Bl}_{p_1,p_2,\ldots,p_r}\mathbb{P}^2_\mathbb{C}$ where $4 \leq r = 9-\deg S \leq 8$ and p_1, p_2, $\ldots$, p_r are general points of $\mathbb{P}^2_\mathbb{C}$. Denote by $\mathbf{e}_0$ the pull-back of the class of a line and by $\mathbf{e}_{p_i}$ the class of the exceptional line E_{p_i} corresponding to the point p_i. The dimension of the Néron-Severi space $\mathrm{NS}(S)\otimes\mathbb{R}$ is $r+1$ and $\mathbf{e}_0$, $\mathbf{e}_{p_1}$, $\mathbf{e}_{p_2}$, $\ldots$, $\mathbf{e}_{p_r}$ is a basis of $\mathrm{NS}(S)\otimes\mathbb{R}$. Note that the equality $\mathbf{e}_{p_i}\cdot\mathbf{e}_{p_i} = -1$ implies that E_{p_i} is the only representative of $\mathbf{e}_{p_i}$ on S.

If $\varphi \in \mathrm{Aut}(S)$ acts as the identity on $\mathrm{NS}(S)\otimes\mathbb{R}$, then φ preserves the exceptional lines E_{p_i} for $1 \leq i \leq r$. Hence φ induces an automorphism of $\mathbb{P}^2_\mathbb{C}$ that fixes p_1, p_2, $\ldots$, p_r. As $r \geq 4$ and as the p_i are in general position, the induced automorphism of $\mathbb{P}^2_\mathbb{C}$ is the identity. The action of $\mathrm{Aut}(S)$ on $\mathrm{NS}(S)\otimes\mathbb{R}$ is thus faithful and we get a faithful representation

$$\mathrm{Aut}(S) \to \mathrm{GL}(r+1,\mathbb{C}).$$

Any element φ of $\mathrm{Aut}(S)$ fixes K_S; as a result, the one-dimensional subspace $\mathbb{R}\cdot K_S$ of $\mathrm{NS}(S)\otimes\mathbb{R}$ is fixed. By projecting the orthogonal complement of K_S in $\mathrm{NS}(S)\otimes\mathbb{R}$ we obtain a faithful representation of $\mathrm{Aut}(S)$ into $\mathrm{GL}(r,\mathbb{C})$. □

A del Pezzo surface of degree 6 is isomorphic to the blow-up of the complex projective plane in three general points, *i.e.*, isomorphic to the surface

$$S_6 = \big\{\big((z_0:z_1:z_2),(a:b:c)\big) \in \mathbb{P}^2_\mathbb{C}\times\mathbb{P}^2_\mathbb{C} \,|\, az_0 = bz_1 = cz_2\big\}.$$

The automorphism group of S_6 is isomorphic to $(\mathbb{C}^*)^2 \rtimes \big(\mathfrak{S}_3 \times \mathbb{Z}/2\mathbb{Z}\big)$ where $\mathfrak{S}_3$ acts by permuting the coordinates of the two factors simultaneously, $\mathbb{Z}/2\mathbb{Z}$ exchanges the two factors and $d \in (\mathbb{C}^*)^2$ acts as follows:

$$d\cdot\big((z_0:z_1:z_2),(a:b:c)\big) = \big(d(z_0:z_1:z_2):d^{-1}(a:b:c)\big).$$

In other words $\mathrm{Aut}(S_6)$ is conjugate to $\big(\mathfrak{S}_3\times \mathbb{Z}/2\mathbb{Z}\big)\ltimes \mathrm{D}_2 \subset \mathrm{GL}(2,\mathbb{Z})\ltimes\mathrm{D}_2$.

LEMMA 3.41 ([**172**]). *The group* $\mathrm{Aut}(S_6)$ *can be embedded in* $\mathrm{GL}(6,\mathbb{C})$.

PROOF. Consider the rational map

$$\phi\colon \mathbb{P}^2_\mathbb{C} \dashrightarrow \mathbb{P}^6_\mathbb{C}, \quad (z_0:z_1:z_2) \dashrightarrow (z_0^2z_1 : z_0^2z_2 : z_0z_1^2 : z_1^2z_2 : z_0z_2^2 : z_1z_2^2 : z_0z_1z_2).$$

The rational action of $(\mathfrak{S}_3\times\mathbb{Z}/2\mathbb{Z})\ltimes\mathrm{D}_2$ on $\phi(\mathbb{P}^2_\mathbb{C})$ extends to a regular action on $\mathbb{P}^6_\mathbb{C}$ that preserves the affine space given by $z_6 \neq 0$. This yields an embedding of $(\mathfrak{S}_3\times\mathbb{Z}/2\mathbb{Z})\ltimes\mathrm{D}_2$ into $\mathrm{GL}(6,\mathbb{C})$. □

3.3.2. Hirzebruch surfaces and their automorphism groups. Let us introduce the Hirzebruch surfaces. Consider the surface $\mathbb{F}_1$ obtained by blowing up $(1:0:0) \in \mathbb{P}^2_{\mathbb{C}}$; it is a compactification of $\mathbb{C}^2$ which has a natural fibration corresponding to the lines $z_1 =$ constant. The divisor at infinity is the union of two rational curves which intersect in one point:

- ◇ one of them is the strict transform of the line at infinity in $\mathbb{P}^2_{\mathbb{C}}$, it is a fiber denoted by f_1;
- ◇ the other one, denoted by s_1, is the exceptional divisor which is a section for the fibration.

Furthermore $f_1^2 = 0$ and $s_1^2 = -1$. More generally for any n, $\mathbb{F}_n$ is a compactification of $\mathbb{C}^2$ with a rational fibration and such that the divisor at infinity is the union of two transversal rational curves: a fiber f_n and a section s_n of self-intersection $-n$. These surfaces are called *Hirzebruch surfaces*. One can go from $\mathbb{F}_n$ to $\mathbb{F}_{n+1}$ as follows. Consider the surface $\mathbb{F}_n$. Set $p = s_n \cap f_n$. Let $\mathfrak{p}_1$ be the blow-up of $p \in \mathbb{F}_n$ and let $\mathfrak{p}_2$ be the contraction of the strict transform $\widetilde{f_n}$ of f_n. One goes from $\mathbb{F}_n$ to $\mathbb{F}_{n+1}$ via $\mathfrak{p}_2 \circ \mathfrak{p}_1^{-1}$. We can also go from $\mathbb{F}_{n+1}$ to $\mathbb{F}_n$ via $\widetilde{\mathfrak{p}_2} \circ \widetilde{\mathfrak{p}_1}^{-1}$ where

- ◇ $\widetilde{\mathfrak{p}_1}$ is the blow-up of a point q such that $q \in f_{n+1}$, $q \notin s_{n+1}$;
- ◇ $\widetilde{\mathfrak{p}_2}$ is the contraction of the strict transform $\widetilde{f_{n+1}}$ of f_{n+1}.

We will say that both $\mathfrak{p}_2 \circ \mathfrak{p}_1^{-1}$ and $\widetilde{\mathfrak{p}_2} \circ \widetilde{\mathfrak{p}_1}^{-1}$ are *elementary transformations.*

The nth Hirzebruch surface $\mathbb{F}_n = \mathbb{P}\big(\mathcal{O}_{\mathbb{P}^1_{\mathbb{C}}} \oplus \mathcal{O}_{\mathbb{P}^1_{\mathbb{C}}}(n)\big)$ is isomorphic to the hypersurface

$$\big\{([x_0, x_1], [y_0, y_1, y_2]) \in \mathbb{P}^1_{\mathbb{C}} \times \mathbb{P}^2_{\mathbb{C}} \,|\, x_0^n y_1 - x_1^n y_2 = 0\big\}$$

of $\mathbb{P}^1_{\mathbb{C}} \times \mathbb{P}^2_{\mathbb{C}}$.

Their automorphism groups are

$$\begin{aligned} \mathrm{Aut}(\mathbb{P}^2_{\mathbb{C}} \times \mathbb{P}^1_{\mathbb{C}}) &= (\mathrm{PGL}(2,\mathbb{C}) \times \mathrm{PGL}(2,\mathbb{C})) \rtimes \langle (z_0, z_1) \mapsto (z_1, z_0) \rangle, \\ \mathrm{Aut}(\mathbb{P}^2_{\mathbb{C}}) &= \mathrm{PGL}(3,\mathbb{C}), \end{aligned}$$

and

$$\begin{aligned} \mathrm{Aut}(\mathbb{F}_n) = \Bigg\{ &(z_0, z_1) \mapsto \left(\frac{az_0 + P(z_1)}{(\gamma z_1 + \delta)^n}, \frac{\alpha z_1 + \beta}{\gamma z_1 + \delta} \right) \,\Big|\, \\ &\begin{pmatrix} \alpha & \beta \\ \gamma & \delta \end{pmatrix} \in \mathrm{PGL}(2,\mathbb{C}),\, a \in \mathbb{C}^*,\, P \in \mathbb{C}[z_1],\, \deg P \leq n \Bigg\}. \end{aligned}$$

In other words as soon as $n \geq 2$, the group $\mathrm{Aut}(\mathbb{F}_n)$ is isomorphic to $\mathbb{C}[z_0, z_1]_n \rtimes \mathrm{GL}(2,\mathbb{C})/\mu_n$ where $\mu_n \subset \mathrm{GL}(2,\mathbb{C})$ is the subgroup of n-torsion elements in the center of $\mathrm{GL}(2,\mathbb{C})$.

LEMMA 3.42 ([**172**]). *If $n \geq 2$ is even, then $\mathrm{GL}(2,\mathbb{C})/\mu_n$ is isomorphic as an algebraic group to* $\mathrm{PGL}(2,\mathbb{C}) \times \mathbb{C}^*$.

If n is odd, then $\mathrm{GL}(2,\mathbb{C})/\mu_n$ is isomorphic as an algebraic group to $\mathrm{PGL}(2,\mathbb{C})$.

In particular all finite subgroups of $\mathrm{Aut}(\mathbb{F}_n)$ *can be embedded into* $\mathrm{PGL}(2,\mathbb{C}) \times \mathrm{PGL}(2,\mathbb{C})$ *as soon as* $n \geq 2$.

3.3.3. Automorphism groups of exceptional conic bundles. An *exceptional conic bundle* S is a conic bundle with singular fiber above $2n$ points in $\mathbb{P}^1_{\mathbb{C}}$ and with two sections s_1 and s_2 of self-intersection $-n$, where $n \geq 2$ (*see* [**26**]).

LEMMA 3.43 ([**26**]). *Let* $\pi\colon S\to\mathbb{P}^1_\mathbb{C}$ *be an exceptional conic bundle. Then* $\mathrm{Aut}(S,\pi)$ *is isomorphic to a subgroup of* $\mathrm{PGL}(2,\mathbb{C})\times\mathrm{PGL}(2,\mathbb{C})$.

3.3.4. $\left(\mathbb{Z}/_{2\mathbb{Z}}\right)^2$**-conic bundles.** A conic bundle $\pi\colon S\to\mathbb{P}^1_\mathbb{C}$ is a $\left(\mathbb{Z}/_{2\mathbb{Z}}\right)^2$-*conic bundle* if

- ◇ the group $\mathrm{Aut}\left(S/_{\mathbb{P}^1_\mathbb{C}}\right)$ is isomorphic to $\left(\mathbb{Z}/_{2\mathbb{Z}}\right)^2$,
- ◇ each of the three involutions of $\mathrm{Aut}\left(S/_{\mathbb{P}^1_\mathbb{C}}\right)$ fixes pointwise an irreducible curve C such that $\pi\colon C\to\mathbb{P}^1_\mathbb{C}$ is a double covering that is ramified over a positive even number of points.

The automorphism group $\mathrm{Aut}(S,\pi)$ of a $\left(\mathbb{Z}/_{2\mathbb{Z}}\right)^2$-conic bundle is finite; its structure is given by the following exact sequence ([**26**])

$$1\longrightarrow V\longrightarrow \mathrm{Aut}(S,\pi)\longrightarrow H_V\longrightarrow 1$$

where $V\simeq\left(\mathbb{Z}/_{2\mathbb{Z}}\right)^2$ and H_V is a finite subgroup of $\mathrm{Aut}(\mathbb{P}^1_\mathbb{C})$.

Note that we also have the following property:

LEMMA 3.44 ([**172**]). *Let* $\mathrm{G}\subset\mathrm{Bir}(\mathbb{P}^2_\mathbb{C})$ *be an infinite torsion group. Assume that for any finitely generated subgroup* $\Gamma\subset\mathrm{G}$ *there exists a* $\left(\mathbb{Z}/_{2\mathbb{Z}}\right)^2$*-conic bundle* $S\to\mathbb{P}^1_\mathbb{C}$ *such that* Γ *is conjugate to a subgroup of* $\mathrm{Aut}(S,\pi)$. *Then any finitely generated subgroup of* G *is isomorphic to a subgroup of* $\mathrm{PGL}(2,\mathbb{C})\times\mathrm{PGL}(2,\mathbb{C})$.

3.3.5. Blanc results. First Blanc proved:

THEOREM 3.45 ([**26**]). *Every algebraic subgroup of* $\mathrm{Bir}(\mathbb{P}^2_\mathbb{C})$ *is contained in a maximal algebraic subgroup of* $\mathrm{Bir}(\mathbb{P}^2_\mathbb{C})$.

The maximal algebraic subgroups of the plane Cremona group are the conjugate of the groups $\mathrm{G}=\mathrm{Aut}(S,\pi)$ *where* S *is a rational surface and* $\pi\colon S\to Y$ *is a morphism such that*

(1) Y *is a point,* $\mathrm{G}=\mathrm{Aut}(S)$ *and* S *is one of the following:*
- ◇ $\mathbb{P}^2_\mathbb{C}$, $\mathbb{P}^1_\mathbb{C}\times\mathbb{P}^1_\mathbb{C}$;
- ◇ *a del Pezzo surface of degree* 1, 4, 5 *or* 6;
- ◇ *a del Pezzo surface of degree* 3 *(resp.* 2*) such that the pair* $(\mathrm{Aut}(S),S)$ *is minimal and such that the fixed points of the action of* $\mathrm{Aut}(S)$ *on* S *are lying on exceptional curves;*

(2) $Y\simeq\mathbb{P}^1_\mathbb{C}$ *and* π *is one of the following conic bundles:*
- ◇ *the fibration by lines of the Hirzebruch surface* $\mathbb{F}_n$ *for* $n\geq 2$;
- ◇ *an exceptional conic bundle with at least* 4 *singular fibers;*
- ◇ *a* $\left(\mathbb{Z}/_{2\mathbb{Z}}\right)^2$*-conic bundle such that* S *is not a del Pezzo surface.*

Moreover, in all these cases, the pair (G,S) is minimal and the fibration $\pi\colon S\to Y$ is a G-Mori fibration which is birationally *superrigid*. This means that two such groups $\mathrm{G}=\mathrm{Aut}(S,\pi)$ and $\mathrm{G}'=\mathrm{Aut}(S',\pi')$ are conjugate if and only if there exists an isomorphism $S\to S'$ which sends fibers of π onto fibers of π'.

Then Blanc described more precisely the structure of these minimal algebraic subgroups of $\mathrm{Bir}(\mathbb{P}^2_\mathbb{C})$. Furthermore he provides algebraic varieties that parameterize the conjugacy classes of these groups:

THEOREM 3.46 ([**26**]). *The maximal algebraic subgroups of* $\mathrm{Bir}(\mathbb{P}^2_{\mathbb{C}})$ *belong up to conjugacy to one of the eleven following families:*

(1) $\mathrm{Aut}(\mathbb{P}^2_{\mathbb{C}}) \simeq \mathrm{PGL}(3,\mathbb{C})$;
(2) $\mathrm{Aut}(\mathbb{P}^1_{\mathbb{C}} \times \mathbb{P}^1_{\mathbb{C}}) \simeq \big(\mathrm{PGL}(2,\mathbb{C})\big)^2 \rtimes \mathbb{Z}/2\mathbb{Z}$;
(3) $\mathrm{Aut}(S) \simeq (\mathbb{C}^*)^2 \rtimes \big(\mathfrak{S}_3 \times \mathbb{Z}/2\mathbb{Z}\big)$ *where S is the del Pezzo surface of degree 6;*
(4) $\mathrm{Aut}(\mathbb{F}_n) \simeq \mathbb{C}^{n+1} \rtimes \mathrm{GL}(2,\mathbb{C})/\mu_n$ *where μ_n is the nth torsion of the center of $\mathrm{GL}(2,\mathbb{C})$ with $n \geq 2$;*
(5) $\mathrm{Aut}(S,\pi)$ *where (S,π) is an exceptional conic bundle with singular fibers over a set $\Delta \subset \mathbb{P}^1_{\mathbb{C}}$ of $2n$ distinct points, $n \geq 2$; the projection of $\mathrm{Aut}(S,\pi)$ onto $\mathrm{PGL}(2,\mathbb{C})$ gives an exact sequence*

$$1 \longrightarrow \mathbb{C}^* \rtimes \mathbb{Z}/2\mathbb{Z} \longrightarrow \mathrm{Aut}(S,\pi) \longrightarrow \mathrm{H}_\Delta \longrightarrow 1$$

where H_Δ is the finite subgroup of $\mathrm{PGL}(2,\mathbb{C})$ formed by elements that preserve Δ;
(6) $\mathrm{Aut}(S) \simeq \mathfrak{S}_5$ *where S is the del Pezzo surface of degree 5;*
(7) $\mathrm{Aut}(S) \simeq \big(\mathbb{Z}/2\mathbb{Z}\big)^4 \rtimes \mathrm{H}_S$ *where S is a del Pezzo surface of degree 4 obtained by blowing up five points in $\mathbb{P}^2_{\mathbb{C}}$ and H_S is the group of automorphisms of $\mathbb{P}^2_{\mathbb{C}}$ that preserve this set of points;*
(8) $\mathrm{Aut}(S)$ *where S is a del Pezzo surface of degree 3 of the following form*
   - ◇ *the triple cover of $\mathbb{P}^2_{\mathbb{C}}$ ramified along a smooth cubic Γ. If S is the Fermat cubic, then $\mathrm{Aut}(S) = \big(\mathbb{Z}/3\mathbb{Z}\big)^3 \rtimes \mathfrak{S}_4$, otherwise we have an exact sequence*

$$1 \longrightarrow \mathbb{Z}/3\mathbb{Z} \longrightarrow \mathrm{Aut}(S) \longrightarrow \mathrm{H}_\Gamma \longrightarrow 1$$

   where H_Γ is the group of automorphisms of $\mathbb{P}^2_{\mathbb{C}}$ that preserve Γ, H_Γ contains a subgroup isomorphic to $\big(\mathbb{Z}/3\mathbb{Z}\big)^2$;
   - ◇ *the Clebsch cubic surface whose automorphism group is isomorphic to $\mathfrak{S}_5$;*
   - ◇ *a cubic surface given by $z_0^3 + z_0(z_1^2 + z_2^2 + z_3^2) + \lambda z_1 z_2 z_3 = 0$ for some $\lambda \in \mathbb{C}$, $9\lambda^3 \neq 8\lambda$, $8\lambda^3 \neq -1$ and whose automorphism group is isomorphic to $\mathfrak{S}_4$;*
(9) $\mathrm{Aut}(S) \simeq \mathbb{Z}/2\mathbb{Z} \rtimes \mathrm{H}_S$ *where S is a del Pezzo surface of degree 2 which is a double cover of a smooth quartic $Q_S \subset \mathbb{P}^2_{\mathbb{C}}$ such that $\mathrm{H}_S = \mathrm{Aut}(Q_S)$ acts without fixed point on the quartic without its bitangent points;*
(10) $\mathrm{Aut}(S)$ *where S is a del Pezzo surface of degree 1, double cover of a quadratic cone Q, ramified along a curve Γ_S of degree 6, complete intersection of Q with a cubic surface of $\mathbb{P}^3_{\mathbb{C}}$. We have the following exact sequence*

$$1 \longrightarrow \mathbb{Z}/2\mathbb{Z} \longrightarrow \mathrm{Aut}(S) \longrightarrow \mathrm{H}_S \longrightarrow 1$$

where H_S denotes the automorphism group of Q preserving the curve Γ_S;
(11) $\mathrm{Aut}(S,\pi)$ *where (S,π) is a $\big(\mathbb{Z}/2\mathbb{Z}\big)^2$-conic bundle such that S is not a del Pezzo surface. The projection of $\mathrm{Aut}(S,\pi)$ onto $\mathrm{PGL}(2,\mathbb{C})$ gives the following exact sequence*

$$1 \longrightarrow V \longrightarrow \mathrm{Aut}(S,\pi) \longrightarrow \mathrm{H}_V \longrightarrow 1$$

where $V \simeq \left(\mathbb{Z}/_{2\mathbb{Z}}\right)^2$ *contains three involutions fixing a hyperelliptic curve ramified over points of* p_1, p_2, $p_3 \subset \mathbb{P}^1_{\mathbb{C}}$ *and* $\mathrm{H}_V \subset \mathrm{Aut}(\mathbb{P}^1_{\mathbb{C}})$ *is the finite subgroup preserving the set* $\{p_1, p_2, p_3\}$.

The eleven families are disjoint and the conjugacy classes in any family are parameterized respectively by

(1), (2), (3), (6) the point;
(4) there is only one conjugacy class for any integer $n \geq 2$;
(5) for any integer $n \geq 2$ the set of $2n$ points of $\mathbb{P}^1_{\mathbb{C}}$ modulo the action of $\mathrm{Aut}(\mathbb{P}^1_{\mathbb{C}}) = \mathrm{PGL}(2, \mathbb{C})$;
(7) the isomorphism classes of del Pezzo surfaces of degree 4;
(8) the isomorphism classes of cubic surfaces given respectively
  - ◇ by the isomorphism classes of elliptic curves;
  - ◇ for the Clebsch surface there is only one isomorphism class;
  - ◇ by the classes of $\{\lambda \in \mathbb{C} \,|\, 9\lambda^3 \neq 8\lambda,\, 8\lambda^3 \neq -1\}$ modulo the equivalence $\lambda \sim -\lambda$.

(9) the isomorphism classes of smoooth quartics of $\mathbb{P}^2_{\mathbb{C}}$ having automorphism groups acting without fixed points on the quartic without its bitangent points;
(10) the isomorphism classes of del Pezzo surfaces of degree 1;
(11) the triplets of ramification $\{p_1, p_2, p_3\} \subset \mathbb{P}^1_{\mathbb{C}}$ that determine $\left(\mathbb{Z}/_{2\mathbb{Z}}\right)^2$ conic bundles on surfaces that are not del Pezzo ones, modulo the action of $\mathbb{P}^1_{\mathbb{C}}$.

The approach of Blanc used the modern viewpoint of Mori's theory and Sarkisov's program, aiming a generalization in higher dimension:

- ◇ he described each maximal algebraic subgroup of the classification as a G-Mori fibration;
- ◇ he then proved that any algebraic subgroup is contained in one of the groups of the classification;
- ◇ he also showed that any group of the classification is a minimal G-fibration that is furthermore superrigid.

Lemmas 3.40, 3.41 and Theorem 3.46 allow us to prove the following statement:

LEMMA 3.47 ([**172**]). *Let* G *be a subgroup of the plane Cremona group. Assume that* G *is conjugate to an automorphism group of a del Pezzo surface* S. *Then* G *can be embedded into* $\mathrm{GL}(8, \mathbb{C})$.

PROOF. If $\deg S \leq 5$, then $\mathrm{Aut}(S)$ is finite and Lemma 3.40 allows us to conclude.

If $\deg S = 6$, then $\mathrm{Aut}(S)$ can be embedded into $\mathrm{GL}(8, \mathbb{C})$ (Lemma 3.41).

If $\deg S = 7$, then $\mathrm{Aut}(S)$ is conjugate to a subgroup of

$$\mathrm{Aut}(\mathbb{P}^1_{\mathbb{C}} \times \mathbb{P}^1_{\mathbb{C}}) \simeq \big(\mathrm{PGL}(2, \mathbb{C}) \times \mathrm{PGL}(2, \mathbb{C})\big) \rtimes \mathbb{Z}/_{2\mathbb{Z}} \subset \mathrm{GL}(6, \mathbb{C}).$$

If $\deg S = 8$, then S is isomorphic either to $\mathbb{F}_0 = \mathbb{P}^1_{\mathbb{C}} \times \mathbb{P}^1_{\mathbb{C}}$ or to $\mathbb{F}_1$. On the one hand

$$\mathrm{Aut}(\mathbb{P}^1_{\mathbb{C}} \times \mathbb{P}^1_{\mathbb{C}}) \simeq \big(\mathrm{PGL}(2, \mathbb{C}) \times \mathrm{PGL}(2, \mathbb{C})\big) \rtimes \mathbb{Z}/_{2\mathbb{Z}} \subset \mathrm{GL}(6, \mathbb{C}).$$

and on the other hand $\mathrm{Aut}(\mathbb{F}_1)$ is not a maximal algebraic subgroup of $\mathrm{Bir}(\mathbb{P}^2_{\mathbb{C}})$ (Theorem 3.46).

If $\deg S = 9$, then $S \simeq \mathbb{P}^2_{\mathbb{C}}$ and $\mathrm{Aut}(S) = \mathrm{PGL}(3,\mathbb{C}) \subset \mathrm{GL}(8,\mathbb{C})$. □

3.4. Closed normal subgroups of the Cremona group

As we have seen, we can endow the Cremona group with a natural Zariski topology induced by morphisms $V \to \mathrm{Bir}(\mathbb{P}^n_{\mathbb{C}})$ where V is an algebraic variety.

In [**47**] Mumford discussed properties of $\mathrm{Bir}(\mathbb{P}^2_{\mathbb{C}})$ and in particular asked if it is a simple group with respect to the Zariski topology, *i.e.*, if every closed normal subgroup of $\mathrm{Bir}(\mathbb{P}^2_{\mathbb{C}})$ is trivial. Blanc and Zimmermann provided an affirmative answer to Mumford question:

THEOREM 3.48 ([**41**]). *Let $n \geq 1$ be an integer. The group $\mathrm{Bir}(\mathbb{P}^n_{\mathbb{C}})$ is topologically simple when endowed with the Zariski topology.*

REMARK 3.49. This statement was proved in dimension 2 by Blanc ([**27**]) using the classical Noether and Castelnuovo Theorem.

Let us mention that $\mathrm{Bir}(\mathbb{P}^n_{\mathbb{C}})$, $n \geq 2$, is not simple as an abstract group (for $n = 2$ *see* [**59**] or §8.6, for $n \geq 3$ *see* [**37**]). Furthermore there is an analogue of Theorem 3.48 when $\mathrm{Bir}(\mathbb{P}^n_{\mathbb{C}})$ is endowed with the Euclidean topology:

THEOREM 3.50 ([**41**]). *Let $n \geq 2$ be an integer. The topological group $\mathrm{Bir}(\mathbb{P}^n_{\mathbb{C}})$ is topologically simple when endowed with the Euclidean topology.*

The proof of Theorem 3.50 is similar to the proof of Theorem 3.48, so we will just focus on this last one.

SKETCH OF THE PROOF OF THEOREM 3.48. Let us first prove the statement for $n = 1$.

LEMMA 3.51. *Let $n \geq 2$ be an integer. The group $\mathrm{PSL}(n,\mathbb{C})$ is dense in $\mathrm{PGL}(n,\mathbb{C})$ with respect to the Zariski topology.*

Moreover every non-trivial normal subgroup of $\mathrm{PGL}(n,\mathbb{C})$ contains $\mathrm{PSL}(n,\mathbb{C})$. In particular $\mathrm{PGL}(n,\mathbb{C})$ does not contain any non-trivial normal strict subgroups closed for the Zariski topology.

PROOF. The group morphism $\det\colon \mathrm{GL}(n,\mathbb{C}) \to \mathbb{C}^*$ yields a group morphism

$$\det\colon \mathrm{PGL}(n,\mathbb{C}) \to \mathbb{C}^*/\{f^n \,|\, f \in \mathbb{C}^*\}$$

whose kernel is the group $\mathrm{PSL}(n,\mathbb{C})$. Denote by id the identity matrix of size $(n-1)\times(n-1)$ and consider the morphism

$$\rho\colon \mathbb{A}^1_{\mathbb{C}} \smallsetminus \{0\} \to \mathrm{PGL}(n,\mathbb{C}), \qquad t \mapsto \begin{pmatrix} t & 0 \\ 0 & \mathrm{id} \end{pmatrix}.$$

Note that $\rho^{-1}(\mathrm{PSL}(n,\mathbb{C}))$ contains $\{t^n \,|\, t \in \mathbb{A}^1_{\mathbb{C}}\}$ which is an infinite subset of $\mathbb{A}^1_{\mathbb{C}}$ and is thus dense in $\mathbb{A}^1_{\mathbb{C}}$. Therefore the closure of $\mathrm{PSL}(n,\mathbb{C})$ contains $\rho(\mathbb{A}^1_{\mathbb{C}} \smallsetminus \{0\})$. Any element of $\mathrm{PGL}(n,\mathbb{C})$ is equal to some $\rho(t)$ modulo $\mathrm{PSL}(n,\mathbb{C})$; hence $\mathrm{PSL}(n,\mathbb{C})$ is dense in $\mathrm{PGL}(n,\mathbb{C})$.

Let N be a non-trivial normal subgroup of $\mathrm{PGL}(n,\mathbb{C})$. Let f be a non-trivial element of N. Let us prove that N contains $\mathrm{PSL}(n,\mathbb{C})$. The center of $\mathrm{PGL}(n,\mathbb{C})$ being trivial, one can replace f by $\alpha\circ f\circ\alpha^{-1}\circ f^{-1}$ where $\alpha \in \mathrm{PGL}(n,\mathbb{C})$ does not commute with f, and assume that f belongs to $\mathrm{N}\cap\mathrm{PSL}(n,\mathbb{C})$. But $\mathrm{PSL}(n,\mathbb{C})$ is a simple group ([**83**, Chapitre II, §2]) so $\mathrm{PSL}(n,\mathbb{C}) \subset \mathrm{N}$.

The first two points imply that $\mathrm{PGL}(n,\mathbb{C})$ does not contain any non-trivial normal strict subgroup which is closed with respect to the Zariski topology. □

We will now focus on $\mathrm{Bir}(\mathbb{P}^n_{\mathbb{C}})$ as soon as $n\geq 2$.

PROPOSITION 3.52 ([**41**]). *Let ϕ be an element of $\mathrm{Bir}(\mathbb{P}^n_{\mathbb{C}})$. Let p be a point of $\mathbb{P}^n_{\mathbb{C}}$ such that ϕ induces a local isomorphism at p, and fixes p. Then there exist morphisms $\nu\colon \mathbb{A}^1_{\mathbb{C}}\smallsetminus\{0\}\to\mathrm{Aut}(\mathbb{P}^n_{\mathbb{C}})$ and $\upsilon\colon\mathbb{A}^1_{\mathbb{C}}\to\mathrm{Bir}(\mathbb{P}^n_{\mathbb{C}})$ such that:*

- ◇ *$\upsilon(t)=\nu(t)^{-1}\circ\phi\circ\nu(t)$ for any $t\in\mathbb{C}$, moreover $\nu(1)=\mathrm{id}$, so $\upsilon(1)=\phi$;*
- ◇ *$\upsilon(0)$ belongs to $\mathrm{Aut}(\mathbb{P}^n_{\mathbb{C}})$ and is the identity if and only if the action of ϕ on the tangent space is trivial.*

PROOF. Up to conjugacy by an element of $\mathrm{Aut}(\mathbb{P}^n_{\mathbb{C}})$ we can assume that $p=(1:0:0:\ldots:0)$. In the affine chart $z_0=1$, one can write ϕ locally as

$$\Big(\frac{p_{1,1}(z_1,\ldots,z_n)+\ldots+p_{1,\ell}(z_1,\ldots,z_n)}{1+q_{1,1}(z_1,\ldots,z_n)+\ldots+q_{1,\ell}(z_1,\ldots,z_n)},\ldots,\\ \frac{p_{n,1}(z_1,\ldots,z_n)+\ldots+p_{n,\ell}(z_1,\ldots,z_n)}{1+q_{n,1}(z_1,\ldots,z_n)+\ldots+q_{n,\ell}(z_1,\ldots,z_n)}\Big)$$

where $p_{i,j}$, $q_{i,j}$ are homogeneous of degree j. For each $t\in\mathbb{C}\smallsetminus\{0\}$ the element

$$\nu_t\colon(z_1,z_2,\ldots,z_n)\mapsto(tz_1,tz_2,\ldots,tz_n)$$

extends to a linear automorphism of $\mathbb{P}^n_{\mathbb{C}}$ that fixes p. Hence the map $t\mapsto\nu_t^{-1}\circ\phi\circ\nu_t$ gives rise to a morphism $\Theta\colon\mathbb{A}^1_{\mathbb{C}}\smallsetminus\{0\}\to\mathrm{Bir}(\mathbb{P}^n_{\mathbb{C}})$ and the image of Θ contains only conjugates of ϕ by linear automorphisms. Note that

$$\Theta\colon t\mapsto\Big(\frac{p_{1,1}(z_1,\ldots,z_n)+tp_{1,2}(z_1,\ldots,z_n)+\ldots+t^{\ell-1}p_{1,\ell}(z_1,\ldots,z_n)}{1+tq_{1,1}(z_1,\ldots,z_n)+t^2q_{1,2}(z_1,\ldots,z_n)+\ldots+t^{\ell}q_{1,\ell}(z_1,\ldots,z_n)},\\ \ldots,\frac{p_{n,1}(z_1,\ldots,z_n)+tp_{n,2}(z_1,\ldots,z_n)+\ldots+t^{\ell-1}p_{n,\ell}(z_1,\ldots,z_n)}{1+tq_{n,1}(z_1,\ldots,z_n)+t^2q_{n,2}(z_1,\ldots,z_n)+\ldots+t^{\ell}q_{n,\ell}(z_1,\ldots,z_n)}\Big)$$

and $\Theta(0)$ corresponds to the linear part of Θ at p which is locally given by

$$\big(p_{1,1}(z_1,\ldots,z_n),\ldots,p_{n,1}(z_1,\ldots,z_n)\big).$$

As ϕ is a local isomorphism at p, this linear part is an automorphism of $\mathbb{P}^n_{\mathbb{C}}$. Furthermore it is trivial if and only if the action of ϕ on the tangent space is trivial. □

Let $\phi\in\mathrm{Bir}(\mathbb{P}^n_{\mathbb{C}})\smallsetminus\{\mathrm{id}\}$; it induces an isomorphism from $\mathcal{U}$ to $\mathcal{V}$ where $\mathcal{U}$, $\mathcal{V}\subset\mathbb{P}^n_{\mathbb{C}}$ are two non-empty open subsets. There exist a point p in $\mathcal{U}$ and two automorphisms α_1, α_2 of $\mathbb{P}^n_{\mathbb{C}}$ such that

- ◇ $\psi=\alpha_1\circ\phi\circ\alpha_2$ fixes p,
- ◇ $\psi=\alpha_1\circ\phi\circ\alpha_2$ is a local isomorphism at p,
- ◇ $D_p\psi$ is not trivial.

According to Proposition 3.52 there exist morphisms $\nu\colon\mathbb{A}^1_{\mathbb{C}}\smallsetminus\{0\}\to\mathrm{Aut}(\mathbb{P}^n_{\mathbb{C}})$ and $\upsilon_1\colon\mathbb{A}^1_{\mathbb{C}}\to\mathrm{Bir}(\mathbb{P}^n_{\mathbb{C}})$ such that

- ◇ $\upsilon_1(t)=\nu(t)^{-1}\circ\psi^{-1}\circ\nu(t)$ for each $t\neq 0$,
- ◇ $\upsilon_1(0)$ is an automorphism of $\mathbb{P}^n_{\mathbb{C}}$.

Consider the morphism $\upsilon_2\colon\mathbb{A}^1_{\mathbb{C}}\to\mathrm{Bir}(\mathbb{P}^n_{\mathbb{C}})$ defined by

$$\upsilon_2(t)=\alpha_1^{-1}\circ\psi\circ\upsilon_1(t)\circ\upsilon_1(0)^{-1}\circ\alpha_2^{-1}.$$

Since α_1, α_2, $\upsilon_1(0)$ and $\nu(t)$ are automorphisms of $\mathbb{P}^n_\mathbb{C}$ for all $t \neq 0$,

$$\upsilon_2(t) = \alpha_1^{-1} \circ \left(\psi \circ \nu(t)^{-1} \circ \psi^{-1}\right) \circ \nu(t) \circ \upsilon_1(0)^{-1} \circ \alpha_2^{-1}$$

belongs for any $t \neq 0$ to the normal subgroup of $\mathrm{Bir}(\mathbb{P}^n_\mathbb{C})$ generated by $\mathrm{Aut}(\mathbb{P}^n_\mathbb{C})$. As a consequence $\phi = \upsilon_2(0)$ belongs to the closure of the normal subgroup of $\mathrm{Bir}(\mathbb{P}^n_\mathbb{C})$ generated by $\mathrm{Aut}(\mathbb{P}^n_\mathbb{C})$. The normal subgroup of $\mathrm{Bir}(\mathbb{P}^n_\mathbb{C})$ generated by $\mathrm{Aut}(\mathbb{P}^n_\mathbb{C})$ is dense in $\mathrm{Bir}(\mathbb{P}^n_\mathbb{C})$ (*see* [**34**]).

In particular $\mathrm{Bir}(\mathbb{P}^n_\mathbb{C})$ does not contain any non-trivial closed normal strict subgroup. Indeed let $\{\mathrm{id}\} \neq \mathrm{N} \subset \mathrm{Bir}(\mathbb{P}^n_\mathbb{C})$ be a closed normal subgroup with respect to the Zariski topology. Then $\mathrm{Aut}(\mathbb{P}^n_\mathbb{C}) \subset \mathrm{N}$ (*see* [**41**, Prop. 3.3, Lemma 3.4]). Since N is closed it contains the closure of the normal subgroup generated by $\mathrm{Aut}(\mathbb{P}^n_\mathbb{C})$ which is equal to $\mathrm{Bir}(\mathbb{P}^n_\mathbb{C})$. □

Furthermore, one has:

THEOREM 3.53 ([**41**]). *If* $n \geq 1$, *the group* $\mathrm{Bir}(\mathbb{P}^n_\mathbb{C})$ *is connected with respect to the Zariski topology.*

If $n \geq 2$, *the group* $\mathrm{Bir}(\mathbb{P}^n_\mathbb{C})$ *is path-connected, and thus connected with respect to the Euclidean topology.*

Let us give an idea of the proof of this statement. We start with an example.

EXAMPLE 6. Let $n \geq 2$ and let α be an element of $\mathbb{C}^*$. Consider the birational self map of $\mathbb{P}^n_\mathbb{C}$ given by

$$\Phi\colon (z_0 : z_1 : \ldots : z_n) \dashrightarrow \left(\frac{z_0(z_1+\alpha z_2)+z_1z_2}{z_1+z_2} : z_1 : z_2 : \ldots : z_n\right).$$

The points $p = (0:1:0:0:\ldots:0)$ and $q = (0:0:1:0:0:\ldots:0)$ are fixed by Φ. Applying Proposition 3.52 to the points p and q we get two morphisms Θ_1, $\Theta_2\colon \mathbb{A}^1_\mathbb{C} \to \mathrm{Bir}(\mathbb{P}^n_\mathbb{C})$ such that

- ◇ $\Theta_1(0)\colon (z_0 : z_1 : \ldots : z_n) \mapsto (z_0 + z_2 : z_1 : z_2 : \ldots : z_n) \in \mathrm{Aut}(\mathbb{P}^n_\mathbb{C})$,
- ◇ $\Theta_2(0)\colon (z_0 : z_1 : \ldots : z_n) \mapsto (\alpha z_0 + z_1 : z_1 : z_2 : \ldots : z_n) \in \mathrm{Aut}(\mathbb{P}^n_\mathbb{C})$,
- ◇ $\Theta_1(1) = \Theta_2(1) = \Phi$.

PROPOSITION 3.54. *Let* $n \geq 2$ *be an integer. For any* ϕ, $\psi \in \mathrm{Bir}(\mathbb{P}^n_\mathbb{C})$ *there is a morphism* $\upsilon\colon \mathbb{P}^1_\mathbb{C} \to \mathrm{Bir}(\mathbb{P}^n_\mathbb{C})$ *such that* $\upsilon(0) = \phi$ *and* $\upsilon(1) = \psi$.

PROOF. Up to composition with ϕ^{-1} one can assume that $\phi = \mathrm{id}$. Let us consider the subset S of $\mathrm{Bir}(\mathbb{P}^n_\mathbb{C})$ given by

$$S = \left\{\varphi \in \mathrm{Bir}(\mathbb{P}^n_\mathbb{C}) \,|\, \exists\, \nu\colon \mathbb{A}^1_\mathbb{C} \to \mathrm{Bir}(\mathbb{P}^n_\mathbb{C}) \text{ morphism such that } \nu(0) = \mathrm{id} \text{ and } \nu(1) = \varphi\right\}.$$

Let ϕ (resp. ψ) be an element of S; denote by ν_ϕ (resp. ν_ψ) the associated morphism. We define a morphism $\nu_{\phi\circ\psi}\colon \mathbb{A}^1_\mathbb{C} \to \mathrm{Bir}(\mathbb{P}^n_\mathbb{C})$ by $\nu_{\phi\circ\psi}(t) = \nu_\phi(t)\circ\nu_\psi(t)$ which satisfies $\nu_{\phi\circ\psi}(0) = \mathrm{id}$ and $\nu_{\phi\circ\psi}(1) = \phi\circ\psi$. For any $\varphi \in \mathrm{Bir}(\mathbb{P}^n_\mathbb{C})$ it is also possible to define a morphism $t \mapsto \varphi \circ \nu_\phi(t) \circ \varphi^{-1}$. Therefore, S is a normal subgroup of $\mathrm{Bir}(\mathbb{P}^n_\mathbb{C})$.

CLAIM 3.55 ([**34**]). *The group* S *contains* $\mathrm{PSL}(n+1,\mathbb{C})$.

Take α, Φ, Θ_1 and Θ_2 as in Example 6; for $i \in \{1, 2\}$ the morphisms

$$t \mapsto \Theta_i(t) \circ (\Theta_i(0))^{-1}$$

show that $g\circ\big(\Theta_1(0)\big)^{-1}$ and $g\circ\big(\Theta_2(0)\big)^{-1}$ belong to S, hence $\Theta_1(0)\circ\big(\Theta_2(0)\big)^{-1}$ belong to S. But $\Theta_1(0)$ belongs to $\mathrm{PSL}(n+1,\mathbb{C})\subset S$, so $\Theta_2(0)$ belongs to S. Thus $\mathrm{Aut}(\mathbb{P}^n_\mathbb{C})=\mathrm{PGL}(n+1,\mathbb{C})$ is contained in S.

Take $\phi\in\mathrm{Bir}(\mathbb{P}^n_\mathbb{C})$ of degree $d\geq 2$. Let p be a point of $\mathbb{P}^n_\mathbb{C}$ such that ϕ induces a local isomorphism at p. Consider an element A of $\mathrm{PSL}(n+1,\mathbb{C})$ such that $A\circ\phi$ fixes p. There exists a morphism $\theta\colon\mathbb{A}^1_\mathbb{C}\to\mathrm{Bir}(\mathbb{P}^n_\mathbb{C})$ such that $\theta(0)$ belongs to $\mathrm{Aut}(\mathbb{P}^n_\mathbb{C})$ and $\theta(1)=A\circ\phi$. Let us define $\theta'\colon\mathbb{A}^1_\mathbb{C}\to\mathrm{Bir}(\mathbb{P}^n_\mathbb{C})$ by $\theta'(t)=\rho(t)\circ\theta(0)^{-1}$. Then $\theta'(1)=A\circ\phi\circ\theta(0)^{-1}$. But A and $\theta(0)$ belong to $\mathrm{Aut}(\mathbb{P}^n_\mathbb{C})\subset S$, so ϕ belongs to S. □

3.5. Regularization of rational group actions

The aim of [**126**] is to give a modern proof of the regularization theorem of Weil which says:

THEOREM 3.56 ([**174**]). *For every rational action of an algebraic group* G *on a variety X there exist a variety Y with a regular action of* G *and a* G*-equivariant birational map $X\dashrightarrow Y$.*

In this section a variety is an algebraic complex variety, and an algebraic group is an algebraic $\mathbb{C}$-group.

A rational map $\phi\colon X\dashrightarrow Y$ is called *biregular* in $p\in X$ if there is an open neighborhood $\mathcal{U}\subset(X\smallsetminus\mathrm{Base}(\phi))$ of p such that $\phi_{|\mathcal{U}}\colon\mathcal{U}\hookrightarrow Y$ is an open immersion. As a result the subset

$$X'=\big\{p\in X\,|\,\phi\text{ is biregular in }p\big\}$$

is open in X, and the induced morphism $\phi\colon X'\hookrightarrow Y$ is an open immersion. We can thus state:

LEMMA 3.57. *Let X and Y be two varieties. Let $\phi\colon X\dashrightarrow Y$ be a birational map. Then the set*

$$\mathrm{Breg}(\phi)=\big\{p\in X\,|\,\phi\textit{ is biregular in }p\big\}$$

is open and dense in X.

3.5.1. Rational group actions. Let X and Z be two varieties. Let us recall that a map $\phi\colon Z\to\mathrm{Bir}(X)$ is a morphism if there exists an open dense set $\mathcal{U}\subset Z\times X$ such that

- ◇ the induced map $\mathcal{U}\to X$, $(q,p)\mapsto\phi(q)(p)$ is a morphism of varieties;
- ◇ for every $q\in Z$ the open set $\mathcal{U}_q=\big\{p\in X\,|\,(q,p)\in\mathcal{U}\big\}$ is dense in X;
- ◇ for every $q\in Z$ the birational map $\phi(q)\colon X\dashrightarrow X$ is defined on $\mathcal{U}_q$.

Equivalently there is a rational map $\phi\colon Z\times X\to X$ such that for every $q\in Z$

- ◇ the open subset $(Z\times X\smallsetminus\mathrm{Base}(\phi))\cap(\{q\}\times X)$ is dense in $\{q\}\times X$;
- ◇ the induced birational map $\phi_q\colon X\dashrightarrow X$, $p\mapsto\phi(q,p)$ is birational.

Recall that this definition allows us to define the Zariski topology on $\mathrm{Bir}(X)$ (*see* §3.1).

We can now define rational group actions on varieties. Let G be an algebraic group, and let X be a variety. A *rational action* of G on X is a morphism $\rho\colon\mathrm{G}\to\mathrm{Bir}(X)$ which is a morphism of groups. In other words there is a rational map still denoted ρ:

$$\rho\colon\mathrm{G}\times X\dashrightarrow X$$

such that

- ◇ the open set $((\mathrm{G} \times X) \setminus \mathrm{Base}(\rho)) \cap (\{g\} \times X)$ is dense in $\{g\} \times X$ for every $g \in \mathrm{G}$;
- ◇ the induced map $\rho_g \colon X \dashrightarrow X$, $p \mapsto \rho(g,p)$ is birational for every $g \in \mathrm{G}$;
- ◇ the map $g \mapsto \rho_g$ is a group morphism.

THEOREM 3.58 ([**126**]). *Let* $\rho\colon \mathrm{G} \to \mathrm{Bir}(X)$ *be a rational action where* X *is affine. Assume that there exists a dense subgroup* $\Gamma \subset \mathrm{G}$ *such that* $\rho(\Gamma) \subset \mathrm{Aut}(X)$. *Then the* G*-action on* X *is regular.*

DEFINITION. Let X and Y be two varieties. Let ρ be a rational G-action on X. Let μ be a rational G-action on Y.

A dominant rational map $\phi\colon X \dashrightarrow Y$ is G-*equivariant* if the following holds: for every $(g,p) \in \mathrm{G} \times X$ such that

- ◇ ρ is defined in (g,p),
- ◇ ϕ is defined in p and in $\rho(g,p)$,
- ◇ μ is defined in $(g, \phi(p))$,

we have $\phi(\rho(g,p)) = \mu(g, \phi(p))$.

REMARK 3.59. The set of $(g,p) \in \mathrm{G} \times X$ satisfying the previous assumptions is open and dense in $\mathrm{G} \times X$ and has the property that it meets all $\{g\} \times X$ in a dense open set.

Let X be a variety with a rational action $\rho\colon \mathrm{G} \times X \dashrightarrow X$ of an algebraic group G. Consider

$$\widetilde{\rho}\colon \mathrm{G} \times X \dashrightarrow \mathrm{G} \times X, \qquad (g,p) \mapsto (g, \rho(g,p)).$$

It is clear that $(\mathrm{G} \times X) \setminus \mathrm{Base}(\widetilde{\rho}) = (\mathrm{G} \times X) \setminus \mathrm{Base}(\rho)$. Furthermore $\widetilde{\rho}$ is birational with inverse $\widetilde{\rho}^{-1}(g,p) = (g, \rho(g^{-1},p))$, that is,

$$\widetilde{\rho}^{-1} = \tau \circ \widetilde{\rho} \circ \tau$$

where τ is the isomorphism

$$\tau\colon \mathrm{G} \times X \to \mathrm{G} \times X, \qquad (g,p) \mapsto (g^{-1},p).$$

DEFINITION. A point $x \in X$ is called G-*regular* for the rational G-action ρ on X if $\mathrm{Breg}(\widetilde{\rho}) \cap (\mathrm{G} \times \{p\})$ is dense in $\mathrm{G} \times \{p\}$.

In other words a point $p \in X$ is called G-regular for the rational G-action ρ on X if $\widetilde{\rho}$ is biregular in (g,p) for all g in a dense open set of G.

Denote by $X_{\mathrm{reg}} \subset X$ the set of G-regular points.

Let $\lambda_g\colon \mathrm{G} \xrightarrow{\sim} \mathrm{G}$ be the left multiplication with $g \in \mathrm{G}$. For any $h \in \mathrm{G}$ the diagram

$$\begin{array}{ccc} \mathrm{G} \times X & \overset{\widetilde{\rho}}{\dashrightarrow} & \mathrm{G} \times X \\ {\scriptstyle \lambda_h \times \mathrm{id}} \downarrow & & \downarrow {\scriptstyle \lambda_h \times \rho_h} \\ \mathrm{G} \times X & \underset{\widetilde{\rho}}{\dashrightarrow} & \mathrm{G} \times X \end{array}$$

commutes. This implies the following statement:

LEMMA 3.60 ([**126**]). *If ρ is defined in (g,p) and if ρ_h is defined in $g\cdot p$, then ρ is defined in (hg,p).*

If $\widetilde{\rho}$ is biregular in (g,p) and if ρ_h is biregular in $g\cdot p$, then $\widetilde{\rho}$ is biregular in (hg,p).

PROPOSITION 3.61 ([**126**]). *The set X_{reg} of* G*-regular points is open and dense in X.*

If p belongs to X_{reg} and if $\widetilde{\rho}$ is biregular in (g,p), then $g\cdot p$ belongs to X_{reg}.

PROOF. Let $\mathrm{G}=\mathrm{G}_0\cup\mathrm{G}_1\cup\ldots\cup\mathrm{G}_n$ be the decomposition into connected components. Then $D_i=\mathrm{Breg}(\rho)\cap(\mathrm{G}_i\times X)$ is open and dense for all i (Lemma 3.57); the same holds for the image $\mathcal{D}_i\subseteq X$ under the projection onto X. Since $X_{\mathrm{reg}}=\bigcap_i\mathcal{D}_i$, the set X_{reg} is open and dense in X.

If $\widetilde{\rho}$ is biregular in (g,p), then $\widetilde{\rho}^{-1}=\tau\circ\widetilde{\rho}\circ\tau$ is biregular in $(g,g\cdot p)$. As a consequence, $\widetilde{\rho}$ is biregular in $\tau(g,g\cdot p)=(g^{-1},g\cdot p)$. If p is G-regular, then ρ_h is biregular in p for all h in a dense open subset G$'$ of G. According to the second assertion of Lemma 3.60 the birational map $\widetilde{\rho}$ is biregular in $(hg^{-1},g\cdot p)$ for all $h\in\mathrm{G}'$. Hence $g\cdot p$ belongs to X_{reg}. □

A consequence of Proposition 3.61 allows us to only consider the case of a rational G-action such that every point is G-regular.

COROLLARY 3.62 ([**126**]). *For the rational* G*-action on X_{reg} every point is* G*-regular.*

LEMMA 3.63 ([**126**]). *Assume that $X=X_{reg}$. If ρ_g is defined in p, i.e., if $p\in X\setminus\mathrm{Base}(\rho_g)$, then ρ_g is biregular in p.*

PROOF. Suppose that ρ_g is defined in $p\in X$. As $X=X_{\mathrm{reg}}$, there exists a dense open subset G$'$ of G such that for all $h\in\mathrm{G}'$

- ◇ ρ_h is biregular in $g\cdot p$,
- ◇ ρ_{hg} is biregular in p.

Since $\rho_{hg}=\rho_h\circ\rho_g$ the map ρ_g is biregular in p. □

Let us recall that if $\phi\colon X\dashrightarrow Y$ is a rational map, its graph $\Gamma(\phi)$ is defined by

$$\Gamma(\phi)=\big\{(x,y)\in X\times Y\,|\,x\in X\setminus\mathrm{Base}(\phi)\text{ and }\phi(x)=y\big\}.$$

In particular $\mathrm{pr}_1(\Gamma(\phi))=X\setminus\mathrm{Base}(\phi)$ and $\mathrm{pr}_2(\Gamma(\phi))=\phi(X\setminus\mathrm{Base}(\phi))$.

LEMMA 3.64 ([**126**]). *Let ρ be a rational* G*-action on a variety X. Suppose that every point of X is* G*-regular, that is, $X=X_{reg}$. Then for every $g\in\mathrm{G}$ the graph $\Gamma(\rho_g)$ of ρ_g is closed in $X\times X$.*

PROOF. Denote by Γ the closure $\overline{\Gamma(\rho_g)}$ of the graph of ρ_g in $X\times X$. Let us prove that for any $(x_0,y_0)\in\Gamma$ the rational map ρ_g is defined in x_0. It is equivalent to prove that the morphism $\mathrm{pr}_{1|\Gamma}\colon\Gamma\to X$ is biregular in (x_0,y_0).

Let h be an element of G such that ρ_{hg} is biregular in x_0, and ρ_h is biregular in y_0. Consider the birational map

$$\phi=(\rho_{hg},\rho_h)\colon X\times X\dashrightarrow X\times X.$$

If ϕ is defined in $(x,y)\in\Gamma(\rho_g)$, $y=g\cdot x$, then $\phi(x,y)=((hg)\cdot x,(hg)\cdot x)$ belongs to the diagonal

$$\Delta(X)=\big\{(x,x)\in X\,|\,x\in X\big\}$$

of $X \times X$. It follows that $\overline{\phi(\Gamma)} \subseteq \Delta(X)$. Since ϕ is biregular in (x_0, y_0), the restriction $\varphi = \phi_{|\Gamma} \colon \Gamma \dashrightarrow \Delta(X)$ of ϕ to Γ is also biregular in (x_0, y_0). By construction

$$\rho_{hg} \circ \mathrm{pr}_{1|\Gamma} = \mathrm{pr}_{1|\Delta(X)} \circ \varphi;$$

indeed

$$\begin{array}{ccc} X \times X & \overset{\rho_{hg} \times \rho_h}{\dashrightarrow} & X \times X \\ \uparrow & & \uparrow \\ \Gamma & \overset{\varphi}{\dashrightarrow} & \Delta(X) \\ {\scriptstyle \mathrm{pr}_{1|\Gamma}} \downarrow & & \downarrow {\scriptstyle \mathrm{pr}_{1|\Delta(X)}} \\ X & \overset{\rho_{hg}}{\dashrightarrow} & X \end{array}$$

But ρ_{hg} is biregular in $\mathrm{pr}_{1|\Gamma}(x_0, x_0)$, φ is biregular in (x_0, y_0), and $\mathrm{pr}_{1|\Delta(X)}$ is an isomorphism, so $\mathrm{pr}_{1|\Gamma}$ is biregular in (x_0, y_0). $\square$

LEMMA 3.65 ([**126**]). *Let ρ be a rational action of* G *on a variety X. Suppose that there is a dense open subset $\mathcal{U}$ of X such that*

$$\widetilde{\rho} \colon \mathrm{G} \times \mathcal{U} \to \mathrm{G} \times X, \qquad (g, p) \mapsto \big(g, \rho(g, p)\big)$$

defines an open immersion. Then the open dense subset $Y = \bigcup_{g \in \mathrm{G}} g \cdot \mathcal{U} \subseteq X$ carries a regular G*-action.*

PROOF. Any ρ_g induces an isomorphism $\mathcal{U} \overset{\sim}{\longrightarrow} g \cdot \mathcal{U}$. Therefore,

$$Y = \bigcup_{g \in \mathrm{G}} g \cdot X \subset X$$

is stable under all ρ_g. By assumption the induced map on $\mathrm{G} \times \mathcal{U}$ is a morphism, so the induced map on $\mathrm{G} \times g \cdot \mathcal{U}$ is a morphism for all $g \in \mathrm{G}$. As a result the induced map $\mathrm{G} \times Y \to Y$ is a morphism. $\square$

3.5.2. Construction of a regular model.

THEOREM 3.66 ([**126**]). *Let X be a variety with a rational action of* G*. Suppose that every point of X is* G*-regular. Then there exists a variety Y with a regular* G*-action and a* G*-equivariant open immersion.*

Assume now that X is a variety with a rational G-action ρ such that $X_{\mathrm{reg}} = X$. Consider a finite subset $S = \{g_0 = e,\, g_1,\, g_2,\, \ldots,\, g_m\}$ of G. Denote by $X^{(0)}$, $X^{(1)}$, $\ldots$, $X^{(m)}$ some copies of X. Consider the disjoint union

$$X(S) = X^{(0)} \cup X^{(1)} \cup \ldots \cup X^{(m)}.$$

Let us define on $X^{(i)}$ the following relations

$$\begin{cases} \forall i \quad p_i \sim p_i' \Longleftrightarrow p_i = p_i', \\ \forall i,\, j \quad i \neq j \quad p_i \sim p_j \Longleftrightarrow \rho_{g_j^{-1} g_i} \text{ is defined in } p_i \text{ and sends } p_i \text{ to } p_j. \end{cases}$$

This defines an equivalence relation (Lemma 3.63 is needed to prove the symmetry). Consider $\widetilde{X}(S) = {}^{X(S)}\!/\!_{\sim}$ the set of equivalence classes endowed with the induced topology.

LEMMA 3.67 ([**126**]). *The maps* $\iota_i\colon X^{(i)} \to \widetilde{X}(S)$ *are open immersions and endow* $\widetilde{X}(S)$ *with the structure of a variety.*

Let us fix the open immersion $\iota_0\colon X = X^{(0)} \hookrightarrow \widetilde{X}(S)$. Then G acts rationally on $\widetilde{X}(S)$ via $\overline{\rho} = \overline{\rho}_S$ such that ι_0 is G-equivariant. Consider any $X^{(i)}$ as the variety X with the rational G-action

$$\rho^{(i)}(g,p) = \rho(g_i g g_i^{-1}, p);$$

by construction of $\widetilde{X}(S)$ the open immersions

$$\iota_i\colon X^{(i)} \hookrightarrow \widetilde{X}(S)$$

are all G-equivariant.

LEMMA 3.68 ([**126**]). *Let* $\widetilde{X}^{(i)}$ *be the image of the open immersion* $\iota_i\colon X^{(i)} \hookrightarrow \widetilde{X}(S)$. *For all* i *the rational map* $\overline{\rho}_{g_i}$ *is defined on* $\widetilde{X}^{(0)}$.

Furthermore $\overline{\rho}_{g_i}\colon \widetilde{X}^{(0)} \xrightarrow{\sim} \widetilde{X}^{(i)}$ *defines an isomorphism.*

PROOF. Consider the open immersion

$$\tau_i = \iota_i \circ \iota_0^{-1}\colon \widetilde{X}^{(0)} \hookrightarrow \widetilde{X}(S).$$

Note that $\operatorname{im}\tau_i = \widetilde{X}^{(i)}$. Let us check that $\tau_i(\overline{p}) = g_i\overline{p}$. It is sufficient to show that it holds on an open dense subset of $\widetilde{X}^{(0)}$. Let $\mathcal{U} \subseteq X$ be the open dense set where $g_i \cdot p$ is defined. Take p in $\mathcal{U}$. On the one hand by definition

$$\iota_0(g_i \cdot p) = \iota_i(p);$$

on the other hand

$$\iota_0(g_i \cdot p) = g_i \cdot \iota_0(p).$$

As a result $g_i \cdot \iota_0(p) = \iota_i(p)$ and

$$\tau_i(\overline{p}) = \iota_i(\iota_0^{-1}(\overline{p})) = g_i \cdot \iota_0(\iota_0^{-1}(\overline{p})) = g_i \cdot \overline{p}$$

for any $\overline{p} \in \iota_0(\mathcal{U})$. □

PROOF OF THEOREM 3.66. Set $D = \mathrm{Breg}(\rho) \cap (\mathrm{G} \times X)$. Since $X_{\mathrm{reg}} = X$ for any $p \in X$ there is an element g in G such that $(g,p) \in D$. As a consequence $\bigcup_{g\in \mathrm{G}} g \cdot D = \mathrm{G} \times X$ where G acts on $\mathrm{G} \times X$ by left-multiplication on G. Hence $\bigcup_i g_i D = \mathrm{G} \times X$ for a suitable finite subset

$$S = \{g_0 = e,\, g_1,\, g_2,\, \ldots,\, g_m\}.$$

Recall that $\widetilde{X}^{(0)} = \mathrm{im}(\iota_0)$. Let $D^{(0)} \subset \mathrm{G} \times \widetilde{X}^{(0)}$ be the image of D. Consider the rational map

$$\widetilde{\rho}_S\colon \mathrm{G} \times \widetilde{X}^{(0)} \dashrightarrow \mathrm{G} \times \widetilde{X}(S).$$

The map $(g,p) \mapsto (g, g\cdot p)$ is the composition of $(g,p) \mapsto (g, (g_i^{-1}g)\cdot p)$ and $(g,y) \mapsto (g, g_i \cdot y)$. The first one is biregular on $g_i \cdot D^{(0)}$ and its image is contained in $\mathrm{G}\times\widetilde{X}^{(0)}$; the second is biregular on $\mathrm{G} \times \widetilde{X}^{(0)}$ (Lemma 3.68). As $\mathrm{G} \times \widetilde{X}^{(0)} = \bigcup_i g_i \cdot D^{(0)}$ the map $\widetilde{\rho}_S$ is biregular. As a consequence the rational action $\overline{\rho}$ of G on $\widetilde{X}(S)$ has the property that

$$\widetilde{\rho}_S \colon \mathrm{G} \times \widetilde{X}^{(0)} \hookrightarrow \mathrm{G} \times \widetilde{X}(S)$$

defines an open immersion. Lemma 3.65 allows us to conclude. □

3.5.3. Proof of Theorem 3.58. Let us start with the following statement:

LEMMA 3.69 ([**126**]). *Let X, Y, Z be varieties. Assume that Z is affine. Let $\phi\colon X \times Y \dashrightarrow Z$ be a rational map. Suppose that*

- ◇ *there exists an open dense subset $\mathcal{U}$ of Y such that ϕ is defined on $X \times \mathcal{U}$;*
- ◇ *there exists a dense subset X' of X such that the induced maps $\phi_p\colon \{p\} \times Y \to Z$ are morphisms for all $p \in X'$.*

Then ϕ is a regular morphism.

Consider a rational action $\rho\colon \mathrm{G} \to \mathrm{Bir}(X)$ of an algebraic group on a variety X. Assume that there is a dense subgroup Γ of G such that $\rho(\Gamma) \subset \mathrm{Aut}(X)$.

- ◇ Let us first prove that the rational G-action on the open dense set $X_{\mathrm{reg}} \subseteq X$ is regular. For every $p \in X_{\mathrm{reg}}$ there is $g \in \Gamma$ such that $\widetilde{\rho}$ is biregular in (g, p). By assumption for any $h \in \Gamma$ the map ρ_h is biregular on X, hence the map $\widetilde{\rho}$ is biregular in (h, p) for any $h \in \Gamma$ (Lemma 3.60). Furthermore $h \cdot p$ belongs to X_{reg} (Proposition 3.61), *i.e.*, X_{reg} is stable under Γ. According to Theorem 3.66 there exists a G-equivariant open immersion

$$X_{\mathrm{reg}} \hookrightarrow Y$$

  where Y is a variety with a regular G-action. The complement $Y \smallsetminus X_{\mathrm{reg}}$ is closed and Γ-stable, so $Y \smallsetminus X_{\mathrm{reg}}$ is stable under $\overline{\Gamma} = \mathrm{G}$.
- ◇ From the previous point the rational map

$$\rho\colon \mathrm{G} \times X \dashrightarrow X$$

  has the following properties:
  - there is a dense open set $X_{\mathrm{reg}} \subseteq X$ such that ρ is regular on $\mathrm{G} \times X_{\mathrm{reg}}$;
  - for every $g \in \Gamma$ the rational map

$$\rho_g\colon X \to X, \qquad\qquad p \mapsto \rho(g, p)$$

    is a regular isomorphism.

  Lemma 3.69 implies that ρ is a regular action in case X is affine.

REMARK 3.70. All the statements in this section hold for an algebraically closed field.

CHAPTER 4

Generators and relations of the Cremona group

As we already say:

THEOREM 4.1 ([**60**]). *The group* $\mathrm{Bir}(\mathbb{P}^2_\mathbb{C})$ *is generated by* $\mathrm{Aut}(\mathbb{P}^2_\mathbb{C}) = \mathrm{PGL}(3,\mathbb{C})$ *and the standard quadratic involution*

$$\sigma_2\colon (z_0 : z_1 : z_2) \dashrightarrow (z_1z_2 : z_0z_2 : z_0z_1).$$

This result is well-known as the theorem of Noether and Castelnuovo. Noether was the first mathematician to state this result at the end of the 19th century. Nevertheless the first exact proof is due to Castelnuovo. Noether's idea was the following. Let us consider a birational self map ϕ of $\mathbb{P}^2_\mathbb{C}$. Take a quadratic birational self map q of $\mathbb{P}^2_\mathbb{C}$ such that the three base-points of q are three base-points of ϕ of highest multiplicity. Then $\deg(\phi\circ q) < \deg\phi$. By induction one gets a birational map of degree 1. But such a quadratic birational map q may not exist. This is, for instance, the case if one starts with the polynomial automorphism

$$(z_0 : z_1 : z_2) \dashrightarrow (z_1^3 - z_0z_2^2 : z_1z_2^2 : z_2^3).$$

In [**2**] Alexander fixes Noether's proof by introducing the notion of complexity of a map: start with a birational self map ϕ of the complex projective plane; one can find a quadratic birational self map q of the complex projective plane such that

- ◇ either the complexity of $\phi\circ q$ is strictly less than the complexity of ϕ;
- ◇ or the complexities of $\phi\circ q$ and ϕ are equal but $\#\mathrm{Base}(\phi\circ q) < \#\mathrm{Base}(\phi)$.

Alexander's proof is a proof by induction on these two integers.

REMARK 4.2. One consequence of the Noether and Castelnuovo theorem is: the Jonquières group and $\mathrm{Aut}(\mathbb{P}^2_\mathbb{C}) = \mathrm{PGL}(3,\mathbb{C})$ generate $\mathrm{Bir}(\mathbb{P}^2_\mathbb{C})$. This result is "weaker"; nevertheless it has the following nice property:

THEOREM 4.3 ([**1**]). *Let* ϕ *be an element of* $\mathrm{Bir}(\mathbb{P}^2_\mathbb{C})$. *There exist* j_1, j_2, $\ldots$, j_k *in* $\mathcal{J}$ *and* A *in* $\mathrm{PGL}(3,\mathbb{C})$ *such that*

- ◇ $\phi = A\circ j_k\circ j_{k-1}\circ\ldots\circ j_2\circ j_1$;
- ◇ *for any* $1\leq i\leq n-1$,

$$\deg(A\circ j_k\circ j_{k-1}\circ\ldots\circ j_{i+1}\circ j_i) > \deg(A\circ j_k\circ j_{k-1}\circ\ldots\circ j_{i+2}\circ j_{i+1}).$$

The first presentation of the plane Cremona group is given by Gizatullin:

THEOREM 4.4 ([**102**]). *The Cremona group* $\mathrm{Bir}(\mathbb{P}^2_\mathbb{C})$ *is generated by the set* $\mathcal{Q}$ *of all quadratic maps.*

The relations in $\mathrm{Bir}(\mathbb{P}^2_\mathbb{C})$ *are consequences of relations of the form* $q_1\circ q_2\circ q_3 = \mathrm{id}$ *where* q_1, q_2, q_3 *are quadratic birational self maps of* $\mathbb{P}^2_\mathbb{C}$. *In other words we have the following presentation:*

$$\mathrm{Bir}(\mathbb{P}^2_\mathbb{C}) = \langle \mathcal{Q}\,|\, q_1\circ q_2\circ q_3 = \mathrm{id}\ \ \forall\, q_1,\, q_2,\, q_3\in\mathcal{Q} \textit{ such that } q_1\circ q_2\circ q_3 = \mathrm{id} \textit{ in } \mathrm{Bir}(\mathbb{P}^2_\mathbb{C})\rangle.$$

Two years later Iskovskikh proved the following statement:

THEOREM 4.5 ([**113,114**]). *The group* $\mathrm{Bir}(\mathbb{P}^1_\mathbb{C}\times\mathbb{P}^1_\mathbb{C})$ *is generated by the group* B *of birational maps preserving the fibration given by the first projection together with* $\tau\colon(z_0,z_1)\mapsto(z_1,z_0)$.

Moreover the following relations form a complete system of relations:

- *relations inside the groups* $\mathrm{Aut}(\mathbb{P}^1_\mathbb{C}\times\mathbb{P}^1_\mathbb{C})$ *and* B*;*
- $\Big(\tau\circ\Big((z_0,z_1)\mapsto\Big(z_0,\frac{z_0}{z_1}\Big)\Big)\Big)^3=\mathrm{id}$*;*
- $\big(\tau\circ\big((z_0,z_1)\mapsto(-z_0,z_1-z_0)\big)\big)^3=\mathrm{id}$.

In 1994 Iskovskikh, Kabdykairov and Tregub presented a list of generators and relations of $\mathrm{Bir}(\mathbb{P}^2_\Bbbk)$ over an arbitrary perfect field $\Bbbk$ (*see* [**116**]).

The group $\mathrm{Bir}(\mathbb{P}^2_\mathbb{C})$ does not have a structure of amalgamated product ([**59**]). Nevertheless a presentation of the plane Cremona group in the form of a generalized amalgam was given by Wright:

THEOREM 4.6 ([**180**]). *The plane Cremona group is the free product of* $\mathrm{PGL}(3,\mathbb{C})$, $\mathrm{Aut}(\mathbb{P}^1_\mathbb{C}\times\mathbb{P}^1_\mathbb{C})$ *and* $\mathcal{J}$ *amalgamated along their pairwise intersections in* $\mathrm{Bir}(\mathbb{P}^2_\mathbb{C})$.

Twenty years later Blanc proved:

THEOREM 4.7 ([**29**]). *The group* $\mathrm{Bir}(\mathbb{P}^2_\mathbb{C})$ *is the amalgamated product of the Jonquières group with the group of automorphisms of the plane along their intersection, divided by the relation* $\sigma_2\circ\tau=\tau\circ\sigma_2$ *where* σ_2 *is the standard involution and* τ *is the involution* $(z_0:z_1:z_2)\mapsto(z_1:z_0:z_2)$.

As we have seen in Chapter 3 there is a Euclidean topology on the Cremona group ([**41**]). With respect to this topology $\mathrm{Bir}(\mathbb{P}^2_\mathbb{C})$ is a Hausdorff topological group. Furthermore the restriction of the Euclidean topology to any algebraic subgroup is the classical Euclidean topology. To show that $\mathrm{Bir}(\mathbb{P}^2_\mathbb{C})$ is compactly presentable with respect to the Euclidean topology Zimmermann established the following statement:

THEOREM 4.8 ([**189**]). *The group* $\mathrm{Bir}(\mathbb{P}^2_\mathbb{C})$ *is isomorphic to the amalgamated product of* $\mathrm{Aut}(\mathbb{P}^2_\mathbb{C})$, $\mathrm{Aut}(\mathbb{F}_2)$, $\mathrm{Aut}(\mathbb{P}^1_\mathbb{C}\times\mathbb{P}^1_\mathbb{C})$ *along their pairwise intersection in* $\mathrm{Bir}(\mathbb{P}^2_\mathbb{C})$ *modulo the relation* $\tau\circ\sigma_2\circ\tau\circ\sigma_2=\mathrm{id}$ *where* σ_2 *is the standard involution and* τ *the involution* $\tau\colon(z_0:z_1:z_2)\mapsto(z_1:z_0:z_2)$.

Urech and Zimmermann got a presentation of the plane Cremona group with respect to the generators given by the theorem of Noether and Castelnuovo:

THEOREM 4.9 ([**173**]). *The Cremona group* $\mathrm{Bir}(\mathbb{P}^2_\mathbb{C})$ *is isomorphic to*

$$\langle\sigma_2,\,\mathrm{PGL}(3,\mathbb{C})\,|\,(\mathcal{R}_1),\,(\mathcal{R}_2),\,(\mathcal{R}_3),\,(\mathcal{R}_4),\,(\mathcal{R}_5)\rangle$$

where

($\mathcal{R}_1$) $g_1 \circ g_2 \circ g_3^{-1} = \mathrm{id}$ *for all* $g_1, g_2, g_3 \in \mathrm{PGL}(3,\mathbb{C})$ *such that* $g_1 \circ g_2 = g_3$;

($\mathcal{R}_2$) $\sigma_2^2 = \mathrm{id}$;

($\mathcal{R}_3$) $\sigma_2 \circ \eta \circ (\eta \circ \sigma_2)^{-1} = \mathrm{id}$ *for all* η *in the symmetric group* $\mathfrak{S}_3 \subset \mathrm{PGL}(3,\mathbb{C})$ *of order* 6 *acting on* $\mathbb{P}^2_\mathbb{C}$ *by coordinate permutations*;

($\mathcal{R}_4$) $\sigma_2 \circ d \circ \sigma_2 \circ d = \mathrm{id}$ *for all diagonal automorphisms* d *in the subgroup* $\mathrm{D}_2 \subset \mathrm{PGL}(3,\mathbb{C})$ *of diagonal automorphisms*;

($\mathcal{R}_5$) $(\sigma_2 \circ h)^3 = \mathrm{id}$ *where* $h\colon (z_0 : z_1 : z_2) \mapsto (z_2 - z_0 : z_2 - z_1 : z_2)$.

REMARKS 4.10.

- ◇ The relations ($\mathcal{R}_2$), ($\mathcal{R}_3$) and ($\mathcal{R}_4$) occur in the group $\mathrm{Aut}(\mathbb{C}^* \times \mathbb{C}^*)$ which is given by the group of monomial maps $\mathrm{GL}(2,\mathbb{Z}) \ltimes \mathrm{D}_2$.
- ◇ ($\mathcal{R}_5$) is a relation from the group $\mathrm{Aut}(\mathbb{P}^1_\mathbb{C} \times \mathbb{P}^1_\mathbb{C})^0 \simeq \mathrm{PGL}(2,\mathbb{C}) \times \mathrm{PGL}(2,\mathbb{C})$ which is considered as a subgroup of $\mathrm{Bir}(\mathbb{P}^2_\mathbb{C})$ by conjugation with the birational equivalence

$$\begin{array}{ccc} \mathbb{P}^1_\mathbb{C} \times \mathbb{P}^1_\mathbb{C} & \dashrightarrow & \mathbb{P}^2_\mathbb{C}, \\ \big((u_0 : u_1), (v_0 : v_1)\big) & \dashrightarrow & (u_1 v_0 : u_0 v_1 : u_1 v_1). \end{array}$$

REMARK 4.11. All the results are stated on $\mathbb{C}$ but indeed

- ◇ [**29**, **59**, **102**, **113**, **114**, **173**, **180**] work for the plane Cremona group over an algebraically closed field,
- ◇ [**189**] works for the plane Cremona group over a locally compact local field.

In the first section we recall the proof of Noether and Castelnuovo due to Alexander.

In the second section we give an outline of the proof of the result of [**59**] that says that the plane Cremona group does not decompose as a non-trivial amalgam. We also recall the proof of Theorem 4.6.

The third section is devoted to generators and relations in the Cremona group. We first give a sketch of the proof of Theorem 4.7. We also give a sketch of the proof of Theorem 4.9. We then explain why there is no Noether and Castelnuovo theorem in higher dimension.

4.1. Noether and Castelnuovo theorem

Let us now deal with the proof of Theorem 4.1 given by Alexander ([**2**]). Recall the two following formulas proved in §1.3. Consider a birational self map ϕ of $\mathbb{P}^2_\mathbb{C}$ of degree ν; denote by $p_1, p_2, \ldots, p_k$ the base-points of ϕ and by m_{p_i} the multiplicity of p_i. Then

$$\sum_{i=0}^{k} m_{p_i} = 3(\nu - 1), \tag{4.1}$$

$$\sum_{i=0}^{k} m_{p_i}^2 = \nu^2 - 1. \tag{4.2}$$

From (4.2) and (4.1) one gets

$$\sum_{i=0}^{k} m_{p_i}(m_{p_i}-1)=(\nu-1)(\nu-2). \tag{4.3}$$

Consider a birational self map of $\mathbb{P}^2_{\mathbb{C}}$ of degree ν. If $\nu=1$, then according to (4.1) the map ϕ is an automorphism of $\mathbb{P}^2_{\mathbb{C}}$. So let us now assume that $\nu>1$. Let Λ_ϕ be the linear system associated to ϕ. Denote by p_1, p_2, $\ldots$, p_k the base-points (in $\mathbb{P}^2_{\mathbb{C}}$ or infinitely near) of ϕ and m_{p_i} their multiplicity. Up to reindexation let us assume that

$$m_{p_0}\geq m_{p_1}\geq\ldots\geq m_{p_k}\geq 1.$$

Alexander introduced the notion of complexity: the *complexity* of Λ_ϕ is the integer $2c=\nu-m_{p_0}$. It is the number of points except p_0 that belong to the intersection of a general line passing through p_0 and a curve of Λ_ϕ.

REMARKS 4.12. One has

- ◇ $2c\geq 0$: the degree of the hypersurfaces of Λ_ϕ is ν, so a point has multiplicity $\leq\nu$;
- ◇ furthermore $2c\geq 1$; indeed if a homogeneous polynomial of degree ν has a point of multiplicity ν, then the hypersurface given by this polynomial is the union of ν lines.

Set

$$C=\big\{p\in\mathrm{Base}(\phi)\smallsetminus\{p_0\}\,|\,m_p>c\big\}$$

and

$$n=\#C.$$

Bezout's theorem implies that the line through p_0 and p_1 intersects any curve of Λ_ϕ in ν points (counted with multiplicity). Furthermore it intersects any curve of Λ_ϕ at p_0 with multiplicity m_{p_0}. Consequently $m_{p_1}\leq\nu-m_{p_0}=2c$ and

$$c<m_{p_k}\leq\ldots\leq m_{p_2}\leq m_{p_1}\leq 2c. \tag{4.4}$$

LEMMA 4.13. *There are at least three base-points of multiplicity* $>c=\frac{\nu-m_{p_0}}{2}$, *i.e.,* $n\geq 2$; *hence* $m_{p_0}>\frac{\nu}{3}$.

Furthermore if $\nu\geq 3$, *then* p_1, p_2, $\ldots$, p_n *are not aligned.*

PROOF. According to (4.2) and (4.3) one has on the one hand

$$c\sum_{i=0}^{k} m_{p_i}(m_{p_i}-1)-(c-1)\sum_{i=0}^{k} m_{p_i}^2=\sum_{i=0}^{k} m_{p_i}(cm_{p_i}-c-cm_{p_i}+m_{p_i})=\sum_{i=0}^{k} m_{p_i}(m_{p_i}-c)$$

and on the other hand

$$(\nu-1)(\nu-2)c-(\nu^2-1)(c-1)=(\nu-1)(\nu c-2c-\nu c+\nu-c+1)=(\nu-1)(\nu-3c+1).$$

As a result

$$\sum_{i=0}^{k} m_{p_i}(m_{p_i}-c)=(\nu-1)(\nu-3c+1). \tag{4.5}$$

Since $m_{p_{n+i}}\leq c$ for any $i>0$ one gets

$$\sum_{i=0}^{n} m_{p_i}(m_{p_i}-c)\geq\sum_{i=0}^{k} m_{p_i}(m_{p_i}-c).$$

According to (4.5)

$$\sum_{i=0}^{n} m_{p_i}(m_{p_i} - c) \geq (\nu - 1)(\nu - 3c + 1) = \nu(\nu - 3c) + 3c - 1.$$

But $3c - 1 \geq \frac{1}{2} > 0$, so

$$\sum_{i=0}^{n} m_{p_i}(m_{p_i} - c) > \nu(\nu - 3c) = \nu(m_{p_0} - c).$$

Consequently

$$\sum_{i=0}^{n} m_{p_i}(m_{p_i} - c) > \nu(m_{p_0} - c) - m_{p_0}(m_{p_0} - c) = (\nu - m_{p_0})(m_{p_0} - c) = 2c(m_{p_0} - c).$$

As $2c \geq m_{p_i}$ for any $i \geq 1$ (see (4.4)) one gets $2c\sum_{i=1}^{n}(m_{p_i} - c) > 2c(m_{p_0} - c)$ and

$$2c\sum_{i=1}^{n}(m_{p_i} - c) > 2c(m_{p_0} - c);$$

that is

$$m_{p_i} - c > m_{p_0} - c \tag{4.6}$$

since $c > 0$. But $m_{p_1} \leq m_{p_0}$, so $n \geq 2$. Therefore, $m_{p_0} + m_{p_1} + m_{p_2} > 3\left(\frac{\nu - m_{p_0}}{2}\right)$ and $m_{p_0} > \frac{\nu}{3}$.

Let us assume that $n \geq 3$; then (4.6) can be rewritten

$$\sum_{i=1}^{n} m_{p_i} - nc > m_{p_0} - c = \nu - 3c$$

and $\sum_{i=1}^{n} m_{p_i} > \nu + (n - 3)c \geq \nu$. □

DEFINITION. A *general quadratic birational self map of* $\mathrm{Bir}(\mathbb{P}^2_{\mathbb{C}})$ *centered at* p, q, r is the map, up to linear automorphism, that blows up the three distinct points p, q, r of $\mathbb{P}^2_{\mathbb{C}}$ and blows down the strict transform of the lines (pq), (qr), and (pr). These lines are thus sent onto points denoted p', q', and r'.

The line $(p'q')$ (resp. $(q'r')$, resp. $(p'r')$) corresponds to the exceptional line of the blow-up of r (resp. p, resp. q).

LEMMA 4.14. *Compose ϕ with a general quadratic birational self map of $\mathbb{P}^2_{\mathbb{C}}$ centered at p_0, q, and r where p_0 is the base-point of ϕ of maximal multiplicity.*

The complexity of the new system is equal to the complexity of the old system if and only if p_0' is of maximal multiplicity.

If it is not the case, then the complexity of the new system is strictly less than the complexity of the old one.

PROOF. The complexity of the new system is $2c' = \nu' - m'_{\max}$ where $m'_{\max}$ denotes the highest multiplicity of the base-points of the new system. Then

$$\begin{aligned}
2c' &= \nu' - m'_{\max} \\
&= 2\nu - m_{p_0} - m_q - m_r - m'_{\max} \\
&= \nu - m_{p_0} + (\nu - m_q - m_r) - m'_{\max} \\
&= \nu - m_{p_0} + m_{p_0'} - m'_{\max} \\
&= 2c + m_{p_0'} - m'_{\max}.
\end{aligned}$$

Hence $c' \leq c$ and $c' = c$ if and only if $m_{p'_0} = m'_{\max}$. □

LEMMA 4.15. *If there exist two points p_i and p_j in $C = \{p_1, p_2, \ldots, p_n\}$ such that*

- ◇ *p_i and p_j are not infinitely near;*
- ◇ *p_i and p_0 are not infinitely near;*
- ◇ *p_j and p_0 are not infinitely near.*

Then there exists a general quadratic birational self map of $\mathbb{P}^2_{\mathbb{C}}$ such that after composition with ϕ

- ◇ *either the complexity of the system decreases,*
- ◇ *or $\# C = n$ decreases by 2.*

PROOF. Suppose that there exist two points p_i and p_j in $C = \{p_1, p_2, \ldots, p_n\}$ such that

- ◇ p_i and p_j are not infinitely near;
- ◇ p_i and p_0 are not infinitely near;
- ◇ p_j and p_0 are not infinitely near.

Let us now compose ϕ with a general quadratic birational self map of $\mathbb{P}^2_{\mathbb{C}}$ centered at p_0, p_i, and p_j. The degree of the new linear system Λ'_ϕ is $\nu' = 2\nu - m_{p_0} - m_{p_i} - m_{p_j}$. Let us remark that

$$\begin{aligned} \nu' &= 2\nu - m_{p_0} - m_{p_i} - m_{p_j} \\ &= \nu + (\nu - m_{p_0} - m_{p_i} - m_{p_j}) \\ &= \nu + (2c - m_{p_i} - m_{p_j}) \\ &< \nu \end{aligned}$$

i.e., the degree has decreased.

The new linear system Λ'_ϕ has complexity c' and we denote by C' the set of points of multiplicity $> c'$.

The points p_0, p_i, and p_j are no more points of indeterminacy; the other base-points and their multiplicity do not change. There are three new base-points which are p'_0, p'_i, and p'_j. By definition the multiplicity of p'_0 (resp. p'_i, resp. p'_j) is equal to the number of intersection points (counted with multiplicity) between the corresponding contracted line and the strict transform of a general curve of the linear system. From Bezout's theorem we thus have

$$\begin{cases} m_{p'_0} = \nu - m_{p_i} - m_{p_j}, \\ m_{p'_i} = \nu - m_{p_0} - m_{p_j}, \\ m_{p'_j} = \nu - m_{p_0} - m_{p_i}. \end{cases}$$

- ◇ If p'_0 is not the point of highest multiplicity, the complexity of the system decreases (Lemma 4.14);
- ◇ otherwise if p'_0 is the point of highest multiplicity, then the complexity remains constant (Lemma 4.14). Furthermore p'_0 belongs to C' (Lemma 4.13). Since $m_{p_i} > c$, $m_{p_j} > c$ and $\nu - m_{p_0} = 2c$, then $m_{p'_i} < c$ and $m_{p'_j} < c$, *i.e.*, $p'_i \notin C'$ and $p'_j \notin C'$. As a consequence $n' = n - 2$. □

LEMMA 4.16. *Assume there exists a base-point p_k in C that is not infinitely near p_0. Then after composition by a general quadratic birational map, one can disperse the points above p_0 and p_k.*

The complexity of the system does not change, the cardinal of C does not change. There is no point infinitely near p_0'.

PROOF. Consider a point q of the complex projective plane such that

- ⋄ the lines (p_0q) and (p_kq) contain no base-point;
- ⋄ there is no point infinitely near p_0 in the direction of the line (p_0q);
- ⋄ there is no point infinitely near p_k in the direction of the line (p_kq).

Compose ϕ with a general quadratic birational map centered at p_0, p_k, and q. The degree of the new linear system is

$$\nu' = 2\nu - m_{p_0} - m_{p_k} = \nu + 2c - m_{p_k} \geq \nu.$$

The point p_0' is the point of highest multiplicity:

$$\begin{cases} m_{p_0'} = \nu - m_{p_k} \geq \nu - m_{p_0} = 2c \geq m_{p_1}, \\ m_{p_k'} = \nu - m_{p_0} = 2c > c, \\ m_{q'} = \nu - m_{p_0} - m_{p_k} = 2c - m_{p_k} < c, \end{cases}$$

hence the complexity remains constant (Lemma 4.14). Note that $\# C' = \# C$.

The assumptions on q allow us to say that a point infinitely near p_k (resp. p_0) is not transformed in a point infinitely near p_0'. Similarly a point infinitely near p_k (resp. p_0) is not transformed in a point infinitely near q'. □

LEMMA 4.17. *Assume that all the points of C are above the point of highest multiplicity p_0. Then one can disperse them with a general quadratic birational self map; in other words there is no base-point of C' infinitely near the point p_0' of highest multiplicity of the new system. The cardinal n increases by 2 but the complexity of the system remains constant.*

PROOF. Take two points q and r in $\mathbb{P}^2_\mathbb{C}$ such that the lines (p_0r), (p_0q), and (rq)

- ⋄ do not contain base-points;
- ⋄ are not in the direction of the points infinitely near p_0.

The degree of the new linear system is $\nu' = 2\nu - m_{p_0} > \nu$. Since the curves of the system do not pass through q and r Bezout's theorem implies that $m_{p_0'} = \nu$; it is thus the point of highest multiplicity. Furthermore

$$2c' = 2\nu - m_{p_0} - \nu = 2c.$$

Any curve of the linear system intersects (p_0r) and (p_0q) in $\nu - m_{p_0} = 2c$ points. As a result $m_{r'} = m_{q'} = 2c > c = c'$. Moreover r' and q' belong to C' and $n' = n + 2$.

The points infinitely near p_0 have been dispersed onto the line $(q'r')$. As there is no base-point on the line (qr) there is no base-point infinitely near p_0'. □

PROOF OF THEOREM 4.1. Let us consider a birational self map ϕ of $\mathbb{P}^2_\mathbb{C}$ of degree ν. Denote by p_0, p_1, ..., p_k its base-points and by Λ_ϕ the linear system associated to ϕ. Let m_{p_i} be the multiplicity of p_i and assume up to reindexation that

$$m_{p_0} \geq m_{p_1} \geq \ldots \geq m_{p_k}.$$

Recall that the complexity of the system Λ_ϕ is c where $2c = \nu - m_{p_0}$, that

$$C = \{p \in \mathrm{Base}(\phi) \smallsetminus \{p_0\} \,|\, m_p > c\}$$

and that $n = \#C$. We will now compose ϕ with a sequence of general quadratic birational maps in order to decrease the complexity until the complexity equals to 1.

Step 1. If all points of C are above p_0, let us apply Lemma 4.17. One gets that p_0' has no more infinitely near base-points and that $n' = n + 2$. Let us now apply Lemma 4.16 until the points of C' are all distinct; note that C' and n' do not change. According to Lemma 4.13 the points of C' are not aligned. Let us take two of these points, denoted by p_i and p_j such that there exist two base-points p_k and p_ℓ with the following property: p_k and p_ℓ do not belong to the lines $(p_0'p_i)$, $(p_0'p_j)$, and (p_ip_j). Apply Lemma 4.15 two times to the points p_k and p_ℓ. If the complexity decreases (the first or the second time, anyway), then let us start this process again; otherwise the first application of Lemma 4.15 yields to $n' = n$ and the second to $n' = n - 2$. Furthermore there is no more base-point of C' infinitely near p_0' and we go to *Step 2.*

Step 2. We distinguish two possibilities:
Step 2i. Either there are two base-points in C that are not infinitely near and one applies Lemma 4.15. If the complexity decreases, come back to *Step 1*, otherwise come back to *Step 2.*
Step 2ii. Or let us apply Lemma 4.16, then there are two base-points that are not infinitely near and one can apply *Step 2i.*

According to Lemma 4.13 if $\nu > 1$, then $\#C \geq 3$. As a result *Step 1* and *Step 2* allow us to decrease the complexity. When the complexity is 1, the point p_0' has the highest multiplicity and from Lemmas 4.15, 4.16 and 4.17 one gets that $\#C$ decreases until 0. In other words our system has at most one base-point. From (4.1) and (4.2) one gets that $\nu = 1$ and that there is no base-point. □

4.2. Amalgamated product and $\mathrm{Bir}(\mathbb{P}^2_\mathbb{C})$

4.2.1. It is not an amalgamated product of two groups. Let us recall that the group $\mathrm{Aut}(\mathbb{A}^2_\mathbb{C}) \subset \mathrm{Bir}(\mathbb{P}^2_\mathbb{C})$ of polynomial automorphisms of the plane is the amalgamated product of the affine group $\mathrm{Aff}_2 = \mathrm{Aut}(\mathbb{P}^2_\mathbb{C}) \cap \mathrm{Aut}(\mathbb{A}^2_\mathbb{C})$ and the group $\mathcal{J}_{\mathbb{A}^2_\mathbb{C}} = \mathcal{J} \cap \mathrm{Aut}(\mathbb{A}^2_\mathbb{C})$ along their intersection. On the contrary, $\mathrm{Bir}(\mathbb{P}^2_\mathbb{C})$ is not the amalgamated product of $\mathrm{Aut}(\mathbb{P}^2_\mathbb{C})$ and $\mathcal{J}$. Indeed there exist elements of $\mathrm{Bir}(\mathbb{P}^2_\mathbb{C})$ of finite order which are neither conjugate to an element of $\mathrm{Aut}(\mathbb{P}^2_\mathbb{C})$, nor to an element of $\mathcal{J}$ (*see* [**28**]), contrary to the case of amalgamated products.

More precisely, Cornulier proved that the plane Cremona group does not decompose as a non-trivial amalgam ([**59**]); we will give a sketch of the proof in this section.

A *graph* Γ consists of two sets X and Y, and two applications

$$Y \to X \times X, \quad y \mapsto (o(y), t(y)) \qquad\qquad Y \to Y, \quad y \mapsto \overline{y}$$

such that:

$$\forall y \in Y \qquad \overline{\overline{y}} = y, \qquad \overline{y} \neq y, \qquad o(y) = t(y).$$

An element of X is a *vertex* of Γ; an element $y \in Y$ is an *oriented edge*, and $\overline{y}$ is the *reversed edge*. The vertex $o(y) = t(\overline{y})$ is the *origin* of y, and the vertex $t(y) = o(\overline{y})$ is the *terminal vertice*. These two vertices are called the *extremities* of y.

An *orientation* of a graph Γ is a part Y_+ of Y such that Y is the disjoint union of Y_+ and $\overline{Y_+}$. An *oriented graph* is defined, up to isomorphism, by the data of two sets X and Y_+, and an application $Y_+ \to X \times X$. The set of edges of the corresponding graph is $Y = Y_+ \bigsqcup \overline{Y_+}$.

A graph is *connected* if two vertices are the extremities of at least one path.

EXAMPLES 4.

◇ Let n be an integer. Let us consider the oriented graph

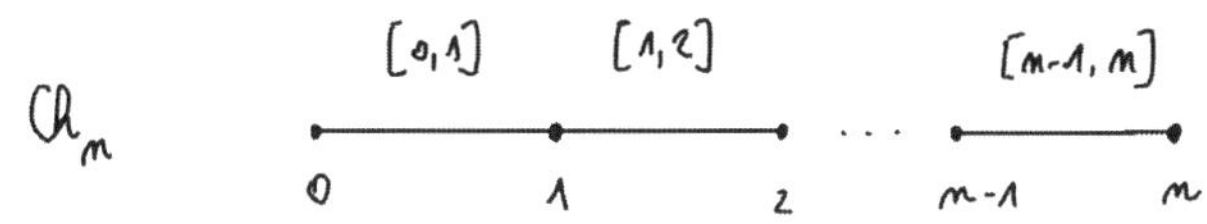

It has $n+1$ vertices $0, 1, \ldots, n$ and the orientation is given by the n egdes $[i, i+1]$, $0 \leq i < n$ with $o([i, i+1]) = i$ and $t([i, i+1]) = i+1$.

◇ Let $n \geq 1$ be an integer. Consider the oriented graph given by

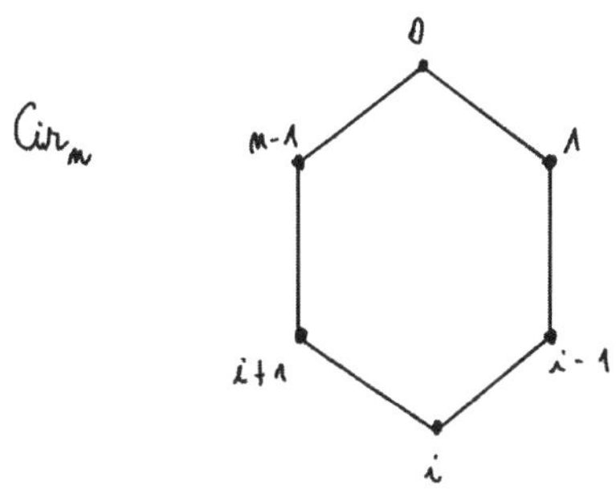

The set of vertices is $\mathbb{Z}/_{n\mathbb{Z}}$, and the orientation is given by the n edges $[i, i+1]$, $i \in \mathbb{Z}/_{n\mathbb{Z}}$, with $o([i, i+1]) = i$ and $t([i, i+1]) = i+1$.

DEFINITIONS. A *path* of length n in a graph Γ is a morphism from Ch_n to Γ.
A *cycle* of length n in a graph is a subgraph isomorphic to Cir_n.
A *tree* is a non-empty, connected graph without cycle.

DEFINITION. A group G is said to have *property (FA)* if every action of G on a tree has a global fixed point.

DEFINITIONS. A *geodesic metric space* is a metric space if given any two points there is a path between them whose length equals the distance between the points.

A *real tree* can be defined in the following equivalent ways ([**63**]):

◇ a geodesic metric space which is 0-hyperbolic in the sense of Gromov;
◇ a uniquely geodesic metric space for which $[a, c] \subset [a, b] \cup [b, c]$ for all a, b and c;
◇ a geodesic metric space with no subspace homeomorphic to the circle.

In a real tree a *ray* is a geodesic embedding of the half line. An *end* is an equivalence class of rays modulo being at bounded distance.

For a group of isometries of a real tree, to *stably fix an end* means to pointwise stabilize a ray modulo eventual coincidence (*i.e.*, it fixes the end as well as the corresponding Busemann function[1]).

DEFINITION. A group has *property (FR)* if for every isometric action on a complete real tree every element has a fixed point.

REMARK 4.18. Property (FR) implies property (FA).

LEMMA 4.19 ([**59**]). *Let* G *be a group. Property (FR) has the following equivalent characterizations:*

- *for every isometric action of* G *on a complete real tree, every finitely generated subgroup has a fixed point;*
- *every isometric action of* G *on a complete real tree either has a fixed point, or stably fixes a point at infinity.*

DEFINITION. A group G decomposes as a *non-trivial amalgam* if $\mathrm{G}\simeq \mathrm{G}_1 *_{\mathrm{H}} \mathrm{G}_2$ with $\mathrm{G}_1\neq \mathrm{H}\neq \mathrm{G}_2$.

THEOREM 4.20 ([**160**], Chapter 6). *A group* G *has property (FA) if and only if it does not decompose as a non-trivial amalgam.*

In the Appendix of [**59**] the author has shown that $\mathrm{Bir}(\mathbb{P}^2_\mathbb{C})$ satisfies the first assertion of Lemma 4.19, hence:

THEOREM 4.21 ([**59**]). *The Cremona group* $\mathrm{Bir}(\mathbb{P}^2_\mathbb{C})$ *has property (FR).*

According to Remark 4.18 the group $\mathrm{Bir}(\mathbb{P}^2_\mathbb{C})$ thus has property (FA). From Theorem 4.20 one gets that:

COROLLARY 4.22 ([**59**]). *The plane Cremona group does not decompose as a non-trivial amalgam.*

Let us give the main steps of the proof of Theorem 4.21. From now on $\mathcal{T}$ is a complete real tree and all actions on $\mathcal{T}$ are isometric.

Step 1. Let $p_0, p_1, \ldots, p_k$ be points of $\mathcal{T}$ and $s\geq 0$. Suppose that the following equality holds:

$$d(p_i,p_j)=s|i-j|$$

for all i, j such that $|i-j|\leq 2$. Then it holds for all i and j.

Step 2. If $d\geq 3$, then $\mathrm{SL}(d,\mathbb{C})$ has property (FR). In particular, if $d\geq 3$, then $\mathrm{PGL}(d,\mathbb{C})$ has property (FR).

Step 3. Let us recall that a *torus* T in a compact Lie group G is a compact, connected, abelian Lie subgroup of G (and therefore isomorphic to the standard torus $\mathbb{T}^n$ for some integer n). Given a torus T, the *Weyl group* of G with respect to T can be defined as the normalizer of T modulo the centralizer of T. A *Cartan subgroup* of an algebraic group is one of the subgroups whose Lie algebra is a Cartan subalgebra. For connected algebraic groups over $\mathbb{C}$ a Cartan subgroup is usually defined as the centralizer of a maximal torus.

Let C be the normalizer of the standard Cartan subgroup of $\mathrm{PGL}(3,\mathbb{C})$, *i.e.*, the semi-direct product of the diagonal matrices by the Weyl group (of order 6).

[1]Let (X,d) be a metric space. Given a ray γ the Busemann function $B_\gamma\colon X\to\mathbb{R}$ is defined by $B_\gamma(x)=\lim\limits_{t\to+\infty}\big(d(\gamma(t),x)-t\big)$.

Set $\varsigma\colon (z_0,z_1)\mapsto(1-z_0,1-z_1)$. The group generated by ς and C coincides with $\mathrm{PGL}(3,\mathbb{C})$:

$$\langle \mathrm{C},\varsigma\rangle=\mathrm{PGL}(3,\mathbb{C}).$$

Step 4. Let $\mathrm{Bir}(\mathbb{P}^2_\mathbb{C})$ act on $\mathcal{T}$ so that $\mathrm{PGL}(3,\mathbb{C})$ has no fixed point and has a (unique) stably fixed end. Then $\mathrm{Bir}(\mathbb{P}^2_\mathbb{C})$ fixes this unique end.

Step 5. Note that $\sigma_2=\big(\varsigma\circ\sigma_2\big)\circ\varsigma\circ\big(\varsigma\circ\sigma_2\big)^{-1}$. Since $\mathrm{Bir}(\mathbb{P}^2_\mathbb{C})=\langle\sigma_2,\,\mathrm{PGL}(3,\mathbb{C})\rangle$ the groups $\mathrm{H}_1=\mathrm{PGL}(3,\mathbb{C})$ and $\mathrm{H}_2=\sigma_2\circ\mathrm{PGL}(3,\mathbb{C})\circ\sigma_2^{-1}$ generate $\mathrm{Bir}(\mathbb{P}^2_\mathbb{C})$. Let us consider an action of $\mathrm{Bir}(\mathbb{P}^2_\mathbb{C})$ on $\mathcal{T}$. By Steps 2 and 4 it is sufficient to consider the case when $\mathrm{PGL}(3,\mathbb{C})$ has a fixed point. Let us prove that $\mathrm{Bir}(\mathbb{P}^2_\mathbb{C})$ has a fixed point; suppose by contradiction that $\mathrm{Bir}(\mathbb{P}^2_\mathbb{C})$ has no fixed point. Denote by $\mathcal{T}_i$ the set of fixed points of H_i, $i=1,\,2$. These two trees are exchanged by σ_2, and as H_1 and H_2 generate $\mathrm{Bir}(\mathbb{P}^2_\mathbb{C})$ they are disjoint. Denote by $\mathcal{S}=[p_1,p_2]$, $p_i\in\mathcal{T}_i$, the minimal segment joining the two trees, and by $s>0$ its length. The segment $\mathcal{S}$ is thus fixed by $\mathrm{C}\subset\mathrm{H}_1\cap\mathrm{H}_2$, and reversed by σ_2. Step 1 implies that for all $k\geq 1$, the distance between the points p_1 and $(\sigma_2\circ\varsigma)^kp_1$ is exactly sk. This contradicts the fact that $(\sigma_2\circ\varsigma)^3=\mathrm{id}$.

4.2.2. It is an amalgamated product of three groups. In [**180**] the author shows that $\mathrm{Bir}(\mathbb{P}^2_\mathbb{C})=\mathrm{Aut}_\mathbb{C}\mathbb{C}(z_0,z_1)$ acts on a two-dimensional simplicial complex C, which has as vertices certain models in the function field $\mathbb{C}(z_0,z_1)$ and whose fundamental domain consists of one face F. This yields a structure description of $\mathrm{Bir}(\mathbb{P}^2_\mathbb{C})$ as an amalgamation of three subgroups along pairwise intersections. The subgroup $\mathrm{Aut}(\mathbb{A}^2_\mathbb{C})$ acts on C by restriction; more precisely the face F has an edge E satisfying the following property: the $\mathrm{Aut}(\mathbb{A}^2_\mathbb{C})$-translates of E form a tree T, and the action of $\mathrm{Aut}(\mathbb{A}^2_\mathbb{C})$ on T yields the well-known structure theory for $\mathrm{Aut}(\mathbb{A}^2_\mathbb{C})$ as an amalgamated product ([**122**]).

Let us give some details. Recall that

$$\mathrm{Aut}(\mathbb{P}^1_\mathbb{C}\times\mathbb{P}^1_\mathbb{C})=\big(\mathrm{PGL}(2,\mathbb{C})\times\mathrm{PGL}(2,\mathbb{C})\big)\rtimes{}^{\mathbb{Z}}\!/_{2\mathbb{Z}}$$

and

$$\mathcal{J}=\mathrm{PGL}(2,\mathbb{C})\ltimes\mathrm{PGL}(2,\mathbb{C}(z_0)).$$

PROOF OF THEOREM 4.6. It is based on Theorem 4.5. Denote by G the group obtained by amalgamating $\mathrm{PGL}(3,\mathbb{C})$, $\mathrm{Aut}(\mathbb{P}^1_\mathbb{C}\times\mathbb{P}^1_\mathbb{C})$, $\mathcal{J}$ along their pairwise intersections in $\mathrm{Bir}(\mathbb{P}^2_\mathbb{C})$. Let τ be the involution $\tau\colon(z_0,z_1)\mapsto(z_1,z_0)$. Consider the group homomorphism $\alpha\colon\mathrm{G}\to\mathrm{Bir}(\mathbb{P}^2_\mathbb{C})$ restricting to the identity on

$$\mathrm{PGL}(3,\mathbb{C})\cup\mathrm{Aut}(\mathbb{P}^1_\mathbb{C}\times\mathbb{P}^1_\mathbb{C})\cup\mathcal{J}.$$

As $\mathrm{im}\,\alpha$ contains $\mathcal{J}$ and $\tau\in\mathrm{Aut}(\mathbb{P}^1_\mathbb{C}\times\mathbb{P}^1_\mathbb{C})$, Theorem 4.5 implies that α is surjective.

Since $\big\{\mathrm{id},\,\tau\big\}\subset\mathrm{Aut}(\mathbb{P}^1_\mathbb{C}\times\mathbb{P}^1_\mathbb{C})$, Theorem 4.5 gives a map $\widetilde{\beta}$ from the free product $\big\{\mathrm{id},\,\tau\big\}*\mathcal{J}$ to G. Since $\mathrm{Aut}(\mathbb{P}^1_\mathbb{C}\times\mathbb{P}^1_\mathbb{C})\subset\mathrm{G}$ the equality

$$\tau\circ(\varphi_0,\varphi_1)\circ\tau=(\varphi_1,\varphi_0)\qquad\forall\,(\varphi_0,\varphi_1)$$

also holds in G. Let us now prove that in G we have $(\tau\circ\varepsilon)^3=\sigma_2$ where $\varepsilon\colon(z_0,z_1)\mapsto\Big(z_0,\frac{z_0}{z_1}\Big)$. First note that the equality $\varepsilon=\rho\circ\sigma_2$, where $\rho\colon(z_0,z_1)\mapsto\Big(\frac{1}{z_0},\frac{z_1}{z_0}\Big)$, holds in $\mathcal{J}$ and so in G. On the one hand σ_2 and ρ commute in $\mathcal{J}$ so in G, on the other hand σ_2 and τ commute in $\mathrm{Aut}(\mathbb{P}^1_\mathbb{C}\times\mathbb{P}^1_\mathbb{C})$ hence in G. Therefore, one has the following equality in G:

$$(\tau\circ\varepsilon)^3=(\tau\circ\rho\circ\sigma_2)^3=(\tau\circ\rho)^3\circ\sigma_2^3.\tag{4.7}$$

The maps τ and ρ belong to $\mathrm{PGL}(3,\mathbb{C})$ and $(\tau\circ\rho)^3 = \mathrm{id}$ in $\mathrm{PGL}(3,\mathbb{C})$; as a consequence $(\tau\circ\rho)^3 = \mathrm{id}$ in G. One has $\sigma_2^3 = \sigma_2$ in $\mathrm{Aut}(\mathbb{P}^1_\mathbb{C}\times\mathbb{P}^1_\mathbb{C})$ so in G. From (4.7) one gets $(\tau\circ\varepsilon)^3 = \sigma_2$ in G. Consequently $\widetilde{\beta}$ induces a map $\beta\colon \mathrm{Bir}(\mathbb{P}^2_\mathbb{C})\to \mathrm{G}$ with the following property: β restricts to the identity on $\mathcal{J}$ and $\{\mathrm{id},\,\tau\}$. According to Theorem 4.5, $\alpha\circ\beta = \mathrm{id}$. The image of β contains $\mathcal{J}\subset \mathrm{G}$ and $\tau\in(\mathrm{PGL}(3,\mathbb{C})\cap \mathrm{Aut}(\mathbb{P}^1_\mathbb{C}\times\mathbb{P}^1_\mathbb{C}))\subset \mathrm{G}$. But both $\mathrm{PGL}(3,\mathbb{C})$ and $\mathrm{Aut}(\mathbb{P}^1_\mathbb{C}\times\mathbb{P}^1_\mathbb{C})$ are generated by their intersection with $\mathcal{J}$ (in $\mathrm{Bir}(\mathbb{P}^2_\mathbb{C})$) together with τ; hence $\mathrm{PGL}(3,\mathbb{C})$ and $\mathrm{Aut}(\mathbb{P}^1_\mathbb{C}\times\mathbb{P}^1_\mathbb{C})$ belong to $\mathrm{im}\,\beta$. As G is generated by $\mathrm{PGL}(3,\mathbb{C})\cup\mathrm{Aut}(\mathbb{P}^1_\mathbb{C}\times\mathbb{P}^1_\mathbb{C})\cup\mathcal{J}$, β is surjective. Therefore, α is an isomorphism ($\alpha^{-1}=\beta$). □

The amalgamated product group structure of Theorem 4.6 reflects the fact that it acts on a simply connected two-dimensional simplicial complex. This follows from a higher-dimensional analogue of Serre's tree theory (*see for instance* [**162, 164**]). Let us detail it.

DEFINITIONS. A *simplicial complex* $\mathcal{K}$ is a finite collection of non-empty finite sets such that if $X\in\mathcal{K}$ and $\emptyset\neq Y\subseteq X$, then $Y\in\mathcal{K}$.

The union of all members of $\mathcal{K}$ is denoted by $V(\mathcal{K})$.

The elements of $V(\mathcal{K})$ are called the *vertices* of $\mathcal{K}$.

The elements of $\mathcal{K}$ are called the *simplices* of $\mathcal{K}$.

The *dimension of a simplex* $S\in\mathcal{K}$ is $\dim S = |S|-1$.

The *dimension* of $\mathcal{K}$ is the maximum dimension of any simplex in $\mathcal{K}$.

Admissible models

A *model* is a reduced, irreducible, separated $\mathbb{C}$-scheme having function field $\mathbb{C}(z_0,z_1)$. Consider the set of models S satisfying one of the three properties

- ◇ $S\simeq\mathbb{P}^2_\mathbb{C}$,
- ◇ $S\simeq\mathbb{P}^1_\mathbb{C}\times\mathbb{P}^1_\mathbb{C}$,
- ◇ $S\simeq\mathbb{P}^1_\Bbbk$ for some subfield $\Bbbk$ of $\mathbb{C}(z_0,z_1)$ necessarily of pure transcendance degree 1 over $\mathbb{C}$.

Such a $\mathbb{C}$-scheme S will be called an *admissible model*. In the first (resp. second, resp. third) case, we say that S is $\mathbb{P}^2_\mathbb{C}$ (resp. S is a $\mathbb{P}^1_\mathbb{C}\times\mathbb{P}^1_\mathbb{C}$, resp. S is a $\mathbb{P}^1_\Bbbk$).

The complex C

It is constructed using as vertices the set of admissible models. The three models S, V and R, where S is a $\mathbb{P}^2_\mathbb{C}$, V is a $\mathbb{P}^1_\mathbb{C}\times\mathbb{P}^1_\mathbb{C}$ and R is a $\mathbb{P}^1_\Bbbk$, determine a face when there exist two distinct points p and q on S such that

- ◇ V is the $\mathbb{P}^1_\mathbb{C}\times\mathbb{P}^1_\mathbb{C}$ ($\simeq\mathbb{F}_0$) obtained by blowing up S at p and q, then blowing down the proper transform of the line in S containing p and q;
- ◇ R is the generic $\mathbb{P}^1_\mathbb{C}$ obtained by blowing up S at p.

If S is the standard $\mathbb{P}^2_\mathbb{C}$, $p=(0:1:0)$ and $q=(1:0:0)$, then V is the standard $\mathbb{P}^1_\mathbb{C}\times\mathbb{P}^1_\mathbb{C}$, and R the standard $\mathbb{P}^1_{\mathbb{C}(z_0)}$. The standard models form a face called the standard face in C.

Fundamental domain

Note that from the construction of C the group $\mathrm{Bir}(\mathbb{P}^2_\mathbb{C})$ acts on C without inverting any edge or rotating any face. A fundamental domain for the action is given by any one face.

If as before we choose S to be the standard $\mathbb{P}^2_\mathbb{C}$, $p=(0:1:0)$ and $q=(1:0:0)$ one gets the standard face. For this choice the centralizer of S, V and R are, respectively, $\mathrm{PGL}(3,\mathbb{C})$, $\mathrm{Aut}(\mathbb{P}^1_\mathbb{C}\times\mathbb{P}^1_\mathbb{C})$, $\mathcal{J}$.

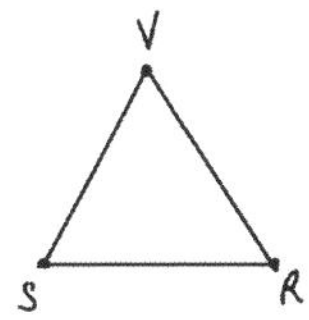

Let us recall that two simplices S_0 and S_n are *k-connected* if there is a sequence of simplices S_0, S_1, S_2, ..., S_n such that any two consecutive ones share a k-face, *i.e.*, they have at least $k+1$ vertices in common. The complex $\mathcal{K}$ is *k-connected* if any two simplices in $\mathcal{K}$ of dimension $\geq k$ are k-connected.

Wright establishes the two following results ([**180**]):

- ◇ the simplicial complex C is 1-connected;
- ◇ the complex C contains the $\mathrm{Aut}(\mathbb{A}^2)$-tree.

4.3. Two presentations of the Cremona group

4.3.1. A simple set of generators and relations for $\mathrm{Bir}(\mathbb{P}^2_\mathbb{C})$**.** In [**29**] Blanc gives a simple set of generators and relations for the plane Cremona group $\mathrm{Bir}(\mathbb{P}^2_\mathbb{C})$. Namely he shows:

THEOREM 4.23 ([**29**]). *The group* $\mathrm{Bir}(\mathbb{P}^2_\mathbb{C})$ *is the amalgamated product of the Jonquières group with the group* $\mathrm{Aut}(\mathbb{P}^2_\mathbb{C})$ *of automorphisms of the plane, divided by the relation* $\sigma_2\circ\tau=\tau\circ\sigma_2$ *where* $\tau\colon(z_0:z_1:z_2)\mapsto(z_1:z_0:z_2)$.

Blanc's proof is inspired by Iskovskikh's proof but Blanc stays on $\mathbb{P}^2_\mathbb{C}$. It is clear that $\sigma_2\circ\tau=\tau\circ\sigma_2$, so it suffices to prove that no other relation holds.

Blanc first establishes the following statement:

LEMMA 4.24 ([**29**]). *Let* φ *be an element of* $\mathcal{J}$ *such that* $\{p_1=(1:0:0),\,q\}\subset\mathrm{Base}(\varphi)$ *where* q *is a proper point of* $\mathbb{P}^2_\mathbb{C}\smallsetminus\{p_1\}$. *If* $\nu\in\mathrm{Aut}(\mathbb{P}^2_\mathbb{C})$ *exchanges* p_1 *and* q, *then*

- ◇ $\psi=\nu\circ\varphi\circ\nu^{-1}$ *belongs to* $\mathcal{J}$,
- ◇ *the relation* $\nu\circ\varphi^{-1}=\psi^{-1}\circ\nu$ *is generated by the relation* $\sigma_2\circ\tau=\tau\circ\sigma_2$ *in the amalgamated product of* $\mathcal{J}$ *and* $\mathrm{Aut}(\mathbb{P}^2_\mathbb{C})$.

Let ϕ be an element of $\mathrm{Aut}(\mathbb{P}^2_\mathbb{C})*_{\mathrm{Aut}(\mathbb{P}^2_\mathbb{C})\cap\mathcal{J}}\mathcal{J}$ modulo the relation $\sigma_2\circ\tau=\tau\circ\sigma_2$. Write ϕ as

$$j_r\circ a_r\circ j_{r-1}\circ a_{r-1}\circ\ldots\circ j_1\circ a_1$$

where $j_i\in\mathcal{J}$ and $a_i\in\mathrm{Aut}(\mathbb{P}^2_\mathbb{C})$ for $i=1,\,\ldots,\,r$. Note that this decomposition is of course not unique.

Let Λ_0 be the linear system of lines of $\mathbb{P}^2_\mathbb{C}$. For any $i=1,\,\ldots,\,r$ let us denote by Λ_i the linear system $(j_i\circ a_i\circ\ldots\circ j_1\circ a_1)(\Lambda_0)$, and by d_i the degree of Λ_i. Set

$$D=\max\big\{d_i\,|\,i=1,\,\ldots,\,r\big\},\qquad n=\max\big\{i\,|\,d_i=D\big\},\qquad k=\sum_{i=1}^{n}\big((\deg j_i)-1\big).$$

Recall that j_i belongs to $\mathcal{J}\subset\mathrm{Bir}(\mathbb{P}^2_\mathbb{C})$ and satisfies the following property:

$$\deg j_i=\deg j_i(\Lambda_0)=\deg j_i^{-1}(\Lambda_0).$$

In particular $\deg j_i=1$ if and only if $j_i\in\mathrm{Aut}(\mathbb{P}^2_\mathbb{C})$.

Let us give an interpretation of k: the number k determines the complexity of the word $j_n \circ a_n \circ j_{n-1} \circ a_{n-1} \circ \ldots \circ j_1 \circ a_1$ which corresponds to the birational self map $j_i \circ a_i \circ \ldots \circ j_1 \circ a_1$ of the highest degree.

Let us now give the strategy of the proof. If $D = 1$, then each j_i is an automorphism of $\mathbb{P}^2_{\mathbb{C}}$ and the word ϕ is equal to an element of $\mathrm{Aut}(\mathbb{P}^2_{\mathbb{C}})$ in the amalgamated product. Since $\mathrm{Aut}(\mathbb{P}^2_{\mathbb{C}}) \hookrightarrow \mathrm{Bir}(\mathbb{P}^2_{\mathbb{C}})$ this eventuality is clear. Assume now that $D > 1$, and prove the result by induction on the pairs (D, k) (we consider the lexicographic order).

Fact. We can suppose that

$$j_{n+1},\, j_n \in \mathcal{J} \smallsetminus \mathrm{Aut}(\mathbb{P}^2_{\mathbb{C}}), \qquad\qquad a_{n+1} \in \mathrm{Aut}(\mathbb{P}^2_{\mathbb{C}}) \smallsetminus \mathcal{J}.$$

Remark. The point $p = (1 : 0 : 0)$ is the base-point of the pencil associated to $\mathcal{J}$. As $a_{n+1} \notin \mathcal{J}$, one has $a_{n+1}(p) \neq p$.

Properties of the Jonquières maps. Since j_n, j_{n+1} do not belong to $\mathrm{Aut}(\mathbb{P}^2_{\mathbb{C}})$, then $\deg j_n > 1$, $\deg j_{n+1} > 1$. Set $D_L = \deg j_{n+1}$, $D_R = \deg j_n$. The maps j_{n+1} and j_n preserve the pencil of lines through p. Furthermore p is a base-point of j_{n+1} (resp. j_n) of multiplicity $D_L - 1$ (resp. $D_R - 1$). Since $j^{\pm}_{n+1}(\Lambda_0)$ (resp. $j^{\pm}_n(\Lambda_0)$) is the image of the system Λ_0, it is a system of rational curves with exactly one free intersection point. The system $j^{\pm}_{n+1}(\Lambda_0)$ (resp. $j^{\pm}_n(\Lambda_0)$) has $2D_L - 2$ (resp. $2D_R - 2$) base-points distinct from p, which all have multiplicity 1.

Set $\Omega_L = (j_{n+1} \circ a_{n+1})^{-1}(\Lambda_0)$ and $\Omega_R = (j_n \circ a_n)(\Lambda_0)$. Since the automorphisms a_{n+1}, a_n are changes of coordinates, the following properties hold:

- ⋄ $\deg \Omega_L = D_L$ and $\ell_0 = a^{-1}_{n+1}(p) \neq p$ is a base-point of Ω_L of multiplicity $D_L - 1$;
- ⋄ $\deg \Omega_R = D_R$ and $r_0 = p$ is a base-point of Ω_R of multiplicity $D_R - 1$.

The author uses these systems to compute the degrees d_{n+1}, resp. d_{n-1}, of the systems $\Lambda_{n+1} = (j_{n+1} \circ a_{n+1})(\Lambda_n)$, resp. $\Lambda_{n-1} = (a_n^{-1} \circ j_n^{-1})(\Lambda_n)$. Indeed for any i the integer d_i coincides with the degree of Λ_i which is, on the one hand, the intersection of Λ_i with a general line, on the other hand, the free intersection of Λ_i with Λ_0. So d_{n+1} (resp. d_{n-1}) is the free intersection of $\Lambda_{n+1} = (j_{n+1} \circ a_{n+1})(\Lambda_n)$ (resp. $\Lambda_{n-1} = (a_n^{-1} \circ j_n^{-1})(\Lambda_n)$) with Λ_0 but also the free intersection of Λ_n with Ω_L (resp. Ω_R).

Denote by $m(q)$ the multiplicity of a point q as a base-point of Λ_n. Let ℓ_1, ..., ℓ_{2D_L-2} (resp. r_1, ..., r_{2D_R-2}) be the base-points of Ω_L (resp. Ω_R). Assume that up to reindexation $m(\ell_i) \geq m(\ell_{i+1})$ (resp. $m(r_i) \geq m(r_{i+1})$) and if ℓ_i (resp. r_i) is infinitely near to ℓ_j (resp. r_j), then $i > j$. The following equalities hold:

$$\left\{\begin{array}{l} d_{n+1} = D_L d_n - (D_L - 1)m(\ell_0) - \displaystyle\sum_{i=1}^{2D_L-2} m(\ell_i) < d_n, \\ d_{n-1} = D_R d_n - (D_R - 1)m(r_0) - \displaystyle\sum_{i=1}^{2D_R-2} m(r_i) < d_n. \end{array}\right. \tag{4.8}$$

Inequalities (4.8) imply

$$\left\{\begin{array}{l} m(\ell_0) + m(\ell_1) + m(\ell_2) > d_n, \\ m(r_0) + m(r_1) + m(r_2) \geq d_n. \end{array}\right.$$

First case: $m(\ell_0) \geq m(\ell_1)$ *and* $m(r_0) \geq m(r_1)$. Let q be a point in $\{\ell_1, \ell_2, r_1, r_2\} \smallsetminus \{\ell_0, r_0\}$ with the maximal multiplicity $m(q)$ and so that q is a proper point of $\mathbb{P}^2_\mathbb{C}$ or infinitely near to ℓ_0 or r_0.

- ◇ Either $\ell_1 = r_0$, $m(q) \geq m(\ell_2)$ and $m(\ell_0)+m(r_0)+m(q) \geq m(\ell_0)+m(\ell_1)+m(\ell_2) > d_n$ by (4.3.1).
- ◇ Or $\ell_1 \neq r_0$, $m(q) \geq m(\ell_1) \geq m(\ell_2)$; hence $m(\ell_0) + m(q) > \frac{2d_n}{3}$. The inequalities $m(r_0) \geq m(r_1) \geq m(r_2)$ imply $m(r_0) \geq \frac{d_n}{3}$ and then $m(\ell_0) + m(r_0) + m(q) > d_n$ holds.

The inequality $m(\ell_0) + m(r_0) + m(q) > d_n$ implies that ℓ_0, r_0 and q are not aligned and there exists an element θ in $\mathcal{J}$ of degree 2 with base points ℓ_0, r_0, q. Note that

$$\deg \theta(\Lambda_n) = 2d_n - m(\ell_0) - m(r_0) - m(q) < d_n.$$

Let us recall that the automorphism a_{n+1} of $\mathbb{P}^2_\mathbb{C}$ sends ℓ_0 onto $r_0 = p$. Take $\nu \in \mathrm{Aut}(\mathbb{P}^2_\mathbb{C}) \cap \mathcal{J}$ such that ν fixes $r_0 = p$ and sends $a_{n+1}(r_0)$ onto ℓ_0. Replace a_{n+1} (resp. j_{n+1}) by $\nu \circ a_{n+1}$ (resp. $j_{n+1} \circ \nu^{-1}$); we can thus assume that a_{n+1} exchanges ℓ_0 and r_0. As a consequence according to Lemma 4.24 and modulo the relation $\sigma_2 \circ \tau = \tau \circ \sigma_2$,

$$j_{n+1} \circ a_{n+1} \circ j_n = j_{n+1} \circ a_{n+1} \circ \theta^{-1} \circ \theta \circ j_n = (j_{n+1} \circ \widetilde{\theta}^{-1}) \circ a_{n+1} \circ (\theta \circ j_n)$$

where $\widetilde{\theta} = a_{n+1} \circ \theta \circ a_{n+1}^{-1} \in \mathcal{J}$. Both $j_{n+1} \circ \widetilde{\theta}^{-1}$ and $\theta \circ j_n$ belong to $\mathcal{J}$, but a_{n+1} belongs to $\mathrm{Aut}(\mathbb{P}^2_\mathbb{C})$. Since $\theta(\Lambda_n) = (\theta \circ j_n)(\Lambda_{n-1})$ has degree $< d_n$ this rewriting decreases the pair (D, k).

Second case: $m(\ell_0) < m(\ell_1)$ *or* $m(r_0) < m(r_1)$. The author comes back to the first case by changing the writing of ϕ in the amalgamated product and modulo the relation $\sigma_2 \circ \tau = \tau \circ \sigma_2$ without changing (D, k) but reversing the inequalities.

4.3.2. An other set of generators and relations for $\mathrm{Bir}(\mathbb{P}^2_\mathbb{C})$**.** The idea of the proof of Theorem 4.9 is the same as in [**29, 113, 114**]. The authors study linear systems of compositions of birational maps of the complex projective plane and use the presentation of $\mathrm{Bir}(\mathbb{P}^2_\mathbb{C})$ given in Theorem 4.7. Before giving the proof of Theorem 4.9 let us state the following:

PROPOSITION 4.25 ([**173**]). *Let* $\phi_1, \phi_2, \ldots, \phi_n$ *be some elements of* $\mathrm{PGL}(3, \mathbb{C}) \cap \mathcal{J}$. *Suppose that* $\phi_n \circ \sigma_2 \circ \phi_{n-1} \circ \sigma_2 \circ \ldots \circ \sigma_2 \circ \phi_1 = \mathrm{id}$ *as maps.*

Then this expression is generated by relations $(\mathcal{R}_1)$–$(\mathcal{R}_5)$.

PROOF OF THEOREM 4.9. Let G be the group generated by σ_2 and $\mathrm{PGL}(3, \mathbb{C})$ divided by the relations $(\mathcal{R}_1)$–$(\mathcal{R}_5)$:

$$\mathrm{G} = \langle \sigma_2, \mathrm{PGL}(3, \mathbb{C}) \,|\, (\mathcal{R}_1) - (\mathcal{R}_5) \rangle.$$

Denote by $\pi\colon \mathrm{G} \to \mathrm{Bir}(\mathbb{P}^2_\mathbb{C})$ the canonical homomorphism that sends generators onto generators. Proposition 4.25 asserts that sending an element of $\mathcal{J}$ onto its corresponding word in G is well defined. Hence there exists a homomorphism $w\colon \mathcal{J} \to \mathrm{G}$ such that

$$\pi \circ w = \mathrm{id}_\mathcal{J}.$$

In particular w is injective.

The universal property of the amalgamated product implies that there exists a unique homomorphism

$$\varphi\colon \mathrm{PGL}(3, \mathbb{C}) *_{\mathrm{PGL}(3,\mathbb{C}) \cap \mathcal{J}} \mathcal{J} \to \mathrm{G}$$

such that the following diagram commutes

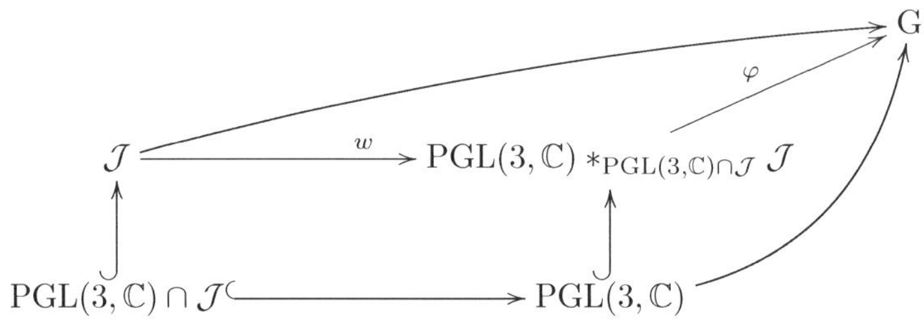

According to Theorem 4.7 the plane Cremona group is isomorphic to

$$\mathrm{PGL}(3,\mathbb{C}) \ast_{\mathrm{PGL}(3,\mathbb{C})\cap\mathcal{J}} \mathcal{J}$$

divided by the relation $\tau\circ\sigma_2\circ\tau\circ\sigma_2$ where $\tau\colon (z_0 : z_1 : z_2) \mapsto (z_1 : z_0 : z_2)$. Note that this relation holds as well in G. As a consequence φ factors through the quotient

$$\mathrm{PGL}(3,\mathbb{C}) \ast_{\mathrm{PGL}(3,\mathbb{C})\cap\mathcal{J}} \mathcal{J}\Big/\langle\tau\circ\sigma_2\circ\tau\circ\sigma_2\rangle.$$

This yields a homomorphism $\overline{\varphi}\colon \mathrm{Bir}(\mathbb{P}^2_\mathbb{C}) \to \mathrm{G}$. More precisely the homomorphisms $\pi\colon \mathrm{G} \to \mathrm{Bir}(\mathbb{P}^2_\mathbb{C})$ and $\overline{\varphi}\colon \mathrm{Bir}(\mathbb{P}^2_\mathbb{C}) \to \mathrm{G}$ both send generators to generators

$$\begin{cases} \pi(\sigma_2) = \sigma_2 \text{ and } \pi(A) = A & \forall\, A \in \mathrm{PGL}(3,\mathbb{C}),\\ \overline{\varphi}(\sigma_2) = \sigma_2 \text{ and } \overline{\varphi}(A) = A & \forall\, A \in \mathrm{PGL}(3,\mathbb{C}).\end{cases}$$

The homomorphisms π and $\overline{\varphi}$ are thus isomorphisms that are inverse to each other. □

Let us give some Lemmas and Remarks that allow us to give a proof of Proposition 4.25.

In [**1**] the author gave a general formula for the degree of a composition of two elements of $\mathrm{Bir}(\mathbb{P}^2_\mathbb{C})$ but the multiplicities of the base-points of the composition is hard to compute in general. If we impose that one of the two maps has degree 2, then it is a rather straightforward computation ([**1**]). Denote by $m_p(\phi)$ the multiplicity of ϕ at the point p. For any $\phi \in \mathcal{J}$ of degree d one has

- ◇ $m_{(1:0:0)}(\phi) = d-1$,
- ◇ $m_p(\phi) = 1\ \forall\, p \in \mathrm{Base}(\phi) \smallsetminus \{(1:0:0)\}$,

so according to [**1**] one has:

LEMMA 4.26 ([**173**]). *Let ϕ, resp. ψ, be a Jonquières map of degree 2, resp. d. Let p_1, p_2 be the base-points of ϕ different from $(1:0:0)$ and q_1, q_2 be the base-points of ϕ^{-1} different from $(1:0:0)$ such that the pencil of lines through p_i is sent by ϕ onto the pencil of lines through q_i.*

Then

- ◇ $\deg(\psi\circ\phi) = d+1-m_{q_1}(\psi)-m_{q_2}(\psi)$,
- ◇ $m_{(1:0:0)}(\psi\circ\phi) = d-m_{q_1}(\psi)-m_{q_2}(\psi) = \deg(\psi\circ\phi)-1$,
- ◇ $m_{p_i}(\psi\circ\phi) = 1-m_{q_j}(\psi)$ *if* $i\neq j$.

REMARK 4.27 ([**173**]). These equalities can translate as follows when Λ_ψ denotes the linear system of ψ:

- ◇ $\deg(\psi\circ\phi) = \deg(\phi^{-1}(\Lambda_\psi)) = d+1-m_{q_1}(\Lambda_\psi)-m_{q_2}(\Lambda_\psi)$,
- ◇ $m_{(1:0:0)}(\phi^{-1}(\Lambda_\psi)) = d-m_{q_1}(\Lambda_\psi)-m_{q_2}(\Lambda_\psi) = \deg\left(\phi^{-1}(\Lambda_\psi)\right)-1$,

$\diamond\ m_{p_i}\big(\phi^{-1}(\Lambda_\psi)\big) = 1 - m_{q_j}(\Lambda_\psi) \qquad i \neq j.$

But the multiplicity of Λ_ψ in a point different from $(1:0:0)$ is 0 or 1 so

$$\begin{cases} \text{either } \deg \phi^{-1}(\Lambda_\psi) = \deg(\Lambda_\psi) + 1 \text{ and } m_{q_1}(\Lambda_\psi) = m_{q_2}(\Lambda_\psi) = 0, \\ \text{or } \deg \phi^{-1}(\Lambda_\psi) = \deg(\Lambda_\psi) \text{ and } m_{q_1}(\Lambda_\psi) + m_{q_2}(\Lambda_\psi) = 1, \\ \text{or } \deg \phi^{-1}(\Lambda_\psi) = \deg(\Lambda_\psi) - 1 \text{ and } m_{q_1}(\Lambda_\psi) = m_{q_2}(\Lambda_\psi) = 1. \end{cases}$$

Furthermore Bezout's theorem implies that $(1:0:0)$ and any other base-points of ψ are not collinear; indeed $(1:0:0)$ is a base-point of multiplicity $d-1$, all other base-points of multiplicity 1 (since ψ belongs to $\mathcal{J}$) and a general member of Λ_ψ intersects a line in d points counted with multiplicity.

LEMMA 4.28 ([**173**]). *Let ϕ be an element of $\mathrm{PGL}(3,\mathbb{C}) \cap \mathcal{J}$. Suppose that $\sigma_2 \circ \phi \circ \sigma_2$ is linear.*

Then $\sigma_2 \circ \phi \circ \sigma_2$ is generated by the relations $(\mathcal{R}_1)$, $(\mathcal{R}_3)$ and $(\mathcal{R}_4)$.

PROOF. By Lemma 4.26 to say that $\sigma_2 \circ \phi \circ \sigma_2$ is linear means that

$$\mathrm{Base}(\sigma_2 \circ \phi) = \mathrm{Base}(\sigma_2) = \big\{(1:0:0),\ (0:1:0),\ (0:0:1)\big\}.$$

Since ϕ belongs to $\mathcal{J}$ it fixes the point $(1:0:0)$, and so permutes $(0:1:0)$ and $(0:0:1)$. As a result there exist φ in $\mathfrak{S}_3 \cap \mathcal{J}$ and d in D_2 such that $\phi = d \circ \varphi$. Hence

$$\sigma_2 \circ \phi \circ \sigma_2 \overset{(1)}{=} \sigma_2 \circ d \circ \varphi \circ \sigma_2 \overset{(3),(4)}{=} d^{-1} \circ \varphi. \qquad \square$$

LEMMA 4.29 ([**173**]). *Let ϕ be an element of $\mathrm{PGL}(3,\mathbb{C}) \cap \mathcal{J}$. Suppose that no three of the base-points of σ_2 and $\sigma_2 \circ \phi$ are collinear.*

Then there exist φ, ψ in $\mathrm{PGL}(3,\mathbb{C}) \cap \mathcal{J}$ such that $\sigma_2 \circ \phi \circ \sigma_2 = \varphi \circ \sigma_2 \circ \psi$. Furthermore this expression is generated by relations $(\mathcal{R}_1)$, $(\mathcal{R}_3)$, $(\mathcal{R}_4)$ and $(\mathcal{R}_5)$.

PROOF. The assumption $\deg(\sigma_2 \circ \phi \circ \sigma_2) = 2$ implies that σ_2 and $\sigma_2 \circ \phi$ have exactly two common base-points (Lemma 4.26), among them $(1:0:0)$ because $\sigma_2 \circ \phi$ and σ_2 belong to $\mathcal{J}$. One can assume up to coordinate permutation that the second point is $(0:1:0)$. More precisely there exist t_1, t_2 in $\mathfrak{S}_3 \cap \mathcal{J}$ such that $t_1 \circ t_2$ fixes $(1:0:0)$ and $(0:1:0)$. As a result

$$t_1 \circ \phi \circ t_2 \colon (z_0 : z_1 : z_2) \mapsto (a_1 z_0 + a_2 z_2 : b_1 z_1 + b_2 z_2 : c z_2)$$

for some complex numbers a_1, a_2, b_1, b_2, c. Since no three of the base-points of σ_2 and $\sigma_2 \circ \phi$ are collinear, $a_2 b_2$ is non-zero. There thus exist d_1, d_2 in D_2 such that

$$t_1 \circ \phi \circ t_2 = d_1 \circ \zeta \circ d_2.$$

We get

$$\begin{aligned} \sigma_2 \circ \phi \circ \sigma_2 &= \sigma_2 \circ t_1^{-1} \circ t_1 \circ \phi \circ t_2 \circ t_2^{-1} \circ \sigma_2 \\ &\overset{(1)}{=} \sigma_2 \circ t_1^{-1} \circ d_1 \circ \zeta \circ d_2 \circ t_2^{-1} \circ \sigma_2 \\ &\overset{(3),(4)}{=} t_1^{-1} \circ d_1^{-1} \circ \sigma_2 \circ \zeta \circ \sigma_2 \circ d_2^{-1} \circ t_2^{-1} \\ &\overset{(5)}{=} t_1^{-1} \circ d_1^{-1} \circ \zeta \circ \sigma_2 \circ \zeta \circ d_2^{-1} \circ t_2^{-1}. \end{aligned}$$

Finally $\varphi = t_1^{-1} \circ d_1^{-1} \circ \zeta$ and $\psi = \zeta \circ d_2^{-1} \circ t_2^{-1}$ suit. $\square$

LEMMA 4.30 ([**173**]). *Let φ_1, φ_2, …, φ_n be elements of* $\mathrm{PGL}(3,\mathbb{C})\cap\mathcal{J}$. *Then there exist ψ_1, ψ_2, …, ψ_n in* $\mathrm{PGL}(3,\mathbb{C})\cap\mathcal{J}$ *and ϕ in $\mathcal{J}$ such that*

$$\phi\circ\varphi_n\circ\sigma_2\circ\varphi_{n-1}\circ\sigma_2\circ\ldots\circ\sigma_2\circ\varphi_1\circ\phi^{-1}=\psi_n\circ\sigma_2\circ\psi_{n-1}\circ\ldots\circ\sigma_2\circ\psi_1,$$

and

- ◇ *the above relation is generated by relations $(\mathcal{R}_1)$–$(\mathcal{R}_5)$,*
- ◇ $\deg(\sigma_2\circ\psi_i\circ\sigma_2\circ\ldots\circ\sigma_2\circ\psi_1)=\deg(\sigma_2\circ\phi_i\circ\sigma_2\circ\ldots\circ\sigma_2\circ\phi_1)$ *for all* $1\leq i\leq n$,
- ◇ $(\sigma_2\circ\psi_i\circ\sigma_2\circ\psi_{i-1}\circ\ldots\circ\sigma_2\circ\psi_1)^{-1}$ *does not have any infinitely near base-points for all* $1\leq i\leq n$.

IDEA OF THE PROOF OF PROPOSITION 4.25. Let us introduce similar notations as in the proof of Theorem 4.7. Let Λ_0 be the complete linear system of lines in $\mathbb{P}^2_{\mathbb{C}}$ and for $1\leq i\leq j$ let Λ_i be the following linear system:

$$\Lambda_i:=\sigma_2\circ\varphi_{i-1}\circ\sigma_2\circ\ldots\circ\sigma_2\circ\varphi_1(\Lambda_0).$$

Set $\delta_i:=\deg\Lambda_i$, $D_i:=\max\big\{\delta_i\,|\,i=1,\,2,\,\ldots,\,j\big\}$, $n:=\max\big\{i\,|\,\delta_i=D\big\}$. Consider the lexicographic order. Let us prove the result by induction on pairs of positive integers (D,n).

If $D=1$, then $j=1$, and there is nothing to prove.

Assume now that $D>1$. We can suppose that for $1\leq i\leq j$ the map

$$\big(\phi_i\circ\sigma_2\circ\phi_{i-1}\circ\sigma_2\circ\ldots\circ\sigma_2\circ\phi_1\big)^{-1}$$

does not have any infinitely near base-points (Lemma 4.30). Furthermore we can do this without increasing the pair (D,n). Hence any Λ_i, $1\leq i\leq j$, does not have any infinitely near base-points.

The maps ϕ_i are Jonquières ones, so fix $(1:0:0)$. The maps $\sigma_2\circ\phi_i$ and σ_2 always have $(1:0:0)$ as common base-points. In particular $\deg(\sigma_2\circ\phi_i\circ\sigma_2)\leq 3$ for any $1\leq i\leq j$ (Lemma 4.26). Let us now deal with the three distinct cases: $\deg(\sigma_2\circ\phi_n\circ\sigma_2)=1$, $\deg(\sigma_2\circ\phi_n\circ\sigma_2)=2$, $\deg(\sigma_2\circ\phi_n\circ\sigma_2)=3$.

- ◇ First case: $\deg(\sigma_2\circ\phi_n\circ\sigma_2)=1$. According to Lemma 4.28 the word $\sigma_2\circ\phi_n\circ\sigma_2$ can be replaced by the linear map $\phi'_n=\sigma_2\circ\phi_n\circ\sigma_2$ using relations $(\mathcal{R}_1)$, $(\mathcal{R}_3)$ and $(\mathcal{R}_4)$. We thus get a new pair (D',n') with $D'\leq D$; moreover if $D=D'$, then $n'<n$.
- ◇ Second case: $\deg(\sigma_2\circ\phi_n\circ\sigma_2)=2$. The maps σ_2 and $\phi_n\circ\sigma_2$ have exactly two common base-points, one of them being $(1:0:0)$. One can assume that the other one is $(0:1:0)$. More precisely there are two coordinate permutations t_1 and t_2 in $\mathfrak{S}_3\cap\mathcal{J}$ such that $t_1\circ\phi_n\circ t_2$ fixes $(1:0:0)$ and $(0:1:0)$, that is

  $$t_1\circ\phi_n\circ t_2\colon(z_0:z_1:z_2)\mapsto(a_1z_0+a_2z_2:b_1z_1+b_2z_2:cz_2)$$

  for some a_1, a_2, b_1, b_2, c in $\mathbb{C}$. Using $(\mathcal{R}_1)$ and $(\mathcal{R}_3)$ we get

  $$\begin{aligned}&\phi_j\circ\sigma_2\circ\ldots\circ\phi_{n+1}\circ\sigma_2\circ t_1^{-1}\circ t_1\circ\phi_n\circ t_2\circ t_2^{-1}\circ\sigma_2\circ\ldots\circ\phi_1\\&\quad=\phi_j\circ\sigma_2\circ\ldots\circ\sigma_2\circ(\phi_{n+1}\circ t_1^{-1})\circ\sigma_2\circ(t_1\circ\sigma_2\circ t_2)\circ\sigma_2\circ(t_2^{-1}\circ\phi_{n-1})\circ\sigma_2\circ\ldots\circ\phi_1.\end{aligned}$$

  The pair (D,n) is unchanged. Let us thus assume that $t_1=t_2=\mathrm{id}$ and

  $$\phi_n\colon(z_0:z_1:z_2)\mapsto(a_1z_0+a_2z_2:b_1z_1+b_2z_2:cz_2).$$

Recall that by assumption for any $1 \leq i \leq n$ the maps $\phi_i \circ \sigma_2 \circ \phi_{i-1} \circ \sigma_2 \circ \ldots \circ \sigma_2 \circ \phi_1$ have no infinitely other base-points. As a result Λ_n has no infinitely near base-points.

CLAIM 4.31 ([**173**]). *The product a_2b_2 is non-zero.*

PROOF. Assume by contradiction that $a_2b_2 = 0$. Then $q := \phi_n^{-1}(0 : 0 : 1)$ is a base-point of $\sigma_2 \circ \phi_n$ that lies on a line contracted by $(\sigma_2 \circ \phi_{n-1})^{-1}$. By Remark 4.27 one has

$$D - 1 = \delta_{n+1} = D + 1 - m_{(0:1:0)}(\Lambda_n) - m_q(\Lambda_n).$$

In particular $m_q(\Lambda_n) = 2 - m_{(0:1:0)}(\Lambda_n) = 1$. As $q \notin \mathrm{Base}(\sigma_2)$ one has: $q \notin \mathrm{Base}\big((\sigma_2 \circ \phi_{n-1})^{-1}\big)$. Its proper image by $(\sigma_2 \circ \phi_{n-1})^{-1}$ is thus a base-point of Λ_{n-1}. But $a_2b_2 = 0$; as a result q is an infinitely near point: contradiction. □

If a_2b_2 is non-zero, then no three of the base-points of σ_2 and $\sigma_2 \circ \phi_n$ are collinear. According to Lemma 4.30 there exist ψ and φ in $\mathrm{PGL}(3, \mathbb{C})$ such that the word $\sigma_2 \circ \phi_n \circ \sigma_2$ can be replaced by the word $\psi \circ \sigma_2 \circ \varphi$ using $(\mathcal{R}_1)$, $(\mathcal{R}_3)$, $(\mathcal{R}_4)$ and $(\mathcal{R}_5)$. We thus get a new pair (D', n') where $D' \leq D$; moreover if $D = D'$, then $n' < n$.

◇ Third case: $\deg(\sigma_2 \circ \phi_n \circ \sigma_2) = 3$. See [**173**]. □

Let us give an application of this new presentation ([**173**]). In [**100**] Gizatullin has considered the following question: can a given group homomorphism $\varphi\colon \mathrm{PGL}(3, \mathbb{C}) \to \mathrm{PGL}(n+1, \mathbb{C})$ be extended to a group homomorphism $\Phi\colon \mathrm{Bir}(\mathbb{P}^2_{\mathbb{C}}) \to \mathrm{Bir}(\mathbb{P}^n_{\mathbb{C}})$? He answers yes when φ is the projective representation induced by the regular action of $\mathrm{PGL}(3, \mathbb{C})$ on the space of plane conics, plane cubics, or plane quartics. To construct these homomorphisms Gizatullin uses the following construction. Denote by $\mathrm{Sym}(n, \mathbb{C})$ the $\mathbb{C}$-algebra of symmetric $n \times n$ matrices. Define $\mathbb{S}(2, n)$ as the quotient $\big(\mathrm{Sym}(n, \mathbb{C})\big)^3 / \mathrm{GL}(n, \mathbb{C})$ where the regular action of $\mathrm{GL}(n, \mathbb{C})$ is given by

$$C \cdot (A_0, A_1, A_2) = \big(CA_0{}^tC,\, CA_1{}^tC,\, CA_2{}^tC\big).$$

LEMMA 4.32 ([**173**]). *The variety $\mathbb{S}(2, n)$ is a rational variety, and*

$$\dim \mathbb{S}(2, n) = \frac{(n+1)(n+2)}{2} - 1.$$

REMARK 4.33. The variety $\mathbb{S}(2, n)$ has thus the same dimension as the space of plane curves of degree n.

An element $A = (A_0, A_1, A_2)$ of $\mathrm{PGL}(3, \mathbb{C})$ induces an automorphism on $(\mathrm{Sym}(n, \mathbb{C}))^3$ by

$$\phi(A_0, A_1, A_2) := \big(\phi_0(A_0, A_1, A_2), \phi_1(A_0, A_1, A_2), \phi_2(A_0, A_1, A_2)\big).$$

This automorphism commutes with the action of $\mathrm{GL}(n, \mathbb{C})$; we thus obtain a regular action of $\mathrm{PGL}(3, \mathbb{C})$ on $\mathbb{S}(2, n)$.

Theorem 4.9 allows us to give a short proof of the following statement:

PROPOSITION 4.34 ([**100**]). *The regular action of* $\mathrm{PGL}(3, \mathbb{C})$ *extends to a rational action of* $\mathrm{Bir}(\mathbb{P}^2_{\mathbb{C}})$.

Proof. Define the birational action of σ_2 on $\mathbb{S}(2,n)$ by

$$(A_0, A_1, A_2) \dashrightarrow (A_0^{-1}, A_1^{-1}, A_2^{-1}).$$

According to Theorem 4.9 to see that this indeed defines a rational action of $\mathrm{Bir}(\mathbb{P}^2_\mathbb{C})$ on $\mathbb{S}(2,n)$ it is sufficient to see that $(\mathcal{R}_1)$–$(\mathcal{R}_5)$ are satisfied, which is the case. □

4.3.3. Why no Noether and Castelnuovo theorem in higher dimension? Let us give an idea of the proof of the fact that there is no Noether and Castelnuovo theorem in higher dimension:

Theorem 4.35 ([**110, 155**]). *Any set of group generators of* $\mathrm{Bir}(\mathbb{P}^n_\mathbb{C})$, $n \geq 3$, *contains uncountably many elements of* $\mathrm{Bir}(\mathbb{P}^n_\mathbb{C}) \smallsetminus \mathrm{PGL}(n+1,\mathbb{C})$.

Let us first recall the following construction of Pan which, given a birational self map of $\mathbb{P}^n_\mathbb{C}$, allows one to construct a birational self map of $\mathbb{P}^{n+1}_\mathbb{C}$. First introduce some notations: let $P \in \mathbb{C}[z_0, z_1, \ldots, z_n]_d$, $Q \in \mathbb{C}[z_0, z_1, \ldots, z_n]_\ell$ and R_0, R_1, …, $R_{n-1} \in \mathbb{C}[z_0, z_1, \ldots, z_n]_{d-\ell}$ be some homogeneous polynomials of degree d, resp. ℓ, resp. $d-\ell$. Consider $\widetilde{\psi}_{P,Q,R}$ and $\widetilde{\psi}_R$ the rational maps given by

$$\widetilde{\psi}_{P,Q,R} \colon (z_0 : z_1 : \ldots : z_n) \dashrightarrow (QR_0 : QR_1 : \ldots : QR_{n-1} : P),$$

$$\widetilde{\psi}_R \colon (z_0 : z_1 : \ldots : z_n) \dashrightarrow (R_0 : R_1 : \ldots : R_{n-1}).$$

Lemma 4.36 ([**155**]). *Let d and ℓ be some integers such that $d \leq \ell + 1 \leq 2$. Take $Q \in \mathbb{C}[z_0, z_1, \ldots, z_n]_\ell$ and $P \in \mathbb{C}[z_0, z_1, \ldots, z_n]_d$ without common factors. Let R_1, R_2, …, R_n be some elements of $\mathbb{C}[z_0, z_1, \ldots, z_{n-1}]_{d-\ell}$. Assume that*

$$P = z_n P_{d-1} + P_d \qquad\qquad Q = z_n Q_{\ell-1} + Q_\ell$$

with P_{d-1}, P_d, $Q_{\ell-1}$, $Q_\ell \in \mathbb{C}[z_0, z_1, \ldots, z_{n-1}]$ of degree $d-1$, resp. d, resp. $\ell-1$, resp. ℓ, and such that $(P_{d-1}, Q_{\ell-1}) \neq (0,0)$.

The map $\widetilde{\psi}_{P,Q,R}$ is birational if and only if $\widetilde{\psi}_R$ is.

This statement allows us to prove that given a hypersurface of $\mathbb{P}^n_\mathbb{C}$ one can construct a birational self map of $\mathbb{P}^n_\mathbb{C}$ that blows down this hypersurface:

Lemma 4.37 ([**155**]). *Let $n \geq 3$. Let S be a hypersurface of $\mathbb{P}^n_\mathbb{C}$ of degree $\ell \geq 1$ having a point p of multiplicity $\geq \ell - 1$.*

Then there exists a birational self map of $\mathbb{P}^n_\mathbb{C}$ of degree $d \geq \ell + 1$ that blows down S onto a point.

Proof. Let us assume without loss of generality that $p = (0 : 0 : \ldots : 0 : 1)$. Suppose that S is given by $(Q = 0)$. Take a generic plane passing through p given by $(H = 0)$. Choose $P = z_n P_{d-1} + P_d$ such that

- ⋄ $P_{d-1} \in \mathbb{C}[z_0, z_1, \ldots, z_{n-1}]$ of degree $d-1$ and $\neq 0$;
- ⋄ $P_d \in \mathbb{C}[z_0, z_1, \ldots, z_{n-1}]$ of degree d;
- ⋄ $\mathrm{pgcd}(P, HQ) = 1$.

Set $\widetilde{Q} = H^{d-\ell-1}q$ and $R_i = z_i$. The statement then follows from Lemma 4.36. □

Proof of Theorem 4.35. Consider the family of hypersurfaces given by

$$Q(z_1, z_2, z_3) = 0$$

where $(Q = 0)$ defines a smooth curve $\mathcal{C}_Q$ of degree ℓ on $\{z_0 = z_4 = z_5 = \ldots = z_n = 0\}$. Note that $(Q = 0)$ is birationally equivalent to $\mathbb{P}^{n-2}_\mathbb{C} \times \mathcal{C}_Q$. Furthermore $(Q = 0)$ and $(Q' = 0)$ are birationally equivalent if and only if $\mathcal{C}_Q$ and $\mathcal{C}_{Q'}$ are isomorphic.

Take $\ell = 2$; the set of isomorphism classes of smooth cubics is a 1-parameter family. For any $\mathcal{C}_Q$ there exists a birational self map of $\mathbb{P}^n_\mathbb{C}$ that blows down $\mathcal{C}_Q$ onto a point (Lemma 4.37). As a result any set of group generators of $\mathrm{Bir}(\mathbb{P}^n_\mathbb{C})$, $n \geq 3$, has to contain uncountably many elements of $\mathrm{Bir}(\mathbb{P}^n_\mathbb{C}) \smallsetminus \mathrm{PGL}(n+1, \mathbb{C})$. □

As we have seen, one consequence of the Noether and Castelnuovo theorem is that the Jonquières group and $\mathrm{Aut}(\mathbb{P}^2_\mathbb{C}) = \mathrm{PGL}(3, \mathbb{C})$ generate $\mathrm{Bir}(\mathbb{P}^2_\mathbb{C})$. This statement also does not hold in higher dimension ([**37**]): let $n \geq 3$, the n-dimensional Cremona group is not generated by $\mathrm{Aut}(\mathbb{P}^n_\mathbb{C})$ and by Jonquières elements, *i.e.*, elements that preserve a family of lines through a given point, which form a subgroup

$$\mathrm{PGL}(2, \mathbb{C}(z_2, z_3, \ldots, z_n)) \rtimes \mathrm{Bir}(\mathbb{P}^{n-1}_\mathbb{C}) \subseteq \mathrm{Bir}(\mathbb{P}^n_\mathbb{C}).$$

A more precise statement has been established in dimension 3 in [**40**]: the 3-dimensional Cremona group is not generated by birational maps preserving a linear fibration $\mathbb{P}^3_\mathbb{C} \dashrightarrow \mathbb{P}^2_\mathbb{C}$.

CHAPTER 5

Algebraic properties of the Cremona group

The group $\mathrm{Bir}(\mathbb{P}^2_\mathbb{C})$ has many properties of linear groups, so we wonder if $\mathrm{Bir}(\mathbb{P}^2_\mathbb{C})$ has a faithful linear representation; in the first section we show that the answer is no ([**62**, **66**]). Still in the first section we give the proof of the following property: the plane Cremona group contains non-linear finitely generated subgroups ([**66**]).

In the second section we give the proof of the facts that

- ◇ the normal subgroup generated by σ_2 in $\mathrm{Bir}(\mathbb{P}^2_\mathbb{C})$ is $\mathrm{Bir}(\mathbb{P}^2_\mathbb{C})$;
- ◇ the normal subgroup, generated by a non-trivial element of $\mathrm{PGL}(3,\mathbb{C}) = \mathrm{Aut}(\mathbb{P}^2_\mathbb{C})$ in $\mathrm{Bir}(\mathbb{P}^2_\mathbb{C})$ is $\mathrm{Bir}(\mathbb{P}^2_\mathbb{C})$.

As a consequence $\mathrm{Bir}(\mathbb{P}^2_\mathbb{C})$ is perfect ([**62**]), that is, $[\mathrm{Bir}(\mathbb{P}^2_\mathbb{C}), \mathrm{Bir}(\mathbb{P}^2_\mathbb{C})] = \mathrm{Bir}(\mathbb{P}^2_\mathbb{C})$.

We finish this chapter with the description of the endomorphisms of the plane Cremona group; as a corollary we get

THEOREM 5.1 ([**78**]). *The plane Cremona group is hopfian, i.e., any surjective endomorphism of* $\mathrm{Bir}(\mathbb{P}^2_\mathbb{C})$ *is an automorphism.*

We use for that the classification of the representations of $\mathrm{SL}(3,\mathbb{Z})$ in $\mathrm{Bir}(\mathbb{P}^2_\mathbb{C})$; we thus recall and establish it in the third section:

THEOREM 5.2 ([**76**]). *Let* Γ *be a finite index subgroup of* $\mathrm{SL}(3,\mathbb{Z})$. *Let* υ *be an injective morphism from* Γ *to* $\mathrm{Bir}(\mathbb{P}^2_\mathbb{C})$. *Then, up to birational conjugacy, either* υ *is the canonical embedding, or* υ *is the involution* $A \mapsto ({}^tA)^{-1}$.

As a result we obtain

COROLLARY 5.3 ([**76**]). *If a morphism from a subgroup of finite index of* $\mathrm{SL}(n,\mathbb{Z})$ *into* $\mathrm{Bir}(\mathbb{P}^2_\mathbb{C})$ *has infinite image, then* $n \leq 3$.

5.1. The group $\mathrm{Bir}(\mathbb{P}^2_\mathbb{C})$ is not linear

Cantat and Lamy proved that $\mathrm{Bir}(\mathbb{P}^2_\mathbb{C})$ is not simple but the non-existence of a faithful representation does not imply the non-existence of a non-trivial representation. So let us deal with the following statement:

PROPOSITION 5.4 ([**62**]). *The plane Cremona group has no faithful linear representation in characteristic zero.*

Before giving the proof let us mention that making an easy refinement of it provides the following stronger result:

PROPOSITION 5.5 ([**66**]). *If* $\mathbb{k}$ *is an algebraically closed field, then there is no non-trivial finite-dimensional linear representation for* $\mathrm{Bir}(\mathbb{P}^2_{\mathbb{k}})$ *over any field.*

Let us recall the following statement due to Birkhoff:

LEMMA 5.6 ([**20**]). *Let* $\mathbb{k}$ *be a field of characteristic zero. Let* A*,* B*, and* C *be three elements of* $\mathrm{GL}(n,\mathbb{k})$ *such that*

- ◇ $[A,B]=C$, $[A,C]=[B,C]=\mathrm{id}$,
- ◇ C *has prime order* p.

Then $p\leq n$.

PROOF. Assume that $\mathbb{k}$ is algebraically closed. Since C is of order p its eigenvalues are p-rooth of unity.

If the eigenvalues of C are all equal to 1, then C is unipotent and $p\leq n$.

Otherwise C admits an eigenvalue $\alpha\neq 1$. Consider the eigenspace

$$E_\alpha=\{v\,|\,Cv=\alpha v\}$$

of C associated to the eigenvalue α. By assumption, A and B commute to C, so E_α is invariant by A and B. From $[A,B]=C$ we get $[A_{|E_\alpha},B_{|E_\alpha}]=C_{|E_\alpha}$; but $C_{|E_\alpha}=\alpha\mathrm{id}_{|E_\alpha}$; hence $[A_{|E_\alpha},B_{|E_\alpha}]=\alpha\mathrm{id}_{|E_\alpha}$, that is $(B^{-1}AB)_{|E_\alpha}=\alpha A_{|E_\alpha}$. Note that $(B^{-1}AB)_{|E_\alpha}$ and $A_{|E_\alpha}$ are conjugate; thus $(B^{-1}AB)_{|E_\alpha}$ and $A_{|E_\alpha}$ have the same eigenvalues. Furthermore these eigenvalues are non-zero. If λ is an eigenvalue of $A_{|E_\alpha}$, then $\alpha\lambda$, $\alpha^2\lambda$, $\ldots$, $\alpha^{p-1}\lambda$ are also eigenvalues of $A_{|E_\alpha}$. As p is prime and α distinct from 1, the numbers α, α^2, $\ldots$, α^{p-1} are distinct, $\dim E_\alpha\geq p$, and $n\geq p$. □

PROOF OF PROPOSITION 5.4. Assume by contradiction that there exists an injective morphism ζ from $\mathrm{Bir}(\mathbb{P}^2_\mathbb{C})$ into $\mathrm{GL}(n,\mathbb{k})$. For any prime p let us consider in the affine chart $z_2=1$ the group generated by the maps

$$(z_0,z_1)\mapsto(\mathrm{e}^{-2\mathrm{i}\pi/p}z_0,z_1),\quad (z_0,z_1)\dashrightarrow(z_0,z_0z_1),\quad (z_0,z_1)\mapsto(z_0,\mathrm{e}^{-2\mathrm{i}\pi/p}z_1).$$

The images of these three elements of $\mathrm{Bir}(\mathbb{P}^2_\mathbb{C})$ satisfy the assumptions of the Birkhoff Lemma; therefore, $p\leq n$ for any prime p: contradiction. □

In [**66**] Cornulier gives an example of a non-linear finitely generated subgroup of the plane Cremona group. The existence of such a subgroup is not new, for instance it follows from an unpublished construction of Cantat. The example in [**66**] has the additional feature of being 3-solvable. To prove its non-linearity Cornulier proves that it contains nilpotent subgroups of arbitrary large nilpotency length.

Let G be a group. Recall that $[g,h]=g\circ h\circ g^{-1}\circ h^{-1}$ denotes the commutator of g and h. If H_1 and H_2 are two subgroups of G, then $[\mathrm{H}_1,\mathrm{H}_2]$ is the subgroup of G generated by the elements of the form $[g,h]$ with $g\in\mathrm{H}_1$ and $h\in\mathrm{H}_2$. We defined the *derived series* of G by setting $\mathrm{G}^{(0)}=\mathrm{G}$ and for all $n\geq 0$

$$\mathrm{G}^{(n+1)}=[\mathrm{G}^{(n)},\mathrm{G}^{(n)}].$$

The *soluble length* $\ell(\mathrm{G})$ of G is defined by

$$\ell(\mathrm{G})=\min\{k\in\mathbb{N}\cup\{0\}\,|\,\mathrm{G}^{(k)}=\{\mathrm{id}\}\}$$

with the convention: $\min\emptyset=\infty$. We say that G is *solvable* if $\ell(\mathrm{G})<\infty$. The *descending central series* of a group G is defined by $C^0\mathrm{G}=\mathrm{G}$ and for all $n\geq 0$

$$C^{n+1}\mathrm{G}=[\mathrm{G},C^n\mathrm{G}].$$

The group G is *nilpotent* if there exists $j \geq 0$ such that $C^j\mathrm{G} = \{\mathrm{id}\}$. If j is the minimum non-negative number with such a property, we say that G is of *nilpotent class* j.

Take f in $\mathbb{C}(z_0)$ and g in $\mathbb{C}(z_0)^*$; define α_f and μ_g by

$$\alpha_f\colon (z_0, z_1) \dashrightarrow \big(z_0, z_1 + f(z_0)\big), \qquad \mu_g\colon (z_0, z_1) \dashrightarrow \big(z_0, z_1 g(z_0)\big).$$

Note that

$$\alpha_{f+f'} = \alpha_f \circ \alpha_{f'} \qquad \mu_{gg'} = \mu_g \circ \mu_{g'} \qquad \mu_g \circ \alpha_f \circ \mu_g^{-1} = \alpha_{fg}. \tag{5.1}$$

Take $t \in \mathbb{C}$ and consider $s_t\colon (z_0, z_1) \mapsto (z_0 + t, z_1)$. The following equalities hold

$$s_t \circ \alpha_{f(z_0)} \circ s_t^{-1} = \alpha_{f(z_0-t)}, \qquad s_t \circ \mu_{g(z_0)} \circ s_t^{-1} = \mu_{g(z_0-t)}. \tag{5.2}$$

Let Γ_n be the subgroup of $\mathrm{Bir}(\mathbb{P}^2_{\mathbb{C}})$ defined for any $n \geq 0$ by

$$\Gamma_n = \langle s_1, \, \alpha_{z_0^n} \rangle.$$

Note that Γ_n is indeed a subgroup of the Jonquières group. It satisfies the following properties:

LEMMA 5.7 ([**66**]). *The nilpotency length of* Γ_n *is exactly* $n+1$*, and* Γ_n *is torsion free.*

PROOF. Let A_n be the abelian subgroup of the Jonquières group consisting of all α_P where P ranges over polynomials of degree at most n. The group A_n is normalized by s_1, and $[s_1, \mathrm{A}_n] \subset \mathrm{A}_{n-1}$ for $n \geq 1$ while $\mathrm{A}_0 = \{\mathrm{id}\}$. Therefore, the largest group generated by s_1 and A_n is nilpotent of class at most $n+1$, and so is Γ_n.

Consider now the n-iterated group commutator given by

$$[s_1, [s_1, \ldots, [s_1, \alpha_{z_0^n}] \ldots]].$$

It coincides with $\alpha_{\Delta^n z_0^n}$ where Δ is the discrete differential operator $\Delta P(z_0) = -P(z_0) + P(z_0 - 1)$. Note that $\Delta^n z_0^n \neq 0$ and Γ_n is not n-nilpotent.

Clearly Γ_n is torsion-free. □

The group

$$\mathrm{G} = \langle s_1, \, \alpha_1, \mu_{z_0} \rangle \subset \mathrm{Bir}(\mathbb{P}^2_{\mathbb{Q}})$$

satisfies the following properties:

PROPOSITION 5.8 ([**66**]). *The group* $\mathrm{G} \subset \mathrm{Bir}(\mathbb{P}^2_{\mathbb{Q}})$ *is solvable of length* 3*, and is not linear over any field.*

A consequence of this statement is Proposition 5.4.

PROOF. Relations (5.1) and (5.2) imply that $\langle s_1, \, \alpha_f, \, \mu_g \,|\, f \in \mathbb{C}(z_0), \, g \in \mathbb{C}(z_0)^* \rangle$ is solvable of length at most three. The subgroup

$$\langle s_1, \, \alpha_f, \, \mu_g \,|\, f \in \mathbb{C}(z_0), \, g = \prod_{n \in \mathbb{Z}} (z_0 - n)^{k_n}, \, k_n \text{ finitely supported } \rangle$$

contains Γ_n, and is torsion free.

As $\mu_{z_0}^n \circ \alpha_1 \circ \mu_{z_0}^{-n} = \alpha_{z_0^n}$, the group Γ_n is contained in G for all n. But Γ_n is nilpotent of length exactly $n+1$, hence G has no linear representation over any field. □

5.2. The Cremona group is perfect

In this section let us prove the following statement

THEOREM 5.9 ([**62**]). *The plane Cremona group is perfect, i.e., the commutator subgroup of* $\mathrm{Bir}(\mathbb{P}^2_\mathbb{C})$ *is* $\mathrm{Bir}(\mathbb{P}^2_\mathbb{C})$*:*

$$\big[\mathrm{Bir}(\mathbb{P}^2_\mathbb{C}),\,\mathrm{Bir}(\mathbb{P}^2_\mathbb{C})\big]=\mathrm{Bir}(\mathbb{P}^2_\mathbb{C}).$$

Let G be a group, and let g be an element of G. We denote by $\ll g\gg_\mathrm{G}$ the *normal subgroup of* G *generated by* g:

$$\ll g\gg_\mathrm{G}=\langle h\circ g\circ h^{-1},\,h\circ g^{-1}\circ h^{-1}\,|\,h\in\mathrm{G}\rangle.$$

Since $\mathrm{PGL}(3,\mathbb{C})$ is simple then

$$\ll A\gg_{\mathrm{PGL}(3,\mathbb{C})}=\mathrm{PGL}(3,\mathbb{C}) \tag{5.3}$$

for any non-trivial element A of $\mathrm{PGL}(3,\mathbb{C})$. Consider now a birational self map ϕ of $\mathrm{Bir}(\mathbb{P}^2_\mathbb{C})$. The Noether and Castelnuovo Theorem implies that

$$\phi=(A_1)\circ\sigma_2\circ A_2\circ\sigma_2\circ A_3\circ\ldots\circ A_n\circ(\sigma_2) \tag{5.4}$$

with $A_i\in\mathrm{PGL}(3,\mathbb{C})$. The relation (5.3) implies that

$$\ll(z_0,z_1)\mapsto(-z_0,-z_1)\gg_{\mathrm{PGL}(3,\mathbb{C})}=\mathrm{PGL}(3,\mathbb{C});$$

thus any A_i in (5.4) can be written

$$B_1\circ\big((z_0,z_1)\mapsto(-z_0,-z_1)\big)\circ B_1^{-1}\circ B_2\circ\big((z_0,z_1)\mapsto(-z_0,-z_1)\big)\circ B_2^{-1}$$
$$\circ\ldots\circ B_n\circ\big((z_0,z_1)\mapsto(-z_0,-z_1)\big)\circ B_n^{-1}$$

with $B_i\in\mathrm{PGL}(3,\mathbb{C})$. The involutions $(z_0,z_1)\mapsto(-z_0,-z_1)$ and σ_2 being conjugate via $(z_0,z_1)\mapsto\left(\frac{z_0+1}{z_0-1},\frac{z_1+1}{z_1-1}\right)\in\mathrm{PGL}(2,\mathbb{C})\times\mathrm{PGL}(2,\mathbb{C})$ any element of $\mathrm{Bir}(\mathbb{P}^2_\mathbb{C})$ can be written as a composition of $\mathrm{Bir}(\mathbb{P}^2_\mathbb{C})$-conjugates of σ_2. As a consequence one has

PROPOSITION 5.10 ([**62**]). *The normal subgroup of* $\mathrm{Bir}(\mathbb{P}^2_\mathbb{C})$ *generated by* σ_2 *in* $\mathrm{Bir}(\mathbb{P}^2_\mathbb{C})$ *is* $\mathrm{Bir}(\mathbb{P}^2_\mathbb{C})$*:*

$$\ll\sigma_2\gg_{\mathrm{Bir}(\mathbb{P}^2_\mathbb{C})}=\mathrm{Bir}(\mathbb{P}^2_\mathbb{C}).$$

Consider now a non-trivial automorphism A of $\mathbb{P}^2_\mathbb{C}$. As $\ll A\gg_{\mathrm{PGL}(3,\mathbb{C})}=\mathrm{PGL}(3,\mathbb{C})$ (*see* (5.3)) the involution $(z_0,z_1)\mapsto(-z_0,-z_1)$ can be written as a composition of $\mathrm{PGL}(3,\mathbb{C})$-conjugates of A. Since $(z_0,z_1)\mapsto(-z_0,-z_1)$ and σ_2 are conjugate via $(z_0,z_1)\mapsto\left(\frac{z_0+1}{z_0-1},\frac{z_1+1}{z_1-1}\right)\in\mathrm{PGL}(2,\mathbb{C})\times\mathrm{PGL}(2,\mathbb{C})$ one gets

$$\sigma_2=\varphi_1\circ A\circ\varphi_1^{-1}\circ\varphi_2\circ A\circ\varphi_2^{-1}\circ\ldots\circ\varphi_n\circ A\circ\varphi_n^{-1}$$

with $\varphi_i\in\mathrm{Bir}(\mathbb{P}^2_\mathbb{C})$. As a consequence the inclusion $\ll\sigma_2\gg_{\mathrm{Bir}(\mathbb{P}^2_\mathbb{C})}\subset\ll A\gg_{\mathrm{Bir}(\mathbb{P}^2_\mathbb{C})}$ holds. But $\ll\sigma_2\gg_{\mathrm{Bir}(\mathbb{P}^2_\mathbb{C})}=\mathrm{Bir}(\mathbb{P}^2_\mathbb{C})$ (Proposition 5.10) hence

PROPOSITION 5.11 ([**62**]). *Let A be a non-trivial automorphism of the complex projective plane. Then*

$$\ll A\gg_{\mathrm{Bir}(\mathbb{P}^2_\mathbb{C})}=\mathrm{Bir}(\mathbb{P}^2_\mathbb{C}).$$

According to (5.4) and Proposition 5.11 one has

COROLLARY 5.12. *Any birational self map of* $\mathbb{P}^2_\mathbb{C}$ *can be written as the composition of* $\mathrm{Bir}(\mathbb{P}^2_\mathbb{C})$*-conjugates of the translation* $(z_0,z_1)\mapsto(z_0,z_1+1)$.

But the translation $(z_0, z_1) \mapsto (z_0, z_1+1)$ is a commutator

$$\Big((z_0,z_1)\mapsto(z_0,z_1+1)\Big) = \left[(z_0,z_1)\mapsto(z_0,3z_1),\,(z_0,z_1)\mapsto\left(z_0,\frac{z_1+1}{2}\right)\right]$$

and Corollary 5.12 thus implies Theorem 5.9.

5.3. Representations of SL($n,\mathbb{Z}$) into Bir($\mathbb{P}^2_\mathbb{C}$) for $n \geq 3$

We will now give a sketch of the proofs of Theorem 5.2 and Corollary 5.3.

Let us introduce some notations. Given $A \in \mathrm{Aut}(\mathbb{P}^2_\mathbb{C}) = \mathrm{PGL}(3,\mathbb{C})$ we denote by ${}^tA$ the linear transpose of A. The involution

$$A \mapsto A^\vee = ({}^tA)^{-1}$$

determines an exterior and algebraic automorphism of the group $\mathrm{Aut}(\mathbb{P}^2_\mathbb{C})$ (*see* [**83**]).

Let us recall some properties about the groups $\mathrm{SL}(n,\mathbb{Z})$ (*see for instance* [**163**]). For any integer q let us introduce the morphism

$$\Theta_q\colon \mathrm{SL}(n,\mathbb{Z}) \to \mathrm{SL}\left(n, {}^{\mathbb{Z}}\!/_{q\mathbb{Z}}\right)$$

induced by the reduction modulo q morphism $\mathbb{Z} \to {}^{\mathbb{Z}}\!/_{q\mathbb{Z}}$. Denote by $\Gamma(n,q)$ the kernel of Θ_q and by $\widetilde{\Gamma}(n,q)$ the reciprocal image of the subgroup of diagonal matrices of $\mathrm{SL}\left(n, {}^{\mathbb{Z}}\!/_{q\mathbb{Z}}\right)$ by Θ_q. The $\Gamma(n,q)$ are normal subgroups called *congruence subgroups*.

THEOREM 5.13 ([**163**]). *Let $n \geq 3$ be an integer. Let Γ be a subgroup of* $\mathrm{SL}(n,\mathbb{Z})$.

If Γ has finite index, there exists an integer q such that the following inclusions hold

$$\Gamma(n,q) \subset \Gamma \subset \widetilde{\Gamma}(n,q).$$

If Γ has infinite index, then Γ is finite.

Take $1 \leq i, j \leq n$, $i \neq j$. Let us denote by δ_{ij} the $n \times n$ Kronecker matrix and set $\mathrm{e}_{ij} = \mathrm{id} + \delta_{i,j}$.

PROPOSITION 5.14 ([**163**]). *The group* $\mathrm{SL}(3,\mathbb{Z})$ *has the following presentation*

$$\langle \mathrm{e}_{ij} \,|\, [\mathrm{e}_{ij}, \mathrm{e}_{k\ell}] = \begin{cases} \mathrm{id} \text{ if } i \neq \ell \text{ and } j \neq k, \\ \mathrm{e}_{i\ell} \text{ if } i \neq \ell \text{ and } j = k, \\ \mathrm{e}_{kj}^{-1} \text{ if } i = \ell \text{ and } j \neq k, \end{cases} \quad (\mathrm{e}_{12}\mathrm{e}_{21}^{-1}\mathrm{e}_{12})^4 = \mathrm{id}\rangle.$$

REMARK 5.15. The e_{ij}^q's generate $\Gamma(3,q)$ and satisfy relations similar to those verified by the e_{ij}'s except $(\mathrm{e}_{12}\mathrm{e}_{21}^{-1}\mathrm{e}_{12})^4 = \mathrm{id}$.

The e_{ij}^q's are called the *standard generators* of $\Gamma(3,q)$.

DEFINITION. Let k be an integer. A *k-Heisenberg group* is a group with the following presentation

$$\mathcal{H}_k = \langle f,\, g,\, h \,|\, [f,g] = h^k,\ [f,h] = [g,h] = \mathrm{id}\rangle.$$

We will say that f, g and h are the *standard generators* of $\mathcal{H}_k$.

REMARKS 5.16.

◇ The subgroup of $\mathcal{H}_k$ generated by f, g and h^k is a subgroup of index k.
◇ The groups $\Gamma(3,q)$ contain a lot of k-Heisenberg groups; for instance if $1\leq i\neq j\neq \ell\leq 3$, then $\langle \mathrm{e}_{ij}^q,\,\mathrm{e}_{i\ell}^q,\,\mathrm{e}_{j\ell}^q\rangle$ is a q-Heisenberg group of $\Gamma(3,q)$.

Let G be a finitely generated group, let $\{a_1,\,a_2,\,\ldots,\,a_n\}$ be a generating set of G, and let g be an element of G. The *length* $||g||$ of g is the smallest integer k for which there exists a sequence $(s_1,\,s_2,\ldots,\,s_k)$ with $s_i\in\{a_1,\,a_2,\,\ldots,\,a_n,\,a_1^{-1},\,a_2^{-1},\,\ldots,\,a_n^{-1}\}$ for any $1\leq i\leq k$, such that

$$g=s_1s_2\ldots s_k.$$

We say that

$$\lim_{k\to+\infty}\frac{||g^k||}{k}$$

is the *stable length* of g. A *distorted* element of G is an element of infinite order of G whose stable length is zero.

LEMMA 5.17 ([**76**]). *Let $\mathcal{H}_k=\langle f,\,g,\,h\,|\,[f,g]=h^k,\,[f,h]=[g,h]=\mathrm{id}\rangle$ be a k-Heisenberg group.*

The element h^k is distorted.

In particular the standard generators of $\Gamma(3,q)$ *are distorted.*

PROOF. Since $[f,g]=[g,h]=\mathrm{id}$, one has $[f^\ell,g^m]=h^{\ell m}$ for any integer ℓ, m. In particular $h^{k\ell^2}=[f^\ell,g^\ell]$. As a result $||h^{k\ell^2}||\leq 4\ell$.

Each standard generator of $\Gamma(3,q)$ satisfies $\mathrm{e}_{ij}^{q^2}=[\mathrm{e}_{i\ell}^q,\mathrm{e}_{\ell j}^q]$. □

LEMMA 5.18 ([**79**]). *Let* G *be a finitely generated group. Let υ be a morphism from* G *to* $\mathrm{Bir}(\mathbb{P}^2_\mathbb{C})$. *Any distorted element g of* G *satisfies $\lambda(\upsilon(g))=1$, i.e., $\upsilon(g)$ is an elliptic map or a parabolic one.*

PROOF. Let $\{a_1,\,a_2,\,\ldots,\,a_n\}$ be a generating set of G. The inequalities

$$\lambda(\upsilon(g))^k\leq \deg(\upsilon(g)^k)\leq \max_i\big(\deg(\upsilon(a_i))\big)^{||g^k||}$$

imply the following ones:

$$0\leq\log\big(\lambda(\upsilon(g))\big)\leq\frac{||g^k||}{k}\log\big(\max_i\big(\deg(\upsilon(a_i))\big)\big).$$

But since g is distorted, $\displaystyle\lim_{k\to+\infty}\frac{||g^k||}{k}=0$ and $\log\big(\lambda(\upsilon(g))\big)=0$. □

REMARK 5.19. We follow the proof of [**76**]; nevertheless it is possible to "simplify it" by using the following result: any distorted element of $\mathrm{Bir}(\mathbb{P}^2_\mathbb{C})$ is algebraic ([**35, 55**]).

According to Corollary 3.34 we thus have: any distorted element of $\mathrm{Bir}(\mathbb{P}^2_\mathbb{C})$ is elliptic.

DEFINITION. Let $\phi_1,\,\phi_2,\,\ldots,\,\phi_k$ be some birational self maps of a rational surface S. Assume that $\phi_1,\,\phi_2,\,\ldots,\,\phi_k$ are virtually isotopic to the identity. We say that $\phi_1,\,\phi_2,\,\ldots,\,\phi_k$ are *simultaneously virtually isotopic to the identity* if there exists a surface $\widetilde{S}$, a birational map $\psi\colon\widetilde{S}\dashrightarrow S$ such that for any $1\leq i\leq k$ the map $\psi^{-1}\circ\phi_i\circ\psi$ belongs to $\mathrm{Aut}(\widetilde{S})$ and $\psi^{-1}\circ\phi_i^\ell\circ\psi$ belongs to $\mathrm{Aut}(\widetilde{S})^0$ for some integer ℓ.

PROPOSITION 5.20 ([**76**]). *Let υ be a representation from*

$$\mathcal{H}_k = \langle f,\, g,\, h \,|\, [f,g] = h^k,\; [f,h] = [g,h] = \mathrm{id}\rangle$$

into $\mathrm{Bir}(\mathbb{P}^2_{\mathbb{C}})$*. Assume that any standard generator $\upsilon(f)$, $\upsilon(g)$ and $\upsilon(h)$ of $\upsilon(\mathcal{H}_k)$ is virtually isotopic to the identity. Then $\upsilon(f)$, $\upsilon(g)$ and $\upsilon(h)$ are simultaneously virtually isotopic to the identity.*

PROOF. According to Proposition 2.12 the maps $\upsilon(f)$ and $\upsilon(g)$ are simultaneously virtually isotopic to the identity. Since g and h commute, $\mathrm{Exc}(\upsilon(g))$ and $\mathrm{Ind}(\upsilon(g))$ are invariant by $\upsilon(h)$. The relation $[f,g] = h^k$ implies that both $\mathrm{Exc}(\upsilon(g))$ and $\mathrm{Ind}(\upsilon(g))$ are invariant by $\upsilon(f)$. A reasoning analogous to that of the proof of Proposition 2.12 and [**85**, Lemma 4.1] allows us to establish the statement. □

The second assertion of Remarks 5.16 leads us to study the representations of Heisenberg k-groups into automorphism groups of minimal rational surfaces. Let us deal with it.

LEMMA 5.21 ([**76**]). *Let υ be a morphism from $\mathcal{H}_k$ into* $\mathrm{Aut}(\mathbb{P}^1_{\mathbb{C}} \times \mathbb{P}^1_{\mathbb{C}})$*.*
The morphism υ is not an embedding.

PROOF. We can assume that $\upsilon(f)$, $\upsilon(g)$ and $\upsilon(h)$ fix the two standard fibrations (if it is not the case we can consider $\mathcal{H}_{2k}$ instead of $\mathcal{H}_k$); in other words we can assume that $\mathrm{im}\,\upsilon$ is contained in $\mathrm{PGL}(2,\mathbb{C}) \times \mathrm{PGL}(2,\mathbb{C})$. Denote by pr_i, $i \in \{1,\, 2\}$, the ith projection. Note that $\mathrm{pr}_i(\upsilon(\mathcal{H}_{2k}))$ is a solvable subgroup of $\mathrm{PGL}(2,\mathbb{C})$. Furthermore $\mathrm{pr}_i(\upsilon(h^k))$ is a commutator. Hence $\mathrm{pr}_i(\upsilon(h^k))$ is conjugate to the translation $z \mapsto z + \beta_i$. Let us prove by contradiction that $\beta_i = 0$; assume $\beta_i \neq 0$. Then both $\mathrm{pr}_i(\upsilon(f))$ and $\mathrm{pr}_i(\upsilon(g))$ are also some translation since they commute with $\mathrm{pr}_i(\upsilon(h^k))$. But then $\mathrm{pr}_i(\upsilon(h^k)) = [\mathrm{pr}_i(\upsilon(f)), \mathrm{pr}_i(\upsilon(g))] = \mathrm{id}$: contradiction with $\beta_i \neq 0$. As a result $\beta_i = 0$ and υ is not an embedding. □

LEMMA 5.22 ([**76**]). *Let υ be a morphism from $\mathcal{H}_k$ into $\mathrm{Aut}(\mathbb{F}_n)$ with $n \geq 1$. Then up to birational conjugacy $\upsilon(\mathcal{H}_k)$ is a subgroup of*

$$\big\{(z_0, z_1) \mapsto (\alpha z_0 + P(z_1), \beta z_1 + \gamma) \,|\, \alpha,\, \beta \in \mathbb{C}^*,\; \gamma \in \mathbb{C},\; P \in \mathbb{C}[z_1]\big\}.$$

Moreover up to birational conjugacy

$$\upsilon(h^{2k})\colon (z_0, z_1) \mapsto (z_0 + P(z_1), z_1)$$

for some $P \in \mathbb{C}[z_1]$.

LEMMA 5.23 ([**76**]). *Let υ be an embedding of $\mathcal{H}_k$ into* $\mathrm{PGL}(3,\mathbb{C})$*. Up to linear conjugacy*

$$\upsilon(f)\colon (z_0, z_1) \mapsto (z_0 + \zeta z_1, z_1 + \beta) \qquad \upsilon(g)\colon (z_0, z_1) \mapsto (z_0 + \gamma z_1, z_1 + \delta)$$
$$\upsilon(h^k)\colon (z_0, z_1) \mapsto (z_0 + k, z_1)$$

where ζ, δ, β γ denote complex numbers such that $\zeta\delta - \beta\gamma = k$.

PROOF. The Zariski closure $\overline{\upsilon(\mathcal{H}_k)}$ of $\upsilon(\mathcal{H}_k)$ is an algebraic unipotent subgroup of $\mathrm{PGL}(3,\mathbb{C})$. By assumption υ is an embedding, so the Lie algebra of $\overline{\upsilon(\mathcal{H}_k)}$ is isomorphic to

$$\mathfrak{h} = \left\{ \begin{pmatrix} 0 & \zeta & \beta \\ 0 & 0 & \gamma \\ 0 & 0 & 0 \end{pmatrix} \,|\, \zeta,\, \beta,\, \gamma \in \mathbb{C} \right\}.$$

Let pr be the canonical projection from $\mathrm{SL}(3,\mathbb{C})$ into $\mathrm{PGL}(3,\mathbb{C})$. The Lie algebra of $\mathrm{pr}^{-1}(\overline{\upsilon(\mathcal{H}_k)})$ coincides with $\mathfrak{h}$ up to conjugacy. Let us recall that the exponential map sends $\mathfrak{h}$ in the group H of upper triangular matrices and that H is a connected algebraic group. As a consequence $\big(\mathrm{pr}^{-1}(\overline{\upsilon(\mathcal{H}_k)})\big)^0=\mathrm{H}$. Any element of $\mathrm{pr}^{-1}(\overline{\upsilon(\mathcal{H}_k)})$ acts by conjugation on H, so it belongs to $\langle \mathrm{H}, \mathbf{j}\cdot\mathrm{id}\,|\,\mathbf{j}^3=1\rangle$. As $\mathrm{pr}(\mathbf{j}\cdot\mathrm{id})=\mathrm{id}$, the restriction $\mathrm{pr}_{|\mathrm{H}}$ of pr to H is surjective on $\overline{\upsilon(\mathcal{H}_k)}$. It is also injective. Hence it is an isomorphism. Therefore, υ can be lifted to a representation $\widetilde{\upsilon}$ from $\mathcal{H}_k$ into H. The map $\widetilde{\upsilon}(h^k)$ can be written as a commutator; it is thus unipotent. The relations satisfied by the generators imply that up to conjugacy in $\mathrm{SL}(3,\mathbb{C})$

$$\upsilon(f)\colon(z_0,z_1)\mapsto(z_0+\zeta z_1,z_1+\beta)\qquad \upsilon(g)\colon(z_0,z_1)\mapsto(z_0+\gamma z_1,z_1+\delta)$$
$$\upsilon(h^k)\colon(z_0,z_1)\mapsto(z_0+k,z_1)$$

with $\zeta\delta-\beta\gamma=k$. □

Let ρ be an embedding of $\Gamma(3,q)$ into $\mathrm{Bir}(\mathbb{P}^2_\mathbb{C})$. According to Lemma 5.17 and Lemma 5.18 for any standard generator e_{ij} of $\mathrm{SL}(3,\mathbb{Z})$ one has $\lambda(\rho(e_{ij}))=1$. Theorem 2.9 implies that

(i) either one of the $\rho(\mathrm{e}_{ij}^q)$ preserves a unique fibration that is rational or elliptic,
(ii) or any standard generator of $\Gamma(3,q)$ is virtually isotopic to the identity.

Let us first assume that (i) holds.

Lemma 5.24 ([**76**]). *Let Γ be a Kazhdan group that is finitely generated. Let ρ be a morphism from Γ into $\mathrm{PGL}(2,\mathbb{C}(z_1))$ (resp. $\mathrm{PGL}(2,\mathbb{C})$). Then ρ has finite image.*

Proof. Denote by γ_i the generators of Γ and by $\begin{pmatrix} a_i(z_1) & b_i(z_1)\\ c_i(z_1) & d_i(z_1)\end{pmatrix}$ their image by ρ. A finitely generated $\mathbb{Q}$-group is isomorphic to a subfield of $\mathbb{C}$. Hence $\mathbb{Q}(a_i(z_0),b_i(z_0),c_i(z_0),d_i(z_0))$ is isomorphic to a subfield of $\mathbb{C}$ and one can assume that $\mathrm{im}\,\rho\subset\mathrm{PGL}(2,\mathbb{C})=\mathrm{Isom}(\mathbb{H}_3)$. As Γ is Kazhdan any continuous action of Γ by isometries of a real or complex hyperbolic space has a fixed point. The image of ρ is thus up to conjugacy a subgroup of $\mathrm{SO}(3,\mathbb{R})$; according to [**72**] the image of ρ is thus finite. □

Proposition 5.25 ([**76**]). *Let ρ be a morphism from $\Gamma(3,q)$ to $\mathrm{Bir}(\mathbb{P}^2_\mathbb{C})$. If one $\rho(\mathrm{e}_{ij}^q)$ preserves a unique fibration, then $\mathrm{im}\,\rho$ is finite.*

Proof. Let us assume without loss of generality that $\rho(\mathrm{e}_{12}^q)$ preserves a unique fibration $\mathcal{F}$. The relations satisfied by the e_{ij}^q imply that $\mathcal{F}$ is invariant by any $\rho(\mathrm{e}_{ij}^{q^2})$. Hence for any $\rho(\mathrm{e}_{ij}^{q^2})$ there exist

- ◇ $F\colon\mathbb{P}^2_\mathbb{C}\to\mathrm{Aut}(\mathbb{P}^1_\mathbb{C})$ that defines $\mathcal{F}$,
- ◇ and $h_{ij}\in\mathrm{PGL}(2,\mathbb{C})$,

such that $F\circ\rho(\mathrm{e}_{ij}^{q^2})=h_{ij}\circ F$.

Let υ be the morphism defined by

$$\upsilon\colon\Gamma(3,q^2)\to\mathrm{PGL}(2,\mathbb{C}),\qquad\qquad \mathrm{e}_{ij}^{q^2}\mapsto h_{ij}.$$

The group $\Gamma(3,q^2)$ is a Kazhdan group, so $\Gamma = \ker \upsilon$ is of finite index (Lemma 5.24); as a consequence Γ is a Kazhdan group.

Note that $\mathcal{F}$ cannot be elliptic; indeed the group of birational maps that preserve fiberwise an elliptic fibration is metabelian and a subgroup of $\Gamma(3,q^2)$ of finite index cannot be metabelian.

Let us assume that $\mathcal{F}$ is a rational fibration. One can assume that $\mathcal{F} = (z_1 = \text{constant})$. The group of birational maps of the complex projective plane that preserves $\mathcal{F}$ is identified with $\mathrm{PGL}(2,\mathbb{C}(z_1)) \rtimes \mathrm{PGL}(2,\mathbb{C})$; hence $\rho_{|\Gamma}\colon \Gamma \to \mathrm{PGL}(2,\mathbb{C}(z_1))$ has finite image (Lemma 5.24). □

Consider now the case (ii), *i.e.*, assume that any standard generator of $\Gamma(3,q)$ is virtually isotopic to the identity.

REMARK 5.26. Two irreducible homologous curves of negative self-intersection coincide. As a consequence an automorphism φ of a surface S isotopic to the identity fixes any curve of negative self-intersection. Furthermore for any sequence of blow-downs ψ from S to a minimal model $\widetilde{S}$ of S the map $\psi\circ\varphi\circ\psi^{-1}$ is an automorphism of $\widetilde{S}$ isotopic to the identity.

According to Remark 5.26, Proposition 5.20, Lemma 5.17 and Lemma 5.18 the maps $\rho(\mathrm{e}_{12}^{qn})$, $\rho(\mathrm{e}_{13}^{qn})$, $\rho(\mathrm{e}_{23}^{qn})$ are, for some integer n, some automorphisms of a minimal rational surface, that is of $\mathbb{P}^2_{\mathbb{C}}$ or of $\mathbb{F}_n$, $n\geq 2$. Let us mention the case $\mathbb{F}_n$, $n\geq 2$ (*see* [**76**] for more details) and detail the case $\mathbb{P}^2_{\mathbb{C}}$.

LEMMA 5.27 ([**76**]). *Let ρ be a morphism from a congruence subgroup $\Gamma(3,q)$ of $\mathrm{SL}(3,\mathbb{Z})$ in the plane Cremona group.*

Assume that $\rho(\mathrm{e}_{12}^{q\ell})$, $\rho(\mathrm{e}_{13}^{q\ell})$ and $\rho(\mathrm{e}_{23}^{q\ell})$ belong to $\mathrm{Aut}(\mathbb{F}_n)$, $n\geq 2$, for some integer ℓ. Then the image of ρ is

- ◇ *either finite,*
- ◇ *or contained in $\mathrm{PGL}(3,\mathbb{C}) = \mathrm{Aut}(\mathbb{P}^2_{\mathbb{C}})$ up to conjugacy.*

LEMMA 5.28 ([**76**]). *Let ρ be an embedding of a congruence subgroup $\Gamma(3,q)$ of $\mathrm{SL}(3,\mathbb{Z})$ into $\mathrm{Bir}(\mathbb{P}^2_{\mathbb{C}})$. If $\rho(\mathrm{e}_{12}^{qn})$, $\rho(\mathrm{e}_{13}^{qn})$ and $\rho(\mathrm{e}_{23}^{qn})$ belong, for some integer n, to $\mathrm{PGL}(3,\mathbb{C}) = \mathrm{Aut}(\mathbb{P}^2_{\mathbb{C}})$, then $\rho(\Gamma(3,q^2n^2))$ is a subgroup of $\mathrm{PGL}(3,\mathbb{C}) = \mathrm{Aut}(\mathbb{P}^2_{\mathbb{C}})$.*

To establish this statement we will need the two following results; the first one was obtained by Cantat and Lamy when they studied the embeddings of lattices from simple Lie groups into the group of polynomial automorphisms $\mathrm{Aut}(\mathbb{A}^2_{\mathbb{C}})$ whereas the second one is a technical one.

THEOREM 5.29 ([**58**]). *Let G be a simple real Lie group. Let Γ be a lattice of G. If there exists an embedding of Γ into $\mathrm{Aut}(\mathbb{A}^2_{\mathbb{C}})$, then G is isomorphic to either $\mathrm{PSO}(1,n)$ or $\mathrm{PSU}(1,n)$ for some integer n.*

LEMMA 5.30 ([**76**]). *Let ϕ be an element of the plane Cremona group. Assume that $\mathrm{Exc}(\phi)$ and $\mathrm{Exc}(\phi^2)$ are non-empty and contained in the line at infinity. If $\mathrm{Ind}(\phi)$ is also contained in the line at infinity, then ϕ is a polynomial automorphism of $\mathbb{A}^2_{\mathbb{C}}$.*

PROOF OF LEMMA 5.28. Lemma 5.23 allows us to assume that

$$\rho(\mathrm{e}_{13}^{qn})\colon (z_0,z_1)\mapsto(z_0+qn,z_1), \qquad \rho(\mathrm{e}_{12}^{qn})\colon (z_0,z_1)\mapsto(z_0+\zeta z_1,z_1+\beta),$$
$$\rho(\mathrm{e}_{23}^{qn})\colon (z_0,z_1)\mapsto(z_0+\gamma z_1,z_1+\delta)$$

where $\zeta\delta-\beta\gamma=q^2n^2$.

- ◇ Let us first suppose that $\beta\delta\neq 0$. Since $[\rho(\mathrm{e}_{13}^{qn}),\rho(\mathrm{e}_{21}^{qn})]=\rho(\mathrm{e}_{23}^{-q^2n^2})$ the curves blown down by $\rho(\mathrm{e}_{21}^{qn})$, if they exist, are of the type $z_1=\text{constant}$. As $\rho(\mathrm{e}_{21}^{qn})$ and $\rho(\mathrm{e}_{23}^{qn})$ commute, the sets $\mathrm{Exc}(\rho(\mathrm{e}_{21}^{qn}))$ and $\mathrm{Ind}(\rho(\mathrm{e}_{21}^{qn}))$ are invariant by $\rho(\mathrm{e}_{23}^{qn})$. As a result $\mathrm{Exc}(\rho(\mathrm{e}_{21}^{qn}))$, $\mathrm{Ind}(\rho(\mathrm{e}_{21}^{qn}))$ and $\mathrm{Exc}((\rho(\mathrm{e}_{21}^{qn}))^2)$ are contained in the line at infinity. Hence $\rho(\mathrm{e}_{21}^{qn})$ belongs to either $\mathrm{PGL}(3,\mathbb{C})$ or $\mathrm{Aut}(\mathbb{A}^2_\mathbb{C})$ (Lemma 5.30). Note that if $\rho(\mathrm{e}_{21}^{qn})$ belongs to $\mathrm{PGL}(3,\mathbb{C})$, then $\rho(\mathrm{e}_{21}^{qn})$ preserves the line at infinity because $\rho(\mathrm{e}_{21}^{qn})$ and $\rho(\mathrm{e}_{23}^{qn})$ commute. In other words $\rho(\mathrm{e}_{21}^{qn})$ also belongs to $\mathrm{Aut}(\mathbb{A}^2_\mathbb{C})$. Using the relations $[\rho(\mathrm{e}_{13}^{qn}),\rho(\mathrm{e}_{32}^{qn})]=\rho(\mathrm{e}_{12}^{q^2n^2})$ and $[\rho(\mathrm{e}_{12}^{qn}),\rho(\mathrm{e}_{32}^{qn})]=\mathrm{id}$ we get that $\rho(\mathrm{e}_{23}^{qn})$ belongs to $\mathrm{Aut}(\mathbb{A}^2_\mathbb{C})$. Finally any $\rho(\mathrm{e}_{ij}^{q^2n^2})$ is a polynomial automorphism of $\mathbb{A}^2_\mathbb{C}$ and ρ is not an embedding (Theorem 5.29).
- ◇ Assume that $\beta\delta=0$. Since $\zeta\delta-\beta\gamma=q^2n^2$ one has $(\beta,\delta)\neq(0,0)$.

  Suppose that $\beta=0$. The conjugacy by

$$(z_0,z_1)\mapsto\left(z_0+\frac{\gamma}{2}z_1-\frac{\gamma}{2\delta}z_1^2,z_1\right)$$

  changes neither $\rho(\mathrm{e}_{13}^{qn})$, nor $\rho(\mathrm{e}_{12}^{qn})$, and sends $\rho(\mathrm{e}_{23}^{qn})$ onto $(z_0,z_1)\mapsto(z_0,z_1+\delta)$. One can thus assume that

$$\rho(\mathrm{e}_{13}^{qn})\colon(z_0,z_1)\mapsto(z_0+qn,z_1),\qquad \rho(\mathrm{e}_{12}^{qn})\colon(z_0,z_1)\mapsto(z_0+\zeta z_1,z_1),$$
$$\rho(\mathrm{e}_{23}^{qn})\colon(z_0,z_1)\mapsto(z_0,z_1+\delta).$$

  The map $\rho(\mathrm{e}_{21}^{qn})$ satisfies the relations $[\rho(\mathrm{e}_{13}^{qn}),\rho(\mathrm{e}_{21}^{qn})]=\rho(\mathrm{e}_{23}^{-q^2n^2})$, and $[\rho(\mathrm{e}_{21}^{qn}),\rho(\mathrm{e}_{23}^{qn})]=\mathrm{id}$, so does the element $\psi\colon(z_0,z_1)\mapsto(z_0,\delta nz_0+z_1)$ of $\mathrm{PGL}(3,\mathbb{C})$. Note that the map $\phi=\rho(\mathrm{e}_{21}^{qn})\circ\psi^{-1}$ commutes to both $\rho(\mathrm{e}_{13}^{qn})$ and $\rho(\mathrm{e}_{23}^{qn})$. As a consequence

$$\phi\colon(z_0,z_1)\mapsto(z_0+a,z_1+b)$$

  for some a, b in $\mathbb{C}$. Finally up to conjugacy by $(z_0,z_1)\mapsto\left(z_0+\frac{b}{\delta},z_1\right)$ one has

$$\rho(\mathrm{e}_{21}^{qn})\colon(z_0,z_1)\mapsto(z_0+a,\delta z_0+z_1);$$

  in particular $\rho(\mathrm{e}_{21}^{qn})$ belongs to $\mathrm{PGL}(3,\mathbb{C})$. Similarly if φ is the map given by

$$(z_0,z_1)\mapsto\left(\frac{z_0}{1+\zeta z_1},\frac{z_1}{1+\zeta z_1}\right)$$

  then the map $\rho(\mathrm{e}_{32}^{qn})\circ\varphi^{-1}$ commutes to both $\rho(\mathrm{e}_{13}^{qn})$ and $\rho(\mathrm{e}_{12}^{qn})$. Therefore

$$\rho(\mathrm{e}_{32}^{qn})\circ\varphi^{-1}\colon(z_0,z_1)\mapsto(z_0+b(z_1),z_1)$$

  and

$$\rho(\mathrm{e}_{32}^{qn})\colon(z_0,z_1)\mapsto\left(\frac{z_0}{1+\zeta z_1}+b\left(\frac{z_1}{1+\zeta z_1}\right),\frac{z_1}{1+\zeta z_1}\right).$$

  Thanks to $[\rho(\mathrm{e}_{23}^{qn}),\rho(\mathrm{e}_{31}^{qn})]=\rho(\mathrm{e}_{21}^{q^2n^2})$, $[\rho(\mathrm{e}_{21}^{qn}),\rho(\mathrm{e}_{31}^{qn})]=\mathrm{id}$ and $[\rho(\mathrm{e}_{12}^{qn}),\rho(\mathrm{e}_{31}^{qn})]=\rho(\mathrm{e}_{32}^{-q^2n^2})$ we get $\rho(\mathrm{e}_{21}^{qn})\colon(z_0,z_1)\mapsto(z_0,\delta z_0+z_1)$. Finally since $\rho(\mathrm{e}_{31}^{qn})$ and $\rho(\mathrm{e}_{32}^{qn})$ commute, $b\equiv 0$ and $\mathrm{im}\,\rho\subset\mathrm{PGL}(3,\mathbb{C})$.

  Assume that $\delta=0$; using a similar reasoning we get a contradiction.

□

Proof of Theorem 5.2. Any $\rho(\mathrm{e}_{ij})$ is virtually isotopic to the identity (Lemma 5.18 and Proposition 5.25). The maps $\rho(\mathrm{e}_{12}^n)$, $\rho(\mathrm{e}_{13}^n)$ and $\rho(\mathrm{e}_{23}^n)$ are, for some integer n, conjugate to automorphisms of a minimal rational surface (Proposition 5.20 and Remark 5.16). Up to conjugacy one can assume that $\rho(\Gamma(3,n^2)) \subset \mathrm{PGL}(3,\mathbb{C})$ (Lemmas 5.21, 5.27 and 5.28). The restriction $\rho_{|\Gamma(3,n^2)}$ of ρ to $\Gamma(3,n^2)$ can be extended to an endomorphism of $\mathrm{PGL}(3,\mathbb{C})$ (*see* [**163**]). But $\mathrm{PGL}(3,\mathbb{C})$ is simple, so this extension is both injective and surjective. The automorphisms of $\mathrm{PGL}(3,\mathbb{C})$ are obtained from inner automorphisms, automorphisms of the field $\mathbb{C}$ and the involution $u \mapsto u^\vee$ (*see* [**83**, Chapter IV]). But automorphisms of the field $\mathbb{C}$ do not act on $\Gamma(3,n^2)$; hence up to linear conjugacy $\rho_{|\Gamma(3,n^2)}$ coincides with the identity or the involution $u \mapsto u^\vee$.

Let ϕ be an element of $\rho(\mathrm{SL}(3,\mathbb{Z})) \smallsetminus \rho(\Gamma(3,n^2))$ that blows down at least one curve $\mathcal{C}$. The group $\Gamma(3,n^2)$ is a normal subgroup of Γ. As a consequence $\mathcal{C}$ is invariant by $\rho(\Gamma(3,n^2))$, and so by $\overline{\rho(\Gamma(3,n^2))} = \mathrm{PGL}(3,\mathbb{C})$ which is impossible. Finally ϕ does not blow-down any curve, and $\rho(\mathrm{SL}(3,\mathbb{Z})) \subset \mathrm{PGL}(3,\mathbb{C})$. □

Proof of Corollary 5.3.

◇ Let Γ be a subgroup of finite index of $\mathrm{SL}(4,\mathbb{Z})$, and let ρ be a morphism from Γ into the plane Cremona group. We will prove that $\mathrm{im}\,\rho$ is finite. To simplify let us suppose that $\Gamma = \mathrm{SL}(4,\mathbb{Z})$. Denote by e_{ij} the standard generators of $\mathrm{SL}(4,\mathbb{Z})$. The morphism ρ induces a faithful representation $\widetilde{\rho}$ from $\mathrm{SL}(3,\mathbb{Z})$ into $\mathrm{Bir}(\mathbb{P}^2_\mathbb{C})$:

$$\mathrm{SL}(4,\mathbb{Z}) \supset \begin{pmatrix} \mathrm{SL}(3,\mathbb{Z}) & 0 \\ 0 & 1 \end{pmatrix} \stackrel{\widetilde{\rho}}{\to} \mathrm{Bir}(\mathbb{P}^2_\mathbb{C}).$$

According to Theorem 5.2 the map $\widetilde{\rho}$ is, up to birational conjugacy, the identity or the involution $u \mapsto u^\vee$.

Let us first assume that up to birational conjugacy $\widetilde{\rho} = \mathrm{id}$. Assume that $\mathrm{Exc}(\rho(\mathrm{e}_{34})) \neq \emptyset$. Since $[\mathrm{e}_{34},\mathrm{e}_{31}] = [\mathrm{e}_{34},\mathrm{e}_{32}] = \mathrm{id}$ the map $\rho(\mathrm{e}_{34})$ commutes with

$$(z_0,z_1,z_2) \mapsto (z_0,z_1,az_0+bz_1+z_2)$$

where a, $b \in \mathbb{C}$ and $\mathrm{Exc}(\rho(\mathrm{e}_{34}))$ is invariant by $(z_0,z_1,z_2) \mapsto (z_0,z_1,az_0 + bz_1 + z_2)$. Moreover e_{34} commutes with e_{12} and e_{21}, in other words e_{34} commutes with the following copy of $\mathrm{SL}(2,\mathbb{Z})$:

$$\mathrm{SL}(4,\mathbb{Z}) \supset \begin{pmatrix} \mathrm{SL}(2,\mathbb{Z}) & 0 & 0 \\ 0 & 1 & 0 \\ 0 & 0 & 1 \end{pmatrix}.$$

The action of $\mathrm{SL}(2,\mathbb{Z})$ on $\mathbb{C}^2$ has no invariant curve, so $\mathrm{Exc}(\rho(\mathrm{e}_{34}))$ is contained in the line at infinity. But the image of this line by $(z_0,z_1,z_2) \mapsto (z_0,z_1,az_0+bz_1+z_2)$ intersects $\mathbb{C}^2$: contradiction. Hence $\mathrm{Exc}(\rho(\mathrm{e}_{34})) = \emptyset$ and $\rho(\mathrm{e}_{34})$ belongs to $\mathrm{PGL}(3,\mathbb{C})$. Similarly we get that $\rho(\mathrm{e}_{43})$ belongs to $\mathrm{PGL}(3,\mathbb{C})$. The relations satisfied by the standard generators thus imply that $\rho(\mathrm{SL}(4,\mathbb{Z}))$ is contained in $\mathrm{PGL}(3,\mathbb{C})$. As a consequence $\mathrm{im}\,\rho$ is finite.

A similar idea allows us to conclude when $\widetilde{\rho}$ is, up to conjugacy, the involution $u \mapsto u^\vee$.

◇ Let $n \geq 4$ be an integer. Consider a subgroup of finite index Γ of $\mathrm{SL}(n,\mathbb{Z})$. Let ρ be a morphism from Γ to $\mathrm{Bir}(\mathbb{P}^2_\mathbb{C})$. According to Theorem 5.13 the group Γ contains a congruence subgroup $\Gamma(n,q)$. The morphism ρ induces a representation $\widetilde{\rho}$ from $\Gamma(4,q)$ to $\mathrm{Bir}(\mathbb{P}^2_\mathbb{C})$. As we just see, the kernel of this representation is infinite, so is $\ker\rho$.

□

5.4. The group $\mathrm{Bir}(\mathbb{P}^2_\mathbb{C})$ is hopfian

Let V be a projective variety defined over a field $\Bbbk\subset\mathbb{C}$. The group $\mathrm{Aut}_\Bbbk(\mathbb{C})$ of automorphisms of the field extension $\mathbb{C}/\Bbbk$ acts on $V(\mathbb{C})$, and on $\mathrm{Bir}(V)$ as follows:

$$^\kappa\psi(p)=(\kappa\circ\psi\circ\kappa^{-1})(p) \tag{5.5}$$

for any $\kappa\in\mathrm{Aut}_\Bbbk(\mathbb{C})$, any $\psi\in\mathrm{Bir}(V)$, and any point $p\in V(\mathbb{C})$ for which both sides of (5.5) are well defined. As a consequence $\mathrm{Aut}_\Bbbk(\mathbb{C})$ acts by automorphisms on $\mathrm{Bir}(V)$. If $\kappa\colon\mathbb{C}\to\mathbb{C}$ is a field morphism, then this construction gives an injective morphism

$$\mathrm{Aut}(\mathbb{P}^n_\mathbb{C})\to\mathrm{Aut}(\mathbb{P}^n_\mathbb{C}), \qquad\qquad g\mapsto {}^\kappa g.$$

Write $\mathbb{C}$ as the algebraic closure of a purely transcendental extension $\mathbb{Q}(x_i,\,i\in I)$ of $\mathbb{Q}$; if $f\colon I\to I$ is an injective map, then there exists a field morphism

$$\kappa\colon\mathbb{C}\to\mathbb{C}, \qquad\qquad x_i\mapsto x_{f(i)}.$$

Such a morphism is surjective if and only if f is onto.

The group $\mathrm{Aut}(\mathrm{Bir}(\mathbb{P}^2_\mathbb{C}))$ has been described in [**77**] and [**76**] via two different methods:

THEOREM 5.31 ([**76, 77**]). *Let φ be an element of $\mathrm{Aut}(\mathrm{Bir}(\mathbb{P}^2_\mathbb{C}))$. Then there exist a birational self map ψ of $\mathbb{P}^2_\mathbb{C}$ and an automorphism κ of the field $\mathbb{C}$ such that*

$$\varphi(\phi)={}^\kappa(\psi\circ\phi\circ\psi^{-1}) \qquad\qquad \forall\,\phi\in\mathrm{Bir}(\mathbb{P}^2_\mathbb{C}).$$

The proof of [**77**] will be dealt with in §7.2. The proof of [**76**] can in fact be used to describe the endomorphisms of the plane Cremona group:

THEOREM 5.32 ([**78**]). *Let φ be a non-trivial endomorphism of $\mathrm{Bir}(\mathbb{P}^2_\mathbb{C})$. Then there exist ψ in $\mathrm{Bir}(\mathbb{P}^2_\mathbb{C})$ and an immersion κ of the field $\mathbb{C}$ such that*

$$\varphi(\phi)={}^\kappa(\psi\circ\phi\circ\psi^{-1}) \qquad\qquad \forall\,\phi\in\mathrm{Bir}(\mathbb{P}^2_\mathbb{C}).$$

Let us work in the affine chart $z_2=1$. The group of translations is

$$\mathrm{T}=\big\{(z_0,z_1)\mapsto(z_0+\alpha,z_1+\beta)\,|\,\alpha,\,\beta\in\mathbb{C}\big\}.$$

LEMMA 5.33 ([**78**]). *Let φ be a birational self map of $\mathbb{P}^2_\mathbb{C}$. Assume that φ commutes with both $(z_0,z_1)\mapsto(z_0+1,z_1)$ and $(z_0,z_1)\mapsto(z_0,z_1+1)$.*

Then φ belongs to T.

PROOF. Let $\varphi=(\varphi_0,\varphi_1)$ be an element of $\mathrm{Bir}(\mathbb{P}^2_\mathbb{C})$ that commutes with both $(z_0,z_1)\mapsto(z_0+1,z_1)$ and $(z_0,z_1)\mapsto(z_0,z_1+1)$. In particular

$$\begin{cases}\varphi_0(z_0+1,z_1)=\varphi_0(z_0,z_1)+1,\\ \varphi_1(z_0+1,z_1)=\varphi_1(z_0,z_1).\end{cases}$$

From $\varphi_1(z_0+1,z_1)=\varphi_1(z_0,z_1)$ we get that $\varphi_1=\varphi_1(z_1)$. The equality $\varphi_0(z_0+1,z_1)=\varphi_0(z_0,z_1)+1$ implies

$$\frac{\partial\varphi_0}{\partial z_0}(z_0+1,z_1)=\frac{\partial\varphi_0}{\partial z_0}(z_0,z_1);$$

as a consequence $\frac{\partial\varphi_0}{\partial z_0}=a(z_1)$ and $\varphi_0=a(z_1)z_0+b(z_1)$ for some a, b in $\mathbb{C}(z_1)$. Then

$$\varphi_0(z_0+1,z_1)=\varphi_0(z_0,z_1)+1$$

yields $a(z_1)=1$. In other words $\varphi\colon(z_0,z_1)\dashrightarrow(z_0+b(z_1),\varphi_1(z_1))$.

Let us now write that $\varphi\circ(z_0,z_1+1)\colon(z_0,z_1)\dashrightarrow(z_0,z_1+1)\circ\varphi$; we get that $\varphi\colon(z_0,z_1)\dashrightarrow(\varphi_0(z_0),z_1+c(z_0))$.

Finally $\varphi\colon(z_0,z_1)\dashrightarrow(z_0+b(z_1),\varphi_1(z_1))$ and $\varphi\colon(z_0,z_1)\dashrightarrow(\varphi_0(z_0),z_1+c(z_0))$ imply that φ belongs to T. □

PROOF OF THEOREM 5.32. Since $\mathrm{PGL}(3,\mathbb{C})$ is simple the restriction $\varphi_{|\mathrm{PGL}(3,\mathbb{C})}$ is either trivial or injective.

Let us first suppose that $\varphi_{|\mathrm{PGL}(3,\mathbb{C})}$ is trivial. Consider the element of $\mathrm{PGL}(3,\mathbb{C})$ given by

$$\ell\colon(z_0,z_1)\mapsto\left(\frac{z_0}{z_0-1},\frac{z_0-z_1}{z_0-1}\right).$$

According to [**102**] one has $(\ell\circ\sigma_2)^3=\mathrm{id}$.

As a result $\varphi((\ell\circ\sigma_2)^3)=\mathrm{id}$. Since $\varphi(\ell)=\ell$ (recall that ℓ belongs to $\mathrm{PGL}(3,\mathbb{C})$) one gets that $\varphi(\sigma_2)=\mathrm{id}$. As the plane Cremona group is generated by $\mathrm{PGL}(3,\mathbb{C})$ and σ_2, one gets that $\varphi=\mathrm{id}$.

Assume now that $\varphi_{|\mathrm{PGL}(3,\mathbb{C})}$ is injective. According to Theorem 5.2 the restriction $\varphi_{|\mathrm{SL}(3,\mathbb{Z})}$ of φ to $\mathrm{SL}(3,\mathbb{Z})$ is, up to inner conjugacy, the canonical embedding or $A\mapsto A^\vee$.

- ◇ Suppose first that $\varphi_{|\mathrm{SL}(3,\mathbb{Z})}$ is the canonical embedding. Denote by $\mathcal{U}$ the group of unipotent upper triangular matrices. Set

$$f_\beta=\varphi(z_0+\beta,z_1),\qquad g_\alpha=\varphi(z_0+\alpha,z_1),\qquad h_\gamma=\varphi(z_0,z_1+\gamma).$$

Since f_β and h_γ commute to both $(z_0,z_1)\mapsto(z_0+1,z_1)$ and $(z_0,z_1)\mapsto(z_0,z_1+1)$ one gets from Lemma 5.33 that

$$f_\beta\colon(z_0,z_1)\mapsto(z_0+\lambda(\beta),z_1+\zeta(\beta))\qquad h_\gamma\colon(z_0,z_1)\mapsto(z_0+\eta(\gamma),z_1+\mu(\gamma))$$

where λ, ζ, η and μ are additive morphisms from $\mathbb{C}$ to $\mathbb{C}$. As g_γ commutes with $(z_0,z_1)\mapsto(z_0+z_1,z_1)$ and $(z_0,z_1)\mapsto(z_0+1,z_1)$, there exists a_α in $\mathbb{C}(y)$ such that

$$g_\gamma\colon(z_0,z_1)\mapsto(z_0+a_\alpha(z_1),z_1).$$

The equality

$$(z_0+\alpha z_1,z_1)\circ(z_0,z_1+\gamma)\circ(z_0+\alpha z_1,z_1)^{-1}\circ(z_0,z_1+\gamma)^{-1}=(z_0+\alpha z_1,z_1)$$

implies that $g_\alpha\circ h_\alpha=f_{\alpha\gamma}\circ h_\gamma\circ g_\alpha$ for any α, γ in $\mathbb{C}$. As a consequence

$$f_\beta\colon(z_0,z_1)\mapsto(z_0+\lambda(\beta),z_1)\qquad g_\alpha\colon(z_0,z_1)\mapsto(z_0+\theta(\alpha)z_1+\zeta(\alpha),z_1)$$

and $\theta(\alpha)\mu(\alpha)=\lambda(\alpha\gamma)$. From

$$\Big[\big((z_0,z_1)\mapsto(z_0+\alpha,z_1)\big),\big((z_0,z_1)\mapsto(z_0,z_1+\beta z_0)\big)\Big]=\big((z_0,z_1)\mapsto(z_0,z_1-\alpha)\big)$$

one gets $h_\gamma\colon(z_0,z_1)\mapsto(z_0,z_1+\mu(\gamma))$. In other words for any α, $\beta\in\mathbb{C}$ one has

$$\varphi(z_0+\alpha,z_1+\beta)=f_\alpha\circ h_\beta=(z_0,z_1)\mapsto(z_0+\lambda(\alpha),z_1+\mu(\beta)).$$

Therefore, $\varphi(\mathrm{T})\subset\mathrm{T}$ and $\varphi(\mathcal{U})\subset\mathcal{U}$. Since $\mathrm{PGL}(3,\mathbb{C})=\langle\mathcal{U},\mathrm{SL}(3,\mathbb{Z})\rangle$ the inclusion $\varphi(\mathrm{PGL}(3,\mathbb{C}))\subset\mathrm{PGL}(3,\mathbb{C})$ holds. According to [**43**] the action of φ on $\mathrm{PGL}(3,\mathbb{C})$ comes, up to inner conjugacy, from an embedding of the field $\mathbb{C}$ into itself.

- ◇ Assume now that $\varphi_{|\mathrm{SL}(3,\mathbb{Z})}$ is $A \mapsto A^\vee$. Similar computations and [**43**] imply that $\varphi_{|\mathrm{PGL}(3,\mathbb{C})}$ comes, up to inner conjugacy, from the composition of $A \mapsto A^\vee$ and an embedding of the field $\mathbb{C}$ into itself.

To finish, let us assume for instance that $\varphi_{|\mathrm{PGL}(3,\mathbb{C})}$ comes, up to inner conjugacy, from the composition of $A \mapsto A^\vee$ and an embedding of the field $\mathbb{C}$ into itself. Set $(\eta_1,\eta_2) = \varphi\Big((z_0,z_1) \mapsto \big(z_0,\frac{1}{z_1}\big)\Big)$. From

$$\left((z_0,z_1) \dashrightarrow \left(z_0,\frac{1}{z_1}\right)\right) \circ ((z_0,z_1) \mapsto (\alpha z_0,\beta z_1)) \circ \left((z_0,z_1) \dashrightarrow \left(z_0,\frac{1}{z_1}\right)\right)$$
$$= \left((z_0,z_1) \mapsto \left(\alpha z_0,\frac{z_1}{\beta}\right)\right)$$

one gets

$$\begin{cases} \eta_1\big(\lambda(\alpha^{-1})z_0,\lambda(\beta^{-1})z_1\big) = \lambda(\alpha^{-1})\eta_1(z_0,z_1), \\ \eta_2\big(\lambda(\alpha^{-1})z_0,\lambda(\beta^{-1})z_1\big) = \lambda(\beta)\eta_2(z_0,z_1). \end{cases}$$

Hence $\varphi\Big((z_0,z_1) \mapsto \big(z_0,\frac{1}{z_1}\big)\Big) = \Big((z_0,z_1) \mapsto \big(\pm z_0,\pm\frac{1}{z_1}\big)\Big)$. But

$$\left(((z_0,z_1) \mapsto (z_1,z_0)) \circ \left((z_0,z_1) \mapsto \left(z_0,\frac{1}{z_1}\right)\right)\right)^2 = \sigma_2,$$

so $\varphi(\sigma_2) = \pm\sigma_2$. Furthermore $\varphi(\ell) = \big((z_0,z_1) \mapsto (-z_0 - z_1 - 1, z_1)\big)$ as $\varphi_{|\mathrm{SL}(3,\mathbb{Z})}$ coincides with $A \mapsto A^\vee$. Then the second component of $\varphi\big(\ell \circ \sigma_2\big)^3$ is $\pm\frac{1}{z_1}$: contradiction with $\varphi\big(\ell \circ \sigma_2\big)^3 = \mathrm{id}$.

If $\varphi_{|\mathrm{PGL}(3,\mathbb{C})}$ comes, up to inner conjugacy, from an embedding of $\mathbb{C}$, similar computations imply that $\varphi(\sigma_2) = \sigma_2$ and one concludes with the Noether and Castelnuovo theorem. □

CHAPTER 6

Finite subgroups of the Cremona group

The classification of finite subgroups of $\mathrm{Bir}(\mathbb{P}^1_\mathbb{C}) = \mathrm{PGL}(2,\mathbb{C})$ is well known and goes back to Klein. It consists of cyclic, dihedral, tetrahedral, octahedral and icosahedral groups. Groups of the same type and same order constitute a unique conjugacy class in $\mathrm{Bir}(\mathbb{P}^1_\mathbb{C})$.

What about the two-dimensional case, *i.e.*, what about the finite subgroups of $\mathrm{Bir}(\mathbb{P}^2_\mathbb{C})$? The story starts a long time ago with Bertini ([**16**]) who classified conjugacy classes of subgroups of order 2 in $\mathrm{Bir}(\mathbb{P}^2_\mathbb{C})$. Already the answer is drastically different from the one-dimensional case. The set of conjugacy classes is parameterized (*see* Theorem 6.3) by a disconnected algebraic variety whose connected components are respectively isomorphic to

- ◇ either the moduli spaces of hyperelliptic curves of genus g,
- ◇ or the moduli space of canonical curves of genus 3,
- ◇ or the moduli space of canonical curves of genus 4 with vanishing theta characteristic.

Bertini's proof is considered to be incomplete; a complete and short proof was published only a few years ago by Bayle and Beauville ([**6**]).

In 1894 Castelnuovo proved that any element of $\mathrm{Bir}(\mathbb{P}^2_\mathbb{C})$ of finite order leaves invariant either a net of lines, or a pencil of lines, or a linear system of cubic curves with $n \leq 8$ base-points ([**60**]). Kantor announced a similar result for arbitrary finite subgroups of $\mathrm{Bir}(\mathbb{P}^2_\mathbb{C})$; his proof relies on a classification of possible groups in each case ([**123**]). Unfortunately Kantor's classification, even with some corrections made by Wiman ([**178**]), is incomplete in the following sense:

- ◇ given some abstract finite group, it is not possible using their list to say whether this group is isomorphic to a subgroup of $\mathrm{Bir}(\mathbb{P}^2_\mathbb{C})$;
- ◇ the possible conjugation between the groups of the list is not considered.

The Russian school has made great progress since the 1960's: Manin and Iskovskikh classified the minimal G-surfaces into automorphisms of del Pezzo surfaces and of conic bundles ([**112, 137**]). Many years afterwards, people came back to this problem. As we already mentioned, Bayle and Beauville classified groups of order 2. It is the first example of a precise description of conjugacy classes; it is shown that the non-rational curves fixed by the groups determine the conjugacy classes. Groups of prime order were also studied ([**9, 69, 183**]). Zhang applies the Bayle and Beauville strategy to the case of birational automorphisms of prime order $p \geq 3$. It turns out that non-linear automorphisms occur only for $p = 3$ and $p = 5$; the author describes them explicitly. The techniques of [**6**] are also generalized by de Fernex to cyclic subgroups of prime order ([**69**]). The list is as precise as one can wish, except for two classes of groups of order 5: the question of their conjugacy is not answered. Beauville and Blanc completed this classification ([**9**]); they proved in particular that a birational self map of the complex projective plane of prime order is not conjugate to a linear automorphism if and only if it fixes some non-rational curve. Beauville classified p-elementary groups ([**8**]). Blanc classified all finite cyclic groups ([**23**]), and all finite abelian groups ([**22**]). The goal of [**89**] is to update the list of Kantor and Wiman. The authors used the modern theory of G-surfaces, the theory of elementary links, and the conjugacy classes of Weyl groups.

In the first section we recall the definitions of Geiser involutions, Bertini involutions, and Jonquières involutions. We give a sketch of the proof of the classification of birational involutions of the complex projective plane due to Bayle and Beauville.

In the second section we deal with finite abelian subgroups of the plane Cremona group. Results due to Dolgachev and Iskovskikh are recalled.

In the last section we state some results of Blanc about finite cyclic subgroups of $\mathrm{Bir}(\mathbb{P}^2_\mathbb{C})$, isomorphism classes of finite abelian subgroups of $\mathrm{Bir}(\mathbb{P}^2_\mathbb{C})$, but also a generalization of a theorem of Castelnuovo which states that an element of finite order which fixes a curve of geometric genus > 1 has order 2, 3, or 4.

6.1. Classification of subgroups of order 2 of $\mathrm{Bir}(\mathbb{P}^2_\mathbb{C})$

6.1.1. Geiser involutions. Let $p_1, p_2, \ldots, p_7$ be seven points of the complex projective plane in general position. Denote by L the linear system of cubics through the p_i's. The linear system L of cubic curves through the p_i's is two-dimensional. Take a general point p, and consider the pencil of curves from L passing through p. A general pencil of cubic curves has nine base-points; let us define $\mathcal{I}_G(p)$ as the ninth base-point of the pencil. The map $\mathcal{I}_G$ is a *Geiser involution* ([**96**]). The algebraic degree of a Geiser involution is equal to 8.

One can also see a Geiser involution as follows. The linear system L defines a rational map of degree 2,

$$\psi\colon \mathbb{P}^2_\mathbb{C} \dashrightarrow |L|^* \simeq \mathbb{P}^2_\mathbb{C}.$$

The points p and $\mathcal{I}_G(p)$ lie in the same fibre. As a consequence, $\mathcal{I}_G$ is a birational deck map of this cover. If we blow-up $p_1, p_2, \ldots, p_7$ we get a del Pezzo surface S of degree 2 and a regular map of degree 2 from S to $\mathbb{P}^2_\mathbb{C}$. Furthermore the Geiser involution becomes an automorphism of S.

Note that the fixed points of $\mathcal{I}_G$ lie on the ramification curve of ψ. It is a curve of degree 6 with double points $p_1, p_2, \ldots, p_7$ and is birationally isomorphic to a canonical curve of genus 3.

A third way to see Geiser involutions is the following. Let S be a del Pezzo surface of degree 2. The linear system $|-K_S|$ defines a double covering $S \to \mathbb{P}^2_\mathbb{C}$, branched along a smooth quartic curve ([**75**]). The involution ι which exchanges the two sheets of this covering is called a Geiser involution; it satisfies

$$\mathrm{Pic}(S)^\iota \otimes \mathbb{Q} \simeq \mathrm{Pic}(\mathbb{P}^2_\mathbb{C}) \otimes \mathbb{Q} = \mathbb{Q}.$$

The exceptional locus of a Geiser involution is the union of seven cubics passing through the seven points of indeterminacy of $\mathcal{I}_G$ and singular at one of these seven points.

6.1.2. Bertini involutions. Let us fix in $\mathbb{P}^2_\mathbb{C}$ eight points $p_1, p_2, \ldots, p_8$ in general position. Consider the pencil of cubic curves through these points. It has a ninth base-point p_9. For any general point p there is a unique cubic curve $\mathcal{C}(p)$ of the pencil passing through p. Take p_9 as the zero of the group law of the cubic $\mathcal{C}(p)$; define $\mathcal{I}_B(p)$ as the negative $-p$ with respect to the group law. The map $\mathcal{I}_B$ is a birational involution called the *Bertini involution* ([**16**]).

The algebraic degree of a Bertini involution is equal to 17. The fixed points of a Bertini involution lie on a canonical curve of genus 4 with vanishing theta characteristic isomorphic to a non-singular intersection of a cubic surface and a quadratic cone in $\mathbb{P}^3_\mathbb{C}$.

Another way to see a Bertini involution is the following. Consider a del Pezzo surface S of degree 1. The map $S \to \mathbb{P}^3_\mathbb{C}$ defined by the linear system $|-2K_S|$ induces a degree 2 morphism of S onto a quadratic cone $Q \subset \mathbb{P}^3_\mathbb{C}$, branched along the vertex of Q and a smooth genus 4 curve ([**75**]). The corresponding involution, the Bertini involution, satisfies $\mathrm{rk}\,\mathrm{Pic}(S)^{\mathcal{I}_B} = 1$.

6.1.3. Jonquières involutions. Let $\mathcal{C}$ be an irreducible curve of degree $\nu \geq 3$. Assume that $\mathcal{C}$ has a unique singular point p and that p is an ordinary multiple point with multiplicity $\nu - 2$. To $(\mathcal{C}, p)$ we associate a birational involution $\mathcal{I}_J$ that fixes pointwise $\mathcal{C}$ and preserves lines through p. Let m be a generic point of $\mathbb{P}^2_\mathbb{C} \smallsetminus \mathcal{C}$. Let r_m, q_m and p be the intersections of the line (mp) and $\mathcal{C}$. The point $\mathcal{I}_J(m)$ is the point such that the cross ratio of m, $\mathcal{I}_J(m)$, q_m and r_m is equal to -1. The map $\mathcal{I}_J$ is a *Jonquières involution* of degree ν centered at p; it preserves $\mathcal{C}$. More precisely its fixed points are the curve $\mathcal{C}$ of genus $\nu - 2$ as soon as $\nu \geq 3$.

If $\nu = 2$, then $\mathcal{C}$ is a smooth conic; the same construction can be done by choosing a point p that does not lie on $\mathcal{C}$.

LEMMA 6.1 ([**89**]). *Let* G *be a finite subgroup of* $\mathrm{Bir}(\mathbb{P}^2_\mathbb{C})$. *Let* $C_1, C_2, \ldots, C_k$ *be non-rational irreducible curves on* $\mathbb{P}^2_\mathbb{C}$ *such that each of them contains an open subset* C_i^0 *whose points are fixed under all* $g \in \mathrm{G}$.

Then the set of birational isomorphism classes of the curves C_i *is an invariant of the conjugacy class of* G *in* $\mathrm{Bir}(\mathbb{P}^2_\mathbb{C})$.

PROOF. Assume that $\mathrm{G} = \psi \circ \mathrm{H} \circ \psi^{-1}$ for some subgroup H of $\mathrm{Bir}(\mathbb{P}^2_\mathbb{C})$ and some birational self map ψ of the complex projective plane. Replacing C_i^0 by a smaller open subset if needed, we assume that $\psi^{-1}(C_i^0)$ is defined and consists of fixed points of H. As C_i is not rational, $\psi^{-1}(C_i^0)$ is not a point. Its Zariski closure is thus a rational irreducible curve C_i' birationally isomorphic to C_i that contains an open subset of fixed points of H. □

COROLLARY 6.2. *Jonquières involutions of degree* ≥ 3 *are not conjugate to each other, not conjugate to projective involutions, not conjugate to Bertini involutions, not conjugate to Geiser involutions.*

Bertini involutions are not conjugate to Geiser involutions, not conjugate to projective involutions.

Geiser involutions are not conjugate to projective involutions.

PROOF. The statement follows from Lemma 6.1 and the above properties:

- ◇ a connected component of the fixed locus of a projective map is a line or a point;

◇ the fixed points of a Geiser involution lie on a curve birationally isomorphic to a canonical curve of genus 3;
◇ the fixed points of a Bertini involution lie on a canonical curve of genus 4 with vanishing theta characteristic;
◇ the set of fixed points of a Jonquières involution of degree $\nu \geq 3$ outside the base locus is a hyperelliptic curve of degree $\nu - 2$. □

We can thus introduce the following definition.

DEFINITION. An involution is of *Jonquières type* if it is birationally conjugate to a Jonquières involution.

An involution is of *Bertini type* if it is birationally conjugate to a Bertini involution.

An involution is of *Geiser type* if it is birationally conjugate to a Geiser involution.

The classification of subgroups of $\mathrm{Bir}(\mathbb{P}^2_\mathbb{C})$ of order 2 is given by the following statement:

THEOREM 6.3 ([**6**]). *A non-trivial birational involution of the complex projective plane is conjugate to one and only one of the following:*
◇ *a Jonquières involution of a given degree* $\nu \geq 2$*;*
◇ *a Geiser involution;*
◇ *a Bertini involution.*

More precisely the parameterization of each conjugacy class is known. Before stating it let us give some definitions.

REMARKS 6.4. Let S, S' be two rational surfaces and let $\iota \in \mathrm{Bir}(S)$, $\iota' \in \mathrm{Bir}(S')$ be two involutions. They are *birationally equivalent* if there exists a birational map $\varphi\colon S \dashrightarrow S'$ such that $\varphi \circ \iota = \iota' \circ \varphi$. Note that in particular two involutions of $\mathrm{Bir}(\mathbb{P}^2_\mathbb{C})$ are equivalent if and only if they are conjugate in $\mathrm{Bir}(\mathbb{P}^2_\mathbb{C})$. Assume that ι fixes a curve C. Then $\iota' = \varphi \circ \iota \circ \varphi^{-1}$ fixes the proper transform of C under φ which is a curve birational to C except possibly if C is rational; indeed, if C is rational it may be contracted to a point. The *normalized fixed curve* of ι is the union of the normalizations of the non-rational curves fixed by ι. This is an invariant of the birational equivalence class of ι.

PROPOSITION 6.5 ([**6**]). *The map which associates to a birational involution of* $\mathbb{P}^2_\mathbb{C}$ *its normalized fixed curve establishes a one-to-one correspondence between*
◇ *conjugacy classes of Jonquières involutions of degree* ν *and isomorphism classes of hyperelliptic curves of genus* $\nu - 2$ $(\nu \geq 3)$*;*
◇ *conjugacy classes of Geiser involutions and isomorphism classes of non-hyperelliptic curves of genus* 3*;*
◇ *conjugacy classes of Bertini involutions and isomorphism classes of non-hyperelliptic curves of genus* 4 *whose canonical model lies on a singular quadric.*

Jonquières involutions of degree 2 *form one conjugacy class.*

The approach of Bayle and Beauville is different from the approach of Castelnuovo. It is based on the following observation: any birational involution of $\mathbb{P}^2_\mathbb{C}$ is conjugate, via an appropriate birational isomorphism $S \stackrel{\sim}{\dashrightarrow} \mathbb{P}^2_\mathbb{C}$ to a biregular involution ι of a rational surface S. Therefore, the authors are reduced to the birational classification of the pairs (S,ι). In [**137**] Manin classified the pairs (S,G) where S is a surface and G a finite group. This question has been simplified by the introduction of Mori theory. This theory allows Bayle and Beauville to show that the minimal pairs (S,ι) fall into two categories, those which admit a ι-invariant base-point free pencil of rational curves, and those with $\mathrm{rk}\,\mathrm{Pic}(S)^\iota=1$. The first case leads to the so-called Jonquières involutions whereas the second one leads to the Geiser and Bertini involutions.

Let us now give some details. By a surface we mean a smooth, projective, connected surface over $\mathbb{C}$. We consider pairs (S,ι) where S is a rational surface and ι a non-trivial biregular involution of S. Recall that the pair (S,ι) is minimal if any birational morphism $\psi\colon S\to S'$ such that there exists a biregular involution ι' of S' with $\psi\circ\iota=\iota'\circ\psi$ is an isomorphism.

LEMMA 6.6 ([**6**]). *The pair (S,ι) is minimal if and only if for any exceptional curve*[1] *E on S the following hold:*

$$\iota(E)\neq E \qquad\qquad E\cap\iota(E)\neq\emptyset.$$

PROOF. Suppose that (S,ι) is not minimal. Then there exist a pair (S',ι') and a birational morphism $\psi\colon S\to S'$ such that $\psi\circ\iota=\iota'\circ\psi$ and ψ contracts some exceptional curve E. Then ψ contracts the divisor $E+\iota(E)$. Therefore, $(E+\iota(E))^2\leq 0$, and so $E\cdot\iota(E)\leq 0$, *i.e.*, $\iota(E)=E$ or $E\cap\iota(E)=\emptyset$.

Conversely assume that there exists an exceptional curve E on S such that $\iota(E)=E$ (resp. $E\cap\iota(E)=\emptyset$). Let S' be the surface obtained by blowing down E (resp. $E\cup\iota(E)$). Then ι induces an involution ι' of S' so that (S,ι) is not minimal. □

The only piece of Mori theory used by Bayle and Beauville is the following one:

LEMMA 6.7 ([**6**]). *Let (S,ι) be a minimal pair with* $\mathrm{rk}\,\mathrm{Pic}(S)^\iota>1$. *Then S admits a base-point free pencil stable under ι.*

It allows them to establish

THEOREM 6.8 ([**6**]). *Let (S,ι) be a minimal pair. One of the following holds:*

(1) *there exists a smooth $\mathbb{P}^1_\mathbb{C}$-fibration $f\colon S\to\mathbb{P}^1_\mathbb{C}$ and a non-trivial involution $\mathcal{I}$ of $\mathbb{P}^1_\mathbb{C}$ such that $f\circ\iota=\mathcal{I}\circ f$;*
(2) *there exists a fibration $f\colon S\to\mathbb{P}^1_\mathbb{C}$ such that $f\circ\iota=f$, the smooth fibres of f are rational curves on which ι induces a non-trivial involution, any singular fibre is the union of two rational curves exchanged by ι, meeting at one point;*
(3) *S is isomorphic to $\mathbb{P}^2_\mathbb{C}$;*
(4) *(S,ι) is isomorphic to $\mathbb{P}^1_\mathbb{C}\times\mathbb{P}^1_\mathbb{C}$ with the involution $(z_0,z_1)\mapsto(z_1,z_0)$;*
(5) *S is a del Pezzo surface of degree 2 and ι is the Geiser involution;*
(6) *S is a del Pezzo surface of degree 1 and ι is the Bertini involution.*

[1] Recall that an exceptional curve E on a surface S is a smooth rational curve with $E^2=-1$.

PROOF.

- Assume $\operatorname{rk}\operatorname{Pic}(S)^\iota = 1$. As $\operatorname{Pic}(S)^\iota$ contains an ample class, $-K_S$ is ample, *i.e.*, S is a del Pezzo surface. If $\operatorname{rk}\operatorname{Pic}(S) = 1$, then one obtains case (3).

  If $\operatorname{rk}\operatorname{Pic}(S) > 1$, then $-\iota$ is the orthogonal reflection with respect to $K_S^\perp$. Such a reflection is of the form

$$x \mapsto x - 2\,\frac{(\alpha\cdot x)}{(\alpha\cdot\alpha)}\,\alpha$$

  with $(\alpha\cdot\alpha)\in\{1,\,2\}$ and K_S proportional to α. If K_S is divisible, then S is isomorphic to $\mathbb{P}^1_\mathbb{C}\times\mathbb{P}^1_\mathbb{C}$ and since ι must act non-trivially on $\operatorname{Pic}(S)$ we get case (4). The only remaining eventualities are $K_S^2\in\{1,\,2\}$. The Geiser and Bertini involutions have the required properties (§6.1.1, §6.1.2). An automorphism φ of S acting trivially on $\operatorname{Pic}(S)$ is the identity; indeed S is the blow-up of $\mathbb{P}^2_\mathbb{C}$ at $9-d$ points in general position, φ induces an automorphism of $\mathbb{P}^2_\mathbb{C}$ which must fix these points. Hence Geiser and Bertini involutions are the only ones to have the required properties.
- Suppose now that $\operatorname{rk}\operatorname{Pic}(S)^\iota > 1$. According to Lemma 6.7 the surface S admits an ι-invariant pencil $|F|$ of rational curves. This defines a fibration $f\colon S\to\mathbb{P}^1_\mathbb{C}$ with fibre F, and an involution $\mathcal{I}$ of $\mathbb{P}^1_\mathbb{C}$ such that $f\circ\iota = \mathcal{I}\circ f$.

  If f is smooth, then this gives (1) or a particular case of (2).

  If f is not smooth, let F_0 be a singular fibre of f. It contains an exceptional divisor E. Since (S,ι) is minimal, then $\iota(E)\neq E$ and $E\cdot\iota(E)\geq 1$. As a result $(E+\iota(E))^2\geq 0$, so $F_0 = E+\iota(E)$ and $E\cdot\iota(E) = 1$. Set $p = E\cap\iota(E)$. The involution induced by ι on T_pS exchanges the directions of E and $\iota(E)$; it thus has eigenvalues 1 and -1. As a consequence, ι fixes a curve passing through p; this curve must be horizontal and $\mathcal{I}$ trivial. Furthermore the fixed curve of ι being smooth, the involution induced by ι on a smooth fibre cannot be trivial. We get case (2). □

Bayle and Beauville precised which pairs in the list of Theorem 6.8 are indeed minimal ([**6**, Proposition 1.7]).

Let us now give the link between biregular involutions of rational surfaces and birational involutions of the complex projective plane:

LEMMA 6.9 ([**6**]). *Let ι be a birational involution of a surface S_1. There exist a birational morphism $\varphi\colon S\to S_1$ and a biregular involution $\mathcal{I}$ of S such that $\varphi\circ\mathcal{I} = \iota\circ\varphi$.*

To prove it we need some results; let us state and prove them.

THEOREM 6.10 (*See for instance* [**7**], Theorem II.7). *Let S be a surface, and let X be a projective variety. Let $\phi\colon S \dashrightarrow X$ be a rational map.*

Then there exist

- *a surface S',*
- *a morphism $\eta\colon S'\to S$ which is the composition of a finite number of blow-ups,*
- *a morphism $\psi\colon S'\to X$*

such that

$$\begin{array}{ccc} & S' & \\ {\scriptstyle\eta}\swarrow & & \searrow{\scriptstyle\psi} \\ S & \overset{}{\underset{\phi}{\dashrightarrow}} & X \end{array}$$

commutes.

PROOF. As X lies in some projective space we may assume that $X = \mathbb{P}^m_\mathbb{C}$. Furthermore we can suppose that $\phi(S)$ lies in no hypersurface of $\mathbb{P}^m_\mathbb{C}$. As a result ϕ corresponds to a linear system $P \subset |D|$ of dimension m on S without fixed component.

If P has no base-point, then ϕ is a morphism and there is nothing to do.

Assume that P has at least one base-point p. Consider the blow-up $\varepsilon\colon \mathrm{Bl}_pS \to S$ at p. Set $S_1 = \mathrm{Bl}_pS$. The exceptional curve E occurs in the fixed part of the linear system $\varepsilon^*P \subset |\varepsilon^*D|$ with some multiplicity $k \geq 1$; that is, the system $P_1 = |\varepsilon^*P - kE| \subset |\varepsilon^*D - kE|$ has no fixed component. It thus defines a rational map $\phi_1 = \phi \circ \varepsilon\colon S_1 \dashrightarrow \mathbb{P}^m_\mathbb{C}$. If ϕ_1 is a morphism, then the result is proved. If not, we repeat the "same step". We get by induction a sequence $\varepsilon_n\colon S_n \to S_{n-1}$ of blow-ups and a linear system $P_n \subset |D_n| = |\varepsilon_n^*D_{n-1} - k_nE_n|$ on S_n with no fixed part. On the one hand $D_n^2 = D_{n-1}^2 - k_n^2 < D_{n-1}^2$; on the other hand P_n has no fixed part, so $D_n^2 \geq 0$ for any n. Consequently the process must end. More precisely after a finite number of blow-ups we obtain a system P_n with no base-points which defines a morphism $\psi\colon S_n \to \mathbb{P}^m_\mathbb{C}$ as required. □

LEMMA 6.11 (*See for instance* [**7**]). *Let S be an irreducible surface. Let S' be a smooth surface. Let $\phi\colon S \to S'$ be a birational morphism. Assume that the rational map ϕ^{-1} is not defined at a point $p \in S'$.*

Then $\phi^{-1}(p)$ is a curve on S.

PROOF. We assume that S is affine so that there is an embedding $j\colon S \hookrightarrow \mathbb{A}^n_\mathbb{C}$. The rational map

$$j \circ \phi^{-1}\colon S' \dashrightarrow \mathbb{A}^n_\mathbb{C}$$

is defined by rational functions $g_1, g_2, \ldots, g_n$. One of them, say, for instance, g_1 is undefined at p, that is, $g_1 \notin \mathcal{O}_{S',p}$. Set $g_1 = \frac{u}{v}$ with $u, v \in \mathcal{O}_{S',p}$, u and v coprime and $v(p) = 0$. Consider the curve D on S given by $\phi^*v = 0$. On $S \subset \mathbb{A}^n_\mathbb{C}$ denote by z_0 the first coordinate function. We have $\phi^*u = z_0\phi^*v$ on S. Hence $\phi^*u = \phi^*v = 0$ on D. Consequently $D = \phi^{-1}(Z)$ where

$$Z = \big\{u = v = 0\big\} \subset S'.$$

By assumption u and v are coprime, so Z is finite. Shrinking S' if necessary we can assume that $Z = \{p\}$. Finally $D = \phi^{-1}(p)$. □

LEMMA 6.12 (*See for instance* [**7**]). *Let S, S' be two surfaces. Let $\phi\colon S \dashrightarrow S'$ be a birational map such that ϕ^{-1} is not defined at $p \in S'$.*

Then there exists a curve C on S such that $\phi(C) = \{p\}$.

PROOF. The map ϕ corresponds to a morphism $\psi\colon \mathcal{U} \to S'$ for some subset $\mathcal{U}$ of S. Denote by

$$\Gamma = \big\{\big(u, \psi(u)\big) \,|\, u \in \mathcal{U}\big\} \subset \mathcal{U} \times S'$$

the graph of ψ. Let $\overline{\Gamma}$ be the closure of Γ in $S \times S'$; it is an irreducible surface, possibly with singularities. The projections

$$\mathrm{pr}_1 \colon \overline{\Gamma} \to S, \qquad\qquad \mathrm{pr}_2 \colon \overline{\Gamma} \to S'$$

are birational morphisms and the diagram

$$\begin{array}{ccc} & \overline{\Gamma} & \\ {}^{\mathrm{pr}_1}\swarrow & & \searrow^{\mathrm{pr}_2} \\ S & \overset{}{\underset{\phi}{\dashrightarrow}} & S' \end{array}$$

is commutative.

By assumption ϕ^{-1} is not defined at $p \in S'$, so does pr_2^{-1}. There is an irreducible curve C' on $\overline{\Gamma}$ such that $\mathrm{pr}_2(C') = \{p\}$ (Lemma 6.11). As $\overline{\Gamma} \subset S \times S'$ the image $\mathrm{pr}_1(C')$ of C' by pr_1 is a curve C in S such that $\phi(C) = \{p\}$. □

PROPOSITION 6.13 (*See for instance* [**129**]). *Let X and S be two surfaces. Let $\phi\colon X \to S$ be a birational morphism of surfaces. Suppose that the rational map ϕ^{-1} is not defined at a point p of S.*

Then

$$\begin{array}{ccc} & \mathrm{Bl}_pS & \\ {}^{\psi}\nearrow & & \searrow^{\varepsilon} \\ X & \underset{\phi}{\longrightarrow} & S \end{array}$$

where $\psi\colon X \to \mathrm{Bl}_pS$ is a birational map and $\varepsilon\colon \mathrm{Bl}_pS \to S$ is the blow-up at p.

PROOF. Set $\psi = \varepsilon^{-1} \circ \phi$. Suppose that ψ is not a morphism, and let m be a point of X such that ψ is not defined at m. On the one hand $\phi(m) = p$ and ϕ is not locally invertible at m; on the other hand there exists a curve in Bl_pS blown down onto m by ψ^{-1} (Lemma 6.12). This curve has to be the exceptional divisor E associated to ε. Let r and q be two distinct points of E at which ψ^{-1} is well defined; consider C, C' two germs of smooth curves transverse to E at r and q respectively. Then $\varepsilon(C)$ and $\varepsilon(C')$ are two germs of smooth curves transverse at p, which are images by ϕ of two germs of curves at m. The differential of ϕ at m has thus rank 2: contradiction with the fact that ϕ is not invertible at m.

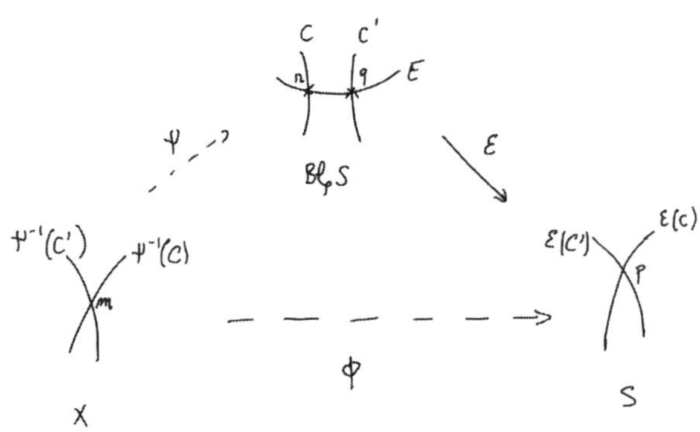

□

Proof of Lemma 6.9. There exists a birational morphism $\varphi\colon S \to S_1$ such that the rational map $\psi = \iota \circ \varphi$ is everywhere defined (Theorem 6.10). Furthermore φ can be written as

$$\varphi = \varepsilon_{n-1} \circ \varepsilon_{n-2} \circ \ldots \circ \varepsilon_1$$

where $\varepsilon_i\colon S_{i+1} \to S_i$, $1 \leq i \leq n-1$, is obtained by blowing up a point $p_i \in S_i$ and $S = S_n$. The map ι is not defined at p_1, so $\psi^{-1} = \varphi^{-1} \circ \iota$ is not defined at p_1. Proposition 6.13 implies that ψ factors as

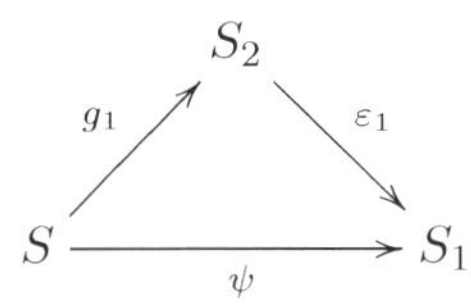

Proceeding by induction we see that ψ factors as $\varphi \circ \mathcal{I}$ where $\mathcal{I}$ is a birational morphism. Since $\varphi \circ \mathcal{I}^2 = \varphi$, the map $\mathcal{I}$ is an involution. □

In other words Lemma 6.9 says that any birational involution of a surface is birationally equivalent to a biregular involution $\iota\colon S \to S$; furthermore (S, ι) can be assumed to be minimal. Therefore, the classification of conjugacy classes of involutions in $\mathrm{Bir}(\mathbb{P}^2_\mathbb{C})$ is equivalent to the classification of minimal pairs (S, ι) up to birational equivalence.

Remark 6.14. Recall that the $\mathbb{P}^1_\mathbb{C}$-bundles over $\mathbb{P}^1_\mathbb{C}$ are of the form

$$\mathbb{F}_n = \mathbb{P}_{\mathbb{P}^1_\mathbb{C}}\big(\mathcal{O}_{\mathbb{P}^1_\mathbb{C}} \oplus \mathcal{O}_{\mathbb{P}^1_\mathbb{C}}(n)\big)$$

for some integers $n \geq 0$ (*see* §3.3.2).

For $n \geq 1$ the fibration

$$f\colon \mathbb{F}_n \to \mathbb{P}^1_\mathbb{C}$$

has a unique section of self-intersection $-n$. Consider a fibre F of f, and a point p of F. Assume that ι is a birational involution of $\mathbb{F}_n$ regular in a neighborhood of F and fixing p. After the elementary transformation at p we get a birational involution of $\mathbb{F}_{n+1}$ regular in a neighborhood of the new fibre.

Proof of Theorem 6.3. The unicity assertion follows from Remark 6.4.

Using Lemma 6.9 we will prove that the involutions of Theorem 6.8 are birationally equivalent to one of Theorem 6.3.

Cases (5) and (6) give by definition the Geiser and Bertini involutions.

An involution of type (4) is birationally equivalent to a Jonquières involution of degree 2. Indeed let Q be a smooth conic in $\mathbb{P}^2_\mathbb{C}$, and let $p \in \mathbb{P}^2_\mathbb{C} \smallsetminus Q$ be a point. Consider the birational involution ι of $\mathbb{P}^2_\mathbb{C}$ that maps a point x to its harmonic conjugate on the line (px) through p and x with respect to the two points of $(px) \cap Q$. This involution is not defined at the following three points: p and the two points q and r where the tangent line to Q passes through p. Set $S = \mathrm{Bl}_{p,q,r}\mathbb{P}^2_\mathbb{C}$. The involution ι extends to a biregular involution $\mathcal{I}$ of S, the Jonquières involution of degree 2.

In case (3) take a point $p \in \mathbb{P}^2_\mathbb{C}$ such that $\iota(p) \neq p$. Let us blow-up p, $\iota(p)$ and then blow-down the proper transform of the line $(p\iota(p))$ which is a ι-invariant exceptional curve. We get a pair (T,ι') with $T\simeq\mathbb{P}^1_\mathbb{C}\times\mathbb{P}^1_\mathbb{C}$ by stereographic projection and $\mathrm{rk}\,\mathrm{Pic}(T)^{\iota'}=1$: we are thus in case (4), so in the case of a Jonquières involution of degree 2.

Let us now deal with case (1). The surface S is isomorphic to $\mathbb{F}_n$ for some $n\geq 0$. The involution ι has two invariant fibres, any of them containing at least two fixed points. One of these points does not belong to s_n (section of self-intersection $-n$ on $\mathbb{F}_n$), hence after a (finite) sequence of elementary transformations we get $n=1$. Let us thus focus on the case $n=1$. Let $\mathbb{F}_1$ be the surface obtained by blowing up a point $p\in\mathbb{P}^2_\mathbb{C}$. Projecting from p defines a $\mathbb{P}^1$-bundle $f\colon\mathbb{F}_1\to\mathbb{P}^1_\mathbb{C}$. Any biregular involution ι of $\mathbb{F}_1$ preserves this fibration; hence it defines a pair $(\mathbb{F}_1,\iota)$ of case (1) or (2). The involution ι preserves the unique exceptional curve E_1 of $\mathbb{F}_1$; the pair $(\mathbb{F}_1,\iota)$ is thus not minimal: ι induces a biregular involution of $\mathbb{P}^2_\mathbb{C}$. We finally get a Jonquières involution of degree 2 as we just see.

We now consider case (2). Let us distinguish two possibilities: denote by F_1, F_2, …, F_s the singular fibres of f and by p_i, $1\leq i\leq s$, the singular point of F_i. The fixed locus of ι is a smooth curve C passing through p_1, p_2, …, p_s. The degree 2 covering $C\to\mathbb{P}^1_\mathbb{C}$ induced by f is ramified at p_1, p_2, …, p_s.

(2a) Either f is smooth, $s=0$ and C is the union of two sections of f which do not intersect;

(2b) or f is not smooth, C is a hyperelliptic curve of genus $g\geq 0$ and $s=2g+2$.

First assume that we are in case (2a). After elementary transformations we can suppose that $S=\mathbb{F}_1$. The fixed locus of ι is the union of E_1 and a section which does not meet E_1. Blowing down E_1 one gets case (4).

Finally let us look at case (2b) for $g\geq 0$. Let us blow-down one of the components in each singular fibre. We thus have a birational involution on a surface $\mathbb{F}_n$, the fixed curve C embedded into $\mathbb{F}_n$. After elementary transformations at general points of C one gets a birational involution on a surface $\mathbb{F}_1$, the fixed curve C embedded into $\mathbb{F}_1$. The genus formula implies that $E_1\cdot C=g$. Suppose that C is tangent to E_1 at some point $q\in\mathbb{F}_1$. After an elementary transformation at q then an elementary transformation at some general point of C the order of contact of C and E_1 at q decreases by 1. Proceeding in this way we arrive at the following situation: E_1 and C meet transversally at g distinct points. Let blow-down E_1 to a point p of $\mathbb{P}^2_\mathbb{C}$; the curve C maps to a plane curve $\overline{C}$ of degree $g+2$ with an ordinary multiple point of multiplicity g at p and no other singularity. This yields to a birational involution of $\mathbb{P}^2_\mathbb{C}$ which preserves the lines through p and admits $\overline{C}$ as fixed curve, *i.e.*, a Jonquières involution with center p and fixed curve $\overline{C}$. □

6.2. Finite abelian subgroups of the Cremona group

Dolgachev and Iskovskikh used a modern approach to the problem initiated in the works of Manin and Iskovskikh who gave a clear understanding of the conjugacy problem via the concept of a G-surface ([**112, 137**]). Let G be a finite group. A *G-surface* is a pair (S,ψ) where S is a non-singular projective surface and ψ is an isomorphism from G to $\mathrm{Aut}(S)$. A *morphism of the pairs* $(S,\psi)\to(S',\psi')$ is defined to be a morphism of surfaces $\phi\colon S\to S'$ such that

$$\psi'(\mathrm{G})=\phi\circ\psi(\mathrm{G})\circ\phi^{-1}.$$

In particular let us note that two subgroups of $\mathrm{Aut}(S)$ define isomorphic G-surfaces if and only if they are conjugate inside $\mathrm{Aut}(S)$.

Let (S,ψ) be a rational G-surface. Take a birational map $\phi\colon S \dashrightarrow \mathbb{P}^2_\mathbb{C}$. For any $g\in \mathrm{G}$ the map $\phi\circ g\circ\phi^{-1}$ belongs to $\mathrm{Bir}(\mathbb{P}^2_\mathbb{C})$. This yields to an injective homomorphism

$$\iota_\phi\colon \mathrm{G}\to\mathrm{Bir}(\mathbb{P}^2_\mathbb{C}).$$

LEMMA 6.15 ([**89**]). *Let (S,ψ) and (S',ψ') be two rational* G*-surfaces. Let $\phi\colon S\dashrightarrow\mathbb{P}^2_\mathbb{C}$ and $\phi'\colon S\dashrightarrow\mathbb{P}^2_\mathbb{C}$ be two birational maps.*

The subgroups $\iota_\phi(\mathrm{G})$ and $\iota_{\phi'}(\mathrm{G})$ are conjugate if and only if there exists a birational map of G*-surfaces $S'\dashrightarrow S$.*

In other words a birational isomorphism class of G-surfaces defines a conjugacy class of subgroups of $\mathrm{Bir}(\mathbb{P}^2_\mathbb{C})$ isomorphic to G. The following result shows that any conjugacy class is obtained in this way:

LEMMA 6.16 ([**89**]). *Let* G *be a finite subgroup of* $\mathrm{Bir}(\mathbb{P}^2_\mathbb{C})$. *There exist a rational* G*-surface (S,ψ) and a birational map $\phi\colon S\dashrightarrow\mathbb{P}^2_\mathbb{C}$ such that*

$$\mathrm{G}=\phi\circ\psi(\mathrm{G})\circ\phi^{-1}.$$

PROOF. If ϕ belongs to G, we denote by $\mathrm{dom}(\phi)$ an open subset on which ϕ is defined. Set $\mathcal{D}=\displaystyle\bigcap_{\phi\in\mathrm{G}}\mathrm{dom}(\phi)$. Then $\mathcal{U}=\displaystyle\bigcap_{\phi\in\mathrm{G}} g(\mathcal{D})$ is an open invariant subset of $\mathbb{P}^2_\mathbb{C}$ on which ϕ acts biregularly. Consider $\mathcal{U}'={}^{\mathcal{U}}\!/_{\mathrm{G}}$ the orbit space; it is a normal algebraic surface. Let us choose any normal projective completion X' of $\mathcal{U}'$. Consider S' the normalization of X' in the field of rational functions of $\mathcal{U}$. It is a normal projective surface on which G acts by biregular transformations. A G-invariant resolution of singularities S of S' suits ([**70**]). □

Hence one has:

THEOREM 6.17 ([**89**]). *There is a natural bijective correspondence between birational isomorphism classes of rational* G*-surfaces and conjugate classes of subgroups of* $\mathrm{Bir}(\mathbb{P}^2_\mathbb{C})$ *isomorphic to* G.

Therefore, the goal of Dolgachev and Iskovskikh is to classify G-surfaces up to birational isomorphism of G-surfaces.

There is a G-equivariant analogue of minimal surfaces:

DEFINITION. A *minimal* G*-surface* is a G-surface (S,ψ) such that any birational morphism of G-surfaces $(S,\psi)\to(S',\psi')$ is an isomorphism.

Note that it is enough to classify minimal rational G-surfaces up to birational isomorphism of G-surfaces. The authors can rely on the following fundamental result:

THEOREM 6.18. *Let S be a minimal rational* G*-surface. Then*

- ◇ *either S admits a structure of a conic bundle with* $\mathrm{Pic}(S)^{\mathrm{G}}\simeq\mathbb{Z}^2$;
- ◇ *or S is isomorphic to a del Pezzo surface with* $\mathrm{Pic}(S)^{\mathrm{G}}\simeq\mathbb{Z}$.

An analogous result from the classical literature is shown by using the method of the termination of adjoints, first introduced for linear systems of plane curves in the work of Castelnuovo. This method is applied to find a G-invariant linear system of curves in the plane in [**123**]; Kantor essentially stated the result above but without the concept of minimality. A first modern proof can be found in [**137**] and [**112**]. Nowadays Theorem 6.18 follows from a G-equivariant version of the Mori theory ([**69**]).

As a result, to complete the classification Dolgachev and Iskovskikh need

(i) to classify all finite groups G that may occur in a minimal G-pair;
(ii) to determine when two minimal G-surfaces are birationally isomorphic.

To achieve (i) the authors computed the full automorphism group of a conic bundle surface on a del Pezzo surface and then made a list of all finite subgroups acting minimally on the surface.

To achieve (ii) the authors used the ideas of Mori theory to decompose a birational map of rational G-surfaces into elementary links.

6.3. Finite cyclic subgroups of $\mathrm{Bir}(\mathbb{P}^2_\mathbb{C})$

In [**23**] the author gave the list of finite cyclic subgroups of the plane Cremona group, up to conjugation. The curves fixed by one element of the group, and the action of the whole group on these curves, are often sufficient to distinguish the conjugacy classes. It was done in [**22**] in many cases, but some remain unsolved. In [**22**] the author completed this classification with the case of abelian non-cyclic groups.

Its classification implies several results we will now mention.

THEOREM 6.19 ([**23**]). *For any integer $n\geq 1$ there are infinitely many conjugacy classes of elements of $\mathrm{Bir}(\mathbb{P}^2_\mathbb{C})$ of order $2n$, that are non-conjugate to a linear automorphism.*

If $n>15$, a birational map of $\mathbb{P}^2_\mathbb{C}$ of order $2n$ is an nth root of a Jonquières involution and preserves a pencil of rational curves.

If an element of $\mathrm{Bir}(\mathbb{P}^2_\mathbb{C})$ is of finite odd order and is not conjugate to a linear automorphism of $\mathbb{P}^2_\mathbb{C}$, then its order is 3, 5, 9 or 15. In particular any birational map of $\mathbb{P}^2_\mathbb{C}$ of odd order >15 is conjugate to a linear automorphism of the plane.

Then Blanc generalized a theorem of Castelnuovo which states that an element of finite order which fixes a curve of geometric genus >1 has order 2, 3 or 4 (*see* [**60**]):

THEOREM 6.20 ([**23**]). *Let G be a finite abelian group which fixes some curve of positive geometric genus.*

Then G is cyclic, of order 2, 3, 4, 5 or 6, and all these cases occur.

If the curve has geometric genus >1, then G is of order 2 or 3.

THEOREM 6.21 ([**23**]). *Let G be a finite abelian subgroup of $\mathrm{Bir}(\mathbb{P}^2_\mathbb{C})$.*

The following assertions are equivalent:

- *any $g\in\mathrm{G}\smallsetminus\{\mathrm{id}\}$ does not fix a curve of positive geometric genus;*
- *the group G is birationally conjugate to a subgroup of $\mathrm{Aut}(\mathbb{P}^2_\mathbb{C})$, or to a subgroup of $\mathrm{Aut}(\mathbb{P}^1_\mathbb{C}\times\mathbb{P}^1_\mathbb{C})$, or to the group isomorphic to $\mathbb{Z}/_{2\mathbb{Z}}\times\mathbb{Z}/_{4\mathbb{Z}}$ generated by the following two elements*

$$(z_0:z_1:z_2)\mapsto\big(z_1z_2:z_0z_1:-z_0z_2\big),$$
$$(z_0:z_1:z_2)\mapsto\big(z_1z_2(z_1-z_2):z_0z_2(z_1+z_2):z_0z_1(z_1+z_2)\big).$$

Furthermore this last group is conjugate neither to a subgroup of $\mathrm{Aut}(\mathbb{P}^2_\mathbb{C})$, nor to a subgroup of $\mathrm{Aut}(\mathbb{P}^1_\mathbb{C}\times\mathbb{P}^1_\mathbb{C})$.

In [**8**] Beauville gave the isomorphism classes of p-elementary subgroups of the plane Cremona group. Blanc generalized it as follows:

THEOREM 6.22 ([**23**]). *The isomorphism classes of finite abelian subgroups of the plane Cremona group are the following:*

- $\mathbb{Z}/_{m\mathbb{Z}} \times \mathbb{Z}/_{n\mathbb{Z}}$ *for any integers* m, $n \geq 1$,
- $\mathbb{Z}/_{2n\mathbb{Z}} \times \left(\mathbb{Z}/_{2\mathbb{Z}}\right)^2$ *for any integer* $n \geq 1$,
- $\left(\mathbb{Z}/_{4\mathbb{Z}}\right)^2 \times \mathbb{Z}/_{2\mathbb{Z}}$,
- $\left(\mathbb{Z}/_{3\mathbb{Z}}\right)^3$,
- $\left(\mathbb{Z}/_{2\mathbb{Z}}\right)^4$.

In [**28**] the author finished the classification of cyclic subgroups of finite order of the Cremona group, up to conjugation. He gave natural parameterizations of conjugacy classes, related to fixed curves of positive genus. The classification of finite cyclic subgroups that are not of Jonquières type was almost achieved in [**89**]. Let us explain what we mean by "almost":

- a list of representative elements is available;
- explicit forms are given;
- the dimension of the varieties which parameterize the conjugacy classes are provided.

What is missing? A finer geometric description of the algebraic variety parameterizing conjugacy classes according to [**89**].

The case of groups conjugate to subgroups of $\mathrm{Aut}(\mathbb{P}^2_\mathbb{C})$ was studied in [**9**]: there is exactly one conjugacy class for each order n, representated by

$$\langle (z_0 : z_1 : z_2) \mapsto (z_0 : z_1 : \mathrm{e}^{2\mathrm{i}\pi/n} z_2)\rangle.$$

Blanc completed the classification of cyclic subgroups of $\mathrm{Bir}(\mathbb{P}^2_\mathbb{C})$ of finite order ([**28**]). For groups of Jonquières type he applied cohomology group theory and algebraic tools to the group $\mathcal{J}$ and got:

THEOREM 6.23 ([**28**]).

- *For any positive integer* m, *there exists a unique conjugacy class of linearisable elements of order* n, *represented by the automorphism*

$$(z_0 : z_1 : z_2) \mapsto (z_0 : z_1 : \mathrm{e}^{2\mathrm{i}\pi/n} z_2).$$

- *Any non-linearisable Jonquières element of finite order of* $\mathrm{Bir}(\mathbb{P}^2_\mathbb{C})$ *has order* $2n$, *for some positive integer* n, *and is conjugate to an element* ϕ, *such that* ϕ *and* ϕ^n *are of the following form:*

$$\phi\colon (z_0, z_1) \dashrightarrow \left(\mathrm{e}^{2\mathrm{i}\pi/n} z_0, \frac{a(z_0)z_1 + (-1)^\delta p(z_0^n)b(z_0)}{b(z_0)z_1 + (-1)^\delta a(z_0)}\right),$$

$$\phi^n\colon (z_0, z_1) \dashrightarrow \left(z_0, \frac{p(z_0^n)}{z_1}\right),$$

where a, b *belongs to* $\mathbb{C}(z_0)$, δ *to* $\{0, \pm 1\}$, *and* $p \in \mathbb{C}[z_0]$ *is a polynomial with simple roots.*

The curve Γ *of equation* $z_1^2 = p(z_0^n)$*, pointwise fixed by* ϕ^n*, is hyperelliptic, of positive geometric genus, and admits a* $(2:1)$*-map* $\phi_1^2\colon \Gamma \to \mathbb{P}^1_{\mathbb{C}}$*. The action of* ϕ *on* Γ *has order* n*, and is not a root of the involution associated to any* ϕ_1^2*.*

Furthermore the above association yields a parameterization of the conjugacy classes of non-linearisable Jonquières elements of order $2n$ *of* $\mathrm{Bir}(\mathbb{P}^2_{\mathbb{C}})$ *by isomorphism classes of pairs* (Γ, ψ)*, where*

- $\diamond$ Γ *is a smooth hyperelliptic curve of positive genus,*
- $\diamond$ $\psi \in \mathrm{Aut}(\Gamma)$ *is an automorphism of order* n*, which preserves the fibres of the* ϕ_1^2 *and is not a root of the involution associated to the* ϕ_1^2*.*

The analogous result for finite Jonquières cyclic groups holds, and follows directly from this statement.

Note that if the curve Γ has geometric genus ≥ 2, the ϕ_1^2 is unique, otherwise it is not.

Blanc also dealt with cyclic subgroups of $\mathrm{Bir}(\mathbb{P}^2_{\mathbb{C}})$ that are not of Jonquières type. Using the classification of [**89**] and some classical tools on surfaces and curves he provided the parameterization of the 29 families of such groups.

The classification is divided in two parts:

- $\diamond$ find representative families and prove that each group is conjugate to one of these;
- $\diamond$ parameterize the conjugacy classes in each family by algebraic varieties.

For cyclic groups of prime order the varieties parameterizing the conjugacy classes are the moduli spaces of the non-rational curves fixed by the groups. Blanc needs to generalize it, by looking for the non-rational curves fixed by the non-trivial elements of the group. Let us give the definition of this invariant which provides a simple way to decide whether two cyclic groups are conjugate. Recall that a birational map of the complex projective plane fixes a curve if it restricts to the identity on the curve.

DEFINITION. Let ϕ be a non-trivial element of $\mathrm{Bir}(\mathbb{P}^2_{\mathbb{C}})$ of finite order.

If no curve of positive geometric genus is (pointwise) fixed by ϕ, then $\mathrm{NFC}(\phi) = \emptyset$; otherwise ϕ fixes exactly one curve of positive genus ([**6**, **69**]), and $\mathrm{NFC}(\phi)$ is then the isomorphism class of the normalization of this curve.

Two involutions ϕ, ψ of $\mathrm{Bir}(\mathbb{P}^2_{\mathbb{C}})$ are conjugate if and only if $\mathrm{NFC}(\phi) = \mathrm{NFC}(\psi)$ (*see* §6.1). If ϕ, ψ are elements of $\mathrm{Bir}(\mathbb{P}^2_{\mathbb{C}})$ of the same prime order, then $\langle\phi\rangle$ and $\langle\psi\rangle$ are conjugate if and only if $\mathrm{NFC}(\phi) = \mathrm{NFC}(\psi)$ (*see* [**9**, **69**]). This is no longer the case for cyclic groups of composite order as observed in [**9**]: the automorphism ϕ of the cubic surface $z_0^3 + z_1^3 + z_2^3 + z_3^3 = 0$ in $\mathbb{P}^3_{\mathbb{C}}$ given by

$$\phi\colon (z_0 : z_1 : z_2 : z_3) \mapsto \big(z_1 : z_0 : z_2 : \zeta z_3\big)$$

where $\zeta^3 = 1$, $\zeta \neq 1$ has only four fixed points while ϕ^2 fixes the elliptic curve $z_3 = 0$.

DEFINITION. Let $\phi \in \mathrm{Bir}(\mathbb{P}^2_{\mathbb{C}})$ be a non-trivial element of finite order n. Then $\mathrm{NFCA}(\phi)$ is the sequence of isomorphism classes of pairs

$$\Big(\mathrm{NFC}(\phi^k), \phi_{|\mathrm{NFC}(\phi^k)}\Big)_{k=1}^{n-1}$$

where $\phi_{|\mathrm{NFC}(\phi^k)}$ is the automorphism induced by ϕ on the curve $\mathrm{NFC}(\phi^k)$ (if $\mathrm{NFC}(\phi^k) = \emptyset$, then ϕ acts trivially on it).

Let us now give a simple way to decide whether two cyclic subgroups of finite order of $\mathrm{Bir}(\mathbb{P}^2_{\mathbb{C}})$ are conjugate:

THEOREM 6.24 ([**28**]). *Let* G *and* H *be two cyclic subgroups of* $\mathrm{Bir}(\mathbb{P}^2_{\mathbb{C}})$ *of the same finite order. Then* G *and* H *are conjugate in* $\mathrm{Bir}(\mathbb{P}^2_{\mathbb{C}})$ *if and only if* $\mathrm{NFCA}(\phi) = \mathrm{NFCA}(\psi)$ *for some generators* ϕ *of* G *and* ψ *of* H.

CHAPTER 7

Uncountable subgroups of the Cremona group

All the results of this chapter have been proved without the construction of the action of the isometric action of $\mathrm{Bir}(\mathbb{P}^2_{\mathbb{C}})$ on the hyperbolic space $\mathbb{H}^\infty$ and we keep this point of view. Different ideas and tools are used in any section: foliations and group theory are the main ingredients.

The study of the automorphism groups starts a long time ago. For instance, for classical groups let us see [**83**]. Consider the automorphism group of the complex projective space $\mathbb{P}^n_{\mathbb{C}}$; it is $\mathrm{PGL}(n+1,\mathbb{C})$. The automorphism group of $\mathrm{PGL}(n+1,\mathbb{C})$ is generated by the inner automorphisms, the involution $M \mapsto M^\vee$ and the action of the field automorphisms of $\mathbb{C}$. In 1963 Whittaker showed that any isomorphism between homeomorphism groups of connex topological varieties is induced by a homeomorphism between the varieties themselves ([**177**]). In 1982 Filipkiewicz proved a similar statement for differentiable varieties.

THEOREM 7.1 ([**94**]). *Let V, W be two connected varieties of class $\mathcal{C}^k$, resp. $\mathcal{C}^j$. Let $\mathrm{Diff}^k(V)$ be the group of $\mathcal{C}^k$-diffeomorphisms of V. Let $\phi\colon \mathrm{Diff}^k(V) \to \mathrm{Diff}^j(V)$ be an isomorphism group. Then $k = j$ and there exists a $\mathcal{C}^k$-difffeomorphism $\psi\colon V \to W$ such that*

$$\phi(\varphi) = \psi \circ \varphi \circ \psi^{-1} \qquad \forall\, \varphi \in \mathrm{Bir}(\mathbb{P}^2_{\mathbb{C}}).$$

The description of uncountable maximal abelian subgroups of the plane Cremona group allows us to characterize the automorphism group of $\mathrm{Bir}(\mathbb{P}^2_{\mathbb{C}})$:

THEOREM 7.2 ([**77**]). *Let φ be an automorphism of $\mathrm{Bir}(\mathbb{P}^2_{\mathbb{C}})$. There exist a birational self map ψ of the complex projective plane and an automorphism κ of the field $\mathbb{C}$ such that*

$$\varphi(\phi) = {}^{\kappa}(\psi \circ \phi \circ \psi^{-1}) \qquad \forall\, \phi \in \mathrm{Bir}(\mathbb{P}^2_{\mathbb{C}}).$$

In other words the non-inner automorphism group of $\mathrm{Bir}(\mathbb{P}^2_{\mathbb{C}})$ can be identified with the automorphisms of the field $\mathbb{C}$.

In the first section we study uncountable maximal abelian subgroups of $\mathrm{Bir}(\mathbb{P}^2_{\mathbb{C}})$; let G be such a group. We give an outline of the proofs of the following results:

- ◇ any element of G preserves at least one singular holomorphic foliation;
- ◇ either no element of G is torsion-free, or G leaves invariant a holomorphic foliation;
- ◇ if G is torsion-free, then G is conjugate to a subgroup of the Jonquières group.

In the second section we describe the automorphism group of $\mathrm{Bir}(\mathbb{P}^2_\mathbb{C})$. A study of the torsion-free maximal abelian subgroups of the Jonquières group shows that the group

$$\mathcal{J}_a = \big\{(z_0, z_1) \dashrightarrow (z_0 + a(z_1), z_1) \,|\, a \in \mathbb{C}(z_1)\big\}$$

is invariant by any automorphism of $\mathrm{Bir}(\mathbb{P}^2_\mathbb{C})$. Some work on special subgroups of $\mathcal{J}_a$ achieves the description of $\mathrm{Aut}(\mathrm{Bir}(\mathbb{P}^2_\mathbb{C}))$.

In a problem session during the International Congress of Mathematicians, Mumford proposed the following ([**145**]):

> "Let $\mathrm{G} = \mathrm{Aut}_\mathbb{C}\mathbb{C}(z_0, z_1)$ be the Cremona group (...) the problem is to topologize G and associate to it a Lie algebra consisting, roughly, of those meromorphic vector fields D on $\mathbb{P}^2_\mathbb{C}$ which "integrate" into an analytic family of Cremona transformations."

In the third section we deal with a contribution in that direction: the description of 1-parameter subgroups of quadratic birational self maps of $\mathbb{P}^2_\mathbb{C}$.

In [**97**] Ghys showed that any nilpotent subgroup of $\mathrm{Diff}^\omega(\mathbb{S}^2)$ is metabelian; as a consequence he got that if Γ is a subgroup of finite index of $\mathrm{SL}(n, \mathbb{Z})$, $n \geq 4$, then any morphism from Γ into $\mathrm{Diff}^\omega(\mathbb{S}^2)$ has finite image. In the same spirit the nilpotent subgroups of the plane Cremona group are described in the fourth section: if Γ is a strongly nilpotent group of length > 1, then either G is metabelian up to finite index, or G is a torsion group. As a consequence as soon as $n \geq 5$ no subgroup of $\mathrm{SL}(n, \mathbb{Z})$ of finite index embeds into $\mathrm{Bir}(\mathbb{P}^2_\mathbb{C})$.

The description of centralizers of discrete dynamical systems is an important problem in real/complex dynamics. Julia ([**120, 121**]), then Ritt ([**157**]), show that the set

$$\mathrm{Cent}(\phi) = \big\{\psi\colon \mathbb{P}^1_\mathbb{C} \to \mathbb{P}^1_\mathbb{C} \,|\, \psi \circ \phi = \phi \circ \psi\big\}$$

of rational functions that commute to a rational function ϕ coincide in general [1] with $\big\{\phi_0^n \,|\, n \in \mathbb{N}\big\}$ where ϕ_0 is an element of $\mathrm{Cent}(\phi)$. In the 1960's Smale considered generic diffeomorphisms ϕ of compact manifolds and asked if its centralizer coincides with $\big\{\phi^n \,|\, n \in \mathbb{Z}\big\}$. Many mathematicians have considered this question (for instance [**42, 152–154**]). The fifth section deals with centralizers of elliptic birational maps, Jonquières twists and Halphen twists.

7.1. Uncountable maximal abelian subgroups of $\mathrm{Bir}(\mathbb{P}^2_\mathbb{C})$

Let S be a complex compact surface. A *foliation* $\mathcal{F}$ on S is given by a family $(\chi_i)_i$ of holomorphic vector fields with isolated zero defined on some open cover $(\mathcal{U}_i)_i$ of S. The vector fields χ_i have to satisfy the following conditions: there exist $g_{ij} \in \mathcal{O}^*(\mathcal{U}_i \cap \mathcal{U}_j)$ such that $\chi_i = g_{ij}\chi_j$ on $\mathcal{U}_i \cap \mathcal{U}_j$. Let us remark that a non-trivial meromorphic vector field on S defines such a foliation.

LEMMA 7.3 ([**77**]). *Let* G *be an uncountable abelian subgroup of* $\mathrm{Bir}(\mathbb{P}^2_\mathbb{C})$. *There exists a rational vector field* χ *such that*

$$\varphi_*\chi = \chi \qquad\qquad \forall\, \varphi \in \mathrm{G}.$$

In particular G *preserves a foliation.*

---

[1] Except monomial maps $z \mapsto z^k$, Tchebychev polynomials, Lattès examples ...

PROOF. Since G is uncountable, there exists an integer d such that

$$\mathrm{G}_d = \mathrm{G} \cap \mathrm{Bir}_d(\mathbb{P}^2_\mathbb{C})$$

is uncountable. Hence the Zariski closure $\overline{\mathrm{G}_d}$ of G_d in $\mathrm{Bir}_{\leq d}(\mathbb{P}^2_\mathbb{C})$ is an algebraic set of dimension ≥ 1. Consider a curve in $\overline{\mathrm{G}_d}$, *i.e.*, a map

$$\eta\colon \mathbb{D} \to \overline{\mathrm{G}_d}, \qquad\qquad t \mapsto \eta(t).$$

Note that elements of $\overline{\mathrm{G}_d}$ are rational maps that commute. Let us define the rational vector field χ at any $m \in \mathbb{P}^2_\mathbb{C} \smallsetminus \mathrm{Ind}(\eta(0)^{-1})$ by

$$\chi(m) = \frac{\partial \eta(s)}{\partial s}\Big|_{s=0}\big(\eta(0)^{-1}(m)\big).$$

Let φ be an element of $\overline{\mathrm{G}_d}$. If we differentiate the equality

$$\varphi\eta(s)\varphi^{-1}(m) = \eta(s)(m)$$

with respect to s, m being fixed, one gets: $\varphi_*\chi = \chi$. In other words χ is invariant by the elements of $\overline{\mathrm{G}_d}$, and so by any element of G. □

As a result for any uncountable abelian subgroup G of $\mathrm{Bir}(\mathbb{P}^2_\mathbb{C})$, there exists a foliation on $\mathbb{P}^2_\mathbb{C}$ invariant by G. Brunella, McQuillan and Mendes have classified, up to birational equivalence, singular holomorphic foliations on projective, compact, complex surfaces ([**48, 141, 142**]). If S is a projective surface endowed with a foliation $\mathcal{F}$, we denote by $\mathrm{Bir}(S,\mathcal{F})$ (resp. $\mathrm{Aut}(S,\mathcal{F})$) the group of birational maps (resp. holomorphic maps) of S preserving the foliation $\mathcal{F}$. In general $\mathrm{Bir}(S,\mathcal{F})$ coincides with $\mathrm{Aut}(S,\mathcal{F})$ and is finite. In [**57**] the authors deal with the opposite case and get a classification.

THEOREM 7.4 ([**57**]). *Let $\mathcal{F}$ be a foliation on S such that $\mathrm{Aut}(X,\varphi^*\mathcal{F}) \subsetneq \mathrm{Bir}(X,\varphi^*\mathcal{F})$ for any birational map $\varphi\colon X \dashrightarrow S$. Then, up to conjugacy, there exists an element of infinite order in $\mathrm{Bir}(S,\mathcal{F})$ and*

- ◇ *either $\mathcal{F}$ is a rational fibration,*
- ◇ *or up to a finite cover there exist some integers p, q, r, s such that*

$$\mathrm{Bir}(\mathbb{P}^2_\mathbb{C},\mathcal{F}) = \big\{(z_0,z_1) \dashrightarrow (z_0^p z_1^q, z_0^r z_1^s),\, (z_0,z_1) \mapsto (\alpha z_0, \beta z_1) \,|\, \alpha,\, \beta \in \mathbb{C}^*\big\}.$$

Before stating the opposite case that $\mathrm{Aut}(S,\mathcal{F})$ is infinite, let us give some definitions. Let Λ be a lattice in $\mathbb{C}^2$; it induces a complex torus $\mathbb{T} = {\mathbb{C}^2}/{\Lambda}$ of dimension 2. For instance, the product of an elliptic curve by itself is a complex torus. An affine map ψ that preserves Λ induces an automorphism of the torus $\mathbb{T}$. If the linear part of ψ is of infinite order, then

- ◇ either the linear part of ψ is hyperbolic and ψ induces an Anosov automorphism that preserves two linear foliations;
- ◇ or the linear part of ψ is unipotent and ψ preserves an elliptic fibration.

Sometimes there is a finite automorphism group of $\mathbb{T}$ normalized by ψ. Denote by $\widetilde{\mathbb{T}/\mathrm{G}}$ the desingularization of $\mathbb{T}/\mathrm{G}$. The automorphism induced by ψ on $\widetilde{\mathbb{T}/\mathrm{G}}$ preserves

- ◇ the foliations induced the stable and unstable foliations preserved by ψ when ψ is hyperbolic;
- ◇ an elliptic fibration when the linear part of ψ is unipotent.

If $\mathrm{G} = \{\mathrm{id},\, (z_0, z_1) \mapsto (-z_0, -z_1)\}$ we say that $\widetilde{\mathbb{T}/_{\mathrm{G}}}$ is a *Kummer surface*; otherwise $\widetilde{\mathbb{T}/_{\mathrm{G}}}$ is a *generalized Kummer surface*.

THEOREM 7.5 ([**57**]). *Let $\mathcal{F}$ be a singular holomorphic foliation on a projective surface S. Assume that* $\mathrm{Aut}(S, \mathcal{F})$ *is infinite. Then* $\mathrm{Aut}(S, \mathcal{F})$ *contains at least one element φ of infinite order and one of the following holds:*

- ⋄ *$\mathcal{F}$ is invariant by a holomorphic vector field;*
- ⋄ *$\mathcal{F}$ is an elliptic fibration;*
- ⋄ *the surface S is a generalized Kummer surface, φ can be lifted to an Anosov automorphism $\widetilde{\varphi}$ of the torus, and $\mathcal{F}$ is the projection on S of the unstable or stable foliation of $\widetilde{\varphi}$.*

REMARKS 7.6.

- ⋄ The foliations invariant by a holomorphic vector field are described in [**57**, Proposition 3.8].
- ⋄ The last two cases are mutually exclusive.

Using these two statements one can prove the following one:

THEOREM 7.7 ([**77**]). *Let* G *be an uncountable maximal abelian subgroup of* $\mathrm{Bir}(\mathbb{P}^2_{\mathbb{C}})$. *Then:*

- ⋄ *either* G *has an element of finite order;*
- ⋄ *or* G *is up to conjugacy a subgroup of the Jonquières group.*

IDEA OF THE PROOF. Assume first that $\mathrm{Aut}(X, \phi^*\mathcal{F}) \subsetneq \mathrm{Bir}(X, \phi^*\mathcal{F})$ for any birational map $\phi\colon X \dashrightarrow S$. Then according to Theorem 7.4 either G preserves a rational fibration, and then G is up to birational conjugacy contained in the Jonquières group; or G is up to conjugacy and a finite cover of a subgroup of

$$\big\{(z_0, z_1) \dashrightarrow (z_0^p z_1^q, z_0^r z_1^s),\, (z_0, z_1) \mapsto (\alpha z_0, \beta z_1) \,|\, \alpha, \beta \in \mathbb{C}^*,\, \alpha = \alpha^p \beta^q,\, \beta = \alpha^r \beta^s\big\}.$$

If G is conjugate to the diagonal group $\mathrm{D} = \big\{(z_0, z_1) \mapsto (\alpha z_0, \beta z_1) \,|\, \alpha, \beta \in \mathbb{C}^*\big\}$, then G contains elements of finite order. Otherwise since G is uncountable it can not be reduced to

$$\langle (z_0, z_1) \dashrightarrow (z_0^p z_1^q, z_0^r z_1^s) \rangle.$$

Therefore, there exists a non-trivial element $(z_0, z_1) \mapsto (\lambda z_0, \mu z_1)$ in G such that $\lambda = \lambda^p \mu^q$ and $\mu = \lambda^r \mu^s$. For any ℓ the map $(z_0, z_1) \mapsto (\lambda^\ell z_0, \mu^\ell z_1)$ satisfies these equalities, so it belongs to G. Consider ℓ such that $\lambda^\ell = \mathbf{i}$; then $\mu^\ell = \mathrm{e}^{\mathbf{i}\pi \frac{1-p}{2q}}$ is also a root of unity and $(z_0, z_1) \mapsto (\lambda^\ell z_0, \mu^\ell z_1)$ is thus an element of finite order of G. More precisely G contains periodic elements of any order.

Suppose now that there exist a surface S and a birational map $\psi\colon S \dashrightarrow \mathbb{P}^2_{\mathbb{C}}$ such that $\mathrm{Aut}(S, \psi^*\mathcal{F}) = \mathrm{Bir}(S, \psi^*\mathcal{F})$. According to Theorem 7.5

- ⋄ either $\psi^*\mathcal{F}$ is invariant by a holomorphic vector field on S;
- ⋄ or $\psi^*\mathcal{F}$ is an elliptic fibration.

Since G is uncountable the last eventuality cannot occur ([**3**]). Let us thus assume that $\psi^*\mathcal{F}$ is invariant by a holomorphic vector field on S. According to [**57**] one can assume up to conjugacy that G is a subgroup of $\mathrm{Aut}(\widetilde{S})$ where $\widetilde{S}$ is a minimal model of S. But minimal rational surfaces are $\mathbb{P}^2_{\mathbb{C}}$, $\mathbb{P}^1_{\mathbb{C}} \times \mathbb{P}^1_{\mathbb{C}}$ and the Hirzebruch surfaces $\mathbb{F}_n$, $n \geq 2$, and their automorphism groups are known (*see* Chapter 3).

The description of the uncountable maximal abelian subgroups of minimal rational surfaces gives:

PROPOSITION 7.8 ([**77**]). *Let S be a minimal rational surface. Let* G *be an uncountable abelian subgroup of* $\mathrm{Aut}(S)$ *maximal in* $\mathrm{Bir}(S)$. *Then:*

- ◇ *either* G *contains an element of finite order,*
- ◇ *or* G *coincides with* $\{(z_0,z_1)\mapsto(z_0+P(z_1),z_1)\,|\,P\in\mathbb{C}[z_1],\,\deg P\leq n\}$,
- ◇ *or* $\mathrm{G}=\{(z_0,z_1)\mapsto(z_0+\alpha,z_1+\beta)\,|\,\alpha,\,\beta\in\mathbb{C}\}$.

A study of the uncountable maximal abelian subgroups of the Jonquières group allows us to refine Theorem 7.7 as follows:

THEOREM 7.9 ([**77**]). *Let* G *be an uncountable maximal abelian subgroup of the plane Cremona group. Then up to conjugacy:*

- ◇ *either* G *contains an element of finite order,*
- ◇ *or* $\mathrm{G}=\{(z_0,z_1)\dashrightarrow(z_0+a(z_1),z_1)\,|\,a\in\mathbb{C}(z_1)\}$,
- ◇ *or* $\mathrm{G}=\{(z_0,z_1)\mapsto(z_0+\alpha,z_1+\beta)\,|\,\alpha,\,\beta\in\mathbb{C}\}$,
- ◇ *or any subgroup of* $\mathrm{Bir}(\mathbb{P}^2_\mathbb{C})$ *acting by conjugacy on* G *is, up to finite index, solvable.* □

7.2. Description of the automorphisms group of the Cremona group

Let us give an idea of the proof of Theorem 7.2. The description of uncountable maximal abelian subgroups of $\mathrm{Bir}(\mathbb{P}^2_\mathbb{C})$ yields to

COROLLARY 7.10 ([**77**]). *Let φ be an automorphism of* $\mathrm{Bir}(\mathbb{P}^2_\mathbb{C})$. *Set*

$$\mathrm{J}_a=\{(z_0,z_1)\dashrightarrow(z_0+a(z_1),z_1)\,|\,a\in\mathbb{C}(z_1)\}.$$

Up to birational conjugacy $\varphi(\mathrm{J}_a)=\mathrm{J}_a$ *and* $(z_0,z_1)\mapsto(z_0+1,z_1)$ *is invariant by φ.*

Let us consider

$$\mathrm{T}_1=\{(z_0,z_1)\mapsto(z_0+\alpha,z_1)\,|\,\alpha\in\mathbb{C}\},\quad \mathrm{T}_2=\{(z_0,z_1)\mapsto(z_0,z_1+\beta)\,|\,\beta\in\mathbb{C}\},$$

and

$$\mathrm{D}_1=\{(z_0,z_1)\mapsto(\alpha z_0,z_1)\,|\,\alpha\in\mathbb{C}^*\},\qquad \mathrm{D}_2=\{(z_0,z_1)\mapsto(z_0,\beta z_1)\,|\,\alpha\in\mathbb{C}\}.$$

PROPOSITION 7.11 ([**77**]). *Let φ be an automorphism of* $\mathrm{Bir}(\mathbb{P}^2_\mathbb{C})$. *Assume that* $\varphi(\mathrm{J}_a)=\mathrm{J}_a$ *and* $(z_0,z_1)\mapsto(z_0+1,z_1)$ *is invariant by φ. Then up to birational conjugacy:*

- ◇ $\varphi(\mathrm{J}_a)=\mathrm{J}_a$,
- ◇ $(z_0,z_1)\mapsto(z_0+1,z_1)$ *is invariant by φ,*
- ◇ $\varphi(\mathrm{T}_1)=\mathrm{T}_1$ *and* $\varphi(\mathrm{T}_2)=\mathrm{T}_2$,
- ◇ $\varphi(\mathrm{D}_1)=\mathrm{D}_1$ *and* $\varphi(\mathrm{D}_2)=\mathrm{D}_2$.

As a consequence an automorphism of $\mathrm{Bir}(\mathbb{P}^2_\mathbb{C})$ induces two automorphisms of the group $\mathrm{Aff}(\mathbb{C})$ of affine maps of the complex line.

LEMMA 7.12. *Let φ be an automorphism of* $\mathrm{Aff}(\mathbb{C})$. *Then φ is the composition of an inner automorphism and an automorphism of the field* $\mathbb{C}$.

SKETCH OF THE PROOF. The maximal abelian subgroups of $\mathrm{Aff}(\mathbb{C})$ are the group of translations

$$\mathrm{T}=\{z\mapsto z+\beta\,|\,\beta\in\mathbb{C}\}$$

and the groups of affine maps that preserve a point

$$\mathrm{D}_{z_0} = \{z \mapsto \alpha(z - z_0) + z_0 \,|\alpha \in \mathbb{C}^*\}.$$

Since T does not contain element of finite order, φ sends T onto T. In other words there exists an additive bijection $\kappa_2 \colon \mathbb{C} \to \mathbb{C}$ such that $\varphi(z+\beta) = z + \kappa_2(\beta)$. Up to conjugacy by an element of T one can assume that $\varphi(\mathrm{D}_0) = \mathrm{D}_0$. In other words there exists a multiplicative bijection $\kappa_1 \colon \mathbb{C}^* \to \mathbb{C}^*$ such that $\varphi(\alpha z) = \kappa_1(\alpha) z$. On the one hand

$$\varphi(z \mapsto \alpha z + \alpha) = \varphi(z \mapsto z + \alpha) \circ \varphi(z \mapsto \alpha z) = (z \mapsto \kappa_1(\alpha)z + \kappa_2(\alpha))$$

and on the other hand

$$\varphi(z \mapsto \alpha z + \alpha) = \varphi(z \mapsto \alpha z) \circ \varphi(z \mapsto z+1) = (z \mapsto \kappa_1(\alpha)z + \kappa_1(\alpha)\kappa_2(1)).$$

Hence for any α the equality $z \mapsto \kappa_1(\alpha)z + \kappa_2(\alpha) = z \mapsto \kappa_1(\alpha)z + \kappa_1(\alpha)\kappa_2(1)$ holds. Since $\mu = \kappa_2(1)$ is non-zero, κ_2 is additive and multiplicative. As a result κ_2 is an isomorphism of the field $\mathbb{C}$ and

$$\begin{aligned}
\varphi(z \mapsto \alpha z + \beta) &= (z \mapsto {}^{\kappa_1}\alpha z + {}^{\kappa_2}\beta) \\
&= (z \mapsto {}^{\kappa_1}\alpha z + \mu^{\kappa_1}\beta) \\
&= (z \mapsto {}^{\kappa_1}(\alpha z + {}^{\kappa_1^{-1}}\mu\beta)) \\
&= (z \mapsto {}^{\kappa_1}(({}^{\kappa_1^{-1}}\mu z) \circ (\alpha z + \beta) \circ ({}^{\kappa_1}\mu z))) \\
&= (z \mapsto {}^{\kappa_1}(({}^{\kappa_1}\mu z)^{-1} \circ (\alpha z + \beta) \circ ({}^{\kappa_1}\mu z))).
\end{aligned}$$

□

SKETCH OF THE PROOF OF THEOREM 7.2. Proposition 7.11 and Lemma 7.12 imply that for any α, β in $\mathbb{C}^*$, for any γ, δ in $\mathbb{C}$ one has

$$\varphi((z_0,z_1) \mapsto (\alpha z_0 + \gamma, \beta z_1 + \delta)) = ((z_0,z_1) \mapsto ({}^{\kappa_1}\alpha z_0 + \mu^{\kappa_1}\gamma, {}^{\kappa_2}\beta z_1 + \eta^{\kappa_2}\delta))$$

where η, μ are two non-zero complex numbers and κ_1, κ_2 are two automorphisms of the field $\mathbb{C}$. Since $(z_0,z_1) \mapsto (z_0+z_1, z_1)$ and $(z_0,z_1) \mapsto (\alpha z_0, \alpha z_1)$ commute their image by φ also, and so $\kappa_1 = \kappa_2$. As a consequence up to conjugacy by an inner automorphism and an automorphism of the field $\mathbb{C}$, the groups

$$\mathrm{T} = \{(z_0,z_1) \mapsto (z_0 + \alpha, z_1 + \beta) \,|\, \alpha,\, \beta \in \mathbb{C}\}$$

and

$$\mathrm{D} = \{(z_0,z_1) \mapsto (\alpha z_0, \beta z_1) \,|\, \alpha,\, \beta \in \mathbb{C}^*\}$$

are pointwise invariant. Then one can check that the involutions $(z_0,z_1) \mapsto \left(z_0, \frac{1}{z_1}\right)$ and $(z_0,z_1) \mapsto (z_1,z_0)$ are invariant by φ. But the group generated by T, D, $(z_0,z_1) \mapsto \left(z_0, \frac{1}{z_1}\right)$ and $(z_0,z_1) \mapsto (z_1,z_0)$ contains $\mathrm{PGL}(3,\mathbb{C})$. Furthermore

$$\sigma_2 = \left(\left((z_0,z_1) \mapsto \left(z_0, \frac{1}{z_1}\right)\right) \circ ((z_0,z_1) \mapsto (z_1,z_0))\right)^2;$$

hence $\varphi(\sigma_2) = \sigma_2$. We conclude thanks to the Noether and Castelnuovo Theorem. □

COROLLARY 7.13 ([**77**]). *An isomorphism of the semi-group of rational self maps of* $\mathbb{P}^2_{\mathbb{C}}$ *is inner up to the action of an automorphism of the field* $\mathbb{C}$.

In the spirit of the result of Filipkiewicz (Theorem 7.1) one has:

COROLLARY 7.14 ([**77**]). *Let S be a complex projective surface. Let φ be an isomorphism between* $\mathrm{Bir}(S)$ *and* $\mathrm{Bir}(\mathbb{P}^2_\mathbb{C})$. *There exist a birational map* $\psi\colon S \dashrightarrow \mathbb{P}^2_\mathbb{C}$ *and an automorphism of the field $\mathbb{C}$ such that*

$$\varphi(\phi) = {}^\kappa(\psi\circ\phi\circ\psi^{-1}) \qquad\qquad \forall\,\phi\in\mathrm{Bir}(S).$$

COROLLARY 7.15 ([**77**]). *The automorphism group of* $\mathbb{C}(z_0, z_1)$ *is isomorphic to the automorphisms group of* $\mathrm{Bir}(\mathbb{P}^2_\mathbb{C})$.

REMARK 7.16. According to [**8**] the groups $\mathrm{Bir}(\mathbb{P}^n_\mathbb{C})$ and $\mathrm{Bir}(\mathbb{P}^2_\mathbb{C})$ are isomorphic if and only if $n = 2$.

Note that there is no description of $\mathrm{Aut}(\mathrm{Bir}(\mathbb{P}^n_\mathbb{C}))$ for $n \geq 3$. Nevertheless there are two results in that direction:

◇ The first one is

THEOREM 7.17 ([**81**]). *Let φ be an automorphism of* $\mathrm{Bir}(\mathbb{P}^n_\mathbb{C})$*; there exist an automorphism κ of the field $\mathbb{C}$, and a birational self map ψ of $\mathbb{P}^n_\mathbb{C}$ such that*

$$\varphi(\phi) = {}^\kappa(\psi\circ\phi\circ\psi^{-1}) \qquad\qquad \forall\,\phi\in\mathrm{G}(n,\mathbb{C}) = \langle\sigma_n,\,\mathrm{PGL}(n+1,\mathbb{C})\rangle.$$

◇ The second one is

THEOREM 7.18 ([**54**]). *Let V be a smooth connected complex projective variety of dimension n. Let r be a positive integer and let* $\rho\colon \mathrm{Aut}(\mathbb{P}^r_\mathbb{C}) \to \mathrm{Bir}(V)$ *be an injective morphism of groups. Then $n \leq r$.*

Furthermore if $n = r$, there exist a field morphism $\kappa\colon \mathbb{C}\to\mathbb{C}$ and a birational map $\psi\colon V \dashrightarrow \mathbb{P}^n_\mathbb{C}$ such that

◇ *either* $\psi\circ\rho(A)\circ\psi^{-1} = {}^\kappa A$ *for all* $A\in\mathrm{Aut}(\mathbb{P}^n_\mathbb{C})$,

◇ *or* $\psi\circ\rho(A)\circ\psi^{-1} = ({}^\kappa A)^\vee$ *for all* $A\in\mathrm{Aut}(\mathbb{P}^n_\mathbb{C})$.

In particular V is rational. Moreover κ is an automorphism of $\mathbb{C}$ if ρ is an isomorphism.

Before giving an idea of the proof of this last result let us state two corollaries of it. The first shows that the Cremona groups $\mathrm{Bir}(\mathbb{P}^n_\mathbb{C})$ are pairwise non-isomorphic, thereby solving an open problem for $n \geq 4$.

COROLLARY 7.19 ([**54**]). *Let n and k be natural integers. The group* $\mathrm{Bir}(\mathbb{P}^n_\mathbb{C})$ *embeds into* $\mathrm{Bir}(\mathbb{P}^k_\mathbb{C})$ *if and only if $n \leq k$.*

In particular $\mathrm{Bir}(\mathbb{P}^n_\mathbb{C})$ *is isomorphic to* $\mathrm{Bir}(\mathbb{P}^k_\mathbb{C})$ *if and only if $n = k$.*

The second characterizes rational varieties V by the structure of $\mathrm{Bir}(V)$, as an abstract group:

COROLLARY 7.20 ([**54**]). *Let V be an irreducible complex projective variety of dimension n. The following properties are equivalent:*

◇ *V is rational,*

◇ $\mathrm{Bir}(V)$ *is isomorphic to* $\mathrm{Bir}(\mathbb{P}^n_\mathbb{C})$ *as an abstract group,*

◇ *there is a non-trivial morphism from* $\mathrm{PGL}(n+1,\mathbb{C})$ *to* $\mathrm{Bir}(V)$.

The strategy that leads to the proof of Theorem 7.18 is similar to the proof of Theorem 7.2 but requires several new ideas:

- ◇ Weil's regularization Theorem (Theorem 3.56), that transforms a group of birational maps of V with uniformly bounded degrees into a group of automorphisms of a new variety by a birational change of variables;
- ◇ Epstein and Thurston work on nilpotent Lie subalgebras in the Lie algebra of smooth vector fields of a compact manifold ([**93**]).

7.3. One-parameter subgroups of $\mathrm{Bir}(\mathbb{P}^2_\mathbb{C})$

7.3.1. Description of 1-parameter subgroups of quadratic birational maps of $\mathbb{P}^2_\mathbb{C}$. A *germ of flow* in $\mathrm{Bir}_{\leq 2}(\mathbb{P}^2_\mathbb{C})$ is a germ of holomorphic application $t \mapsto \phi_t \in \mathrm{Bir}_{\leq 2}(\mathbb{P}^2_\mathbb{C})$ such that

$$\begin{cases} \phi_{t+s} = \phi_t \circ \phi_s, \\ \phi_0 = \mathrm{id}. \end{cases}$$

Since a germ of flow can be generalized we speak about *flow*. The set of lines blown down by the flow ϕ_t is a germ of analytic sets in the Grassmaniann of lines in $\mathbb{P}^2_\mathbb{C}$, *i.e.*, in the dual space $(\mathbb{P}^2_\mathbb{C})^\vee$. Similarly the set of indeterminacy points of the ϕ_t is a germ of analytic sets of $\mathbb{P}^2_\mathbb{C}$.

We call a *family of contracted curves* a continuous map (indeed an analytic one) defined over a germ of closed sector Δ of vertex 0 in $\mathbb{C}$:

$$\mathcal{D}\colon \Delta \to (\mathbb{P}^2_\mathbb{C})^\vee$$

such that for any $t \in \Delta$ the lines $\mathcal{D}_t$ coincide with a line $\mathcal{D}(t)$ blown down by ϕ_t.

Similarly a *family of indeterminacy points* is a continuous map $t \mapsto p_t$ defined on a sector Δ such that any p_t is an indeterminacy point of ϕ_t.

Let ϕ_t be a flow. Let $\mathcal{D}_t$ (resp. p_t) be a family of curves blown down by ϕ_t (resp. a family of indeterminacy points of ϕ_t). If $\mathcal{D}_t$ (resp. p_t) is independent of t, the family is *unmobile*, otherwise it is *mobile*.

A rational vector field χ on $\mathbb{P}^2_\mathbb{C}$ is *rationally integrable* if its flow is a flow of birational maps.

A germ of flow in $\mathrm{Bir}_2(\mathbb{P}^2_\mathbb{C})$ is the flow of a rationally integrable vector field $\chi = \frac{\partial \phi_t}{\partial t}\Big|_{t=0}$ called an *infinitesimal generator* of ϕ_t. To this vector field is associated a foliation whose leaves are "grosso modo" the trajectories of χ. Recall that a *fibration by lines* $\mathcal{L}$ of $\mathbb{P}^2_\mathbb{C}$ is given by

$$\lambda\ell_1 + \mu\ell_2 = 0$$

where ℓ_1, ℓ_2 are linear forms that are not proportional. The *base-point* is the intersection point p of all these lines. We also say that $\mathcal{L}$ is a *pencil of lines* through p, or $\mathcal{L}$ is a *foliation by lines* singular at p. Recall that a birational self map of $\mathbb{P}^2_\mathbb{C}$ that preserves a rational fibration belongs up to birational conjugacy to $\mathcal{J}$.

Let ϕ_t be a germ of flow in $\mathrm{Bir}_2(\mathbb{P}^2_\mathbb{C})$. Then the following properties hold:

- ◇ assume that ϕ_t blows down a mobile line, then ϕ_t preserves a fibration by lines, more precisely the family of contracted lines belongs to a fibration invariant by any element of the flow ([**62**, Proposition 2.5, Remark 2.6]);
- ◇ there is at most one unmobile line blown down by ϕ_t (*see* [**62**, Lemma 2.10]);

- ◇ if ϕ_t blows down a unique line that is moreover unmobile, then there exists an invariant affine chart $\mathbb{C}^2$ such that $\phi_{t|\mathbb{C}^2}\colon \mathbb{C}^2 \to \mathbb{C}^2$ is polynomial for any t (*see* [**62**, Proposition 2.12]);
- ◇ assume that there exists an invariant affine chart $\mathbb{C}^2$ such that $\phi_{t|\mathbb{C}^2}\colon \mathbb{C}^2 \to \mathbb{C}^2$ is polynomial for any t. Then ϕ_t preserves a pencil of lines. Furthermore either ϕ_t is affine, or there exists a normal form for ϕ_t up to linear conjugacy ([**62**, Proposition 2.15]).

Combining all these properties one can state the following result:

THEOREM 7.21 ([**62**]). *A germ of flow in* $\mathrm{Bir}_2(\mathbb{P}^2_\mathbb{C})$ *preserves a fibration by lines.*

Let ϕ_t be a quadratic birational flow, and let χ be its infinitesimal generator. A *strong symmetry* Y of χ is a rationally integrable vector field of flow ψ_s such that

- ◇ ϕ_t and ψ_s commute, *i.e.*, $[\chi, Y] = 0$,
- ◇ $\psi_s \in \mathrm{Bir}_2(\mathbb{P}^2_\mathbb{C})$ for all s,
- ◇ χ and Y are not $\mathbb{C}$-colinear.

Let ϕ_t be a flow in $\mathrm{Bir}_2(\mathbb{P}^2_\mathbb{C})$, and let χ (resp. $\mathcal{F}_\chi$) be the associated vector field (resp. foliation). We denote by $\overline{\langle\phi_t\rangle}^Z \subset \mathrm{Bir}_2(\mathbb{P}^2_\mathbb{C})$ the Zariski closure of $\langle\phi_t\rangle$ in $\mathrm{Bir}_2(\mathbb{P}^2_\mathbb{C})$. Let $\mathrm{G}(\chi)$ be the maximal algebraic abelian subgroup of $\mathrm{Bir}_2(\mathbb{P}^2_\mathbb{C})$ that contains $\overline{\langle\phi_t\rangle}^Z$.

THEOREM 7.22 ([**62**]). *Let* ϕ_t *be a germ of flow in* $\mathrm{Bir}_2(\mathbb{P}^2_\mathbb{C})$, *and let* χ *be its infinitesimal generator.*

- ◇ *If* $\dim \mathrm{G}(\chi) = 1$, *then* $\mathcal{F}_\chi$ *is a rational fibration.*
- ◇ *If* $\dim \mathrm{G}(\chi) \geq 2$, *then* $\mathcal{F}_\chi$ *has a strong symmetry.*

In both cases $\mathcal{F}_\chi$ *is defined by a rational closed* 1*-form.*

PROOF. Let us prove the first assertion. If $\dim \mathrm{G}(\chi) = 1$, then $\overline{\langle\phi_t\rangle}^Z$ is the component of $\mathrm{G}(\chi)$ that contains the identity. This group viewed as a Lie group is isomorphic to $\mathbb{C}$, or $\mathbb{C}^*$, or ${}^{\mathbb{C}}\!/_{\Lambda}$. According to Theorem 7.21 the group $\overline{\langle\phi_t\rangle}^Z$ preserves a fibration by lines; let us assume that this fibration is given by $z_1 =$ constant. One yields a morphism

$$\pi\colon \overline{\langle\phi_t\rangle}^Z \to \mathrm{PGL}(2,\mathbb{C})$$

that describes the action of ϕ_t on the fibers.

If π is trivial (*i.e.*, if the fibration is preserved fiberwise), then $\mathcal{F}_\chi = \big\{z_1 = \text{constant}\big\}$ and the result holds.

Otherwise $\overline{\langle\phi_t\rangle}^Z$ is not isomorphic to ${}^{\mathbb{C}}\!/_{\Lambda}$ because there is no ${}^{\mathbb{C}}\!/_{\Lambda}$ among the subgroup of $\mathrm{PGL}(2,\mathbb{C})$. Hence the topological closure of $\overline{\langle\phi_t\rangle}^Z$ in $\mathbb{P}^{17}_\mathbb{C} \simeq \mathrm{Rat}_2$ is a rational curve. But according to Darboux a foliation of $\mathbb{P}^2_\mathbb{C}$ whose closure of all leaves are algebraic curves has a non-constant rational first integral ([**119**]). In our case the curves are rational, so $\mathcal{F}_\chi$ is a rational fibration.

Let us now prove the second assertion. Assume $\dim \mathrm{G}(\chi) \geq 2$. One can find a germ of a 1-parameter group ψ_s in $\mathrm{G}(\chi)$ not contained in $\langle\phi_t\rangle$. Let Y be the infinitesimal generator of ψ_s. The vector fields χ and Y commute and are not $\mathbb{C}$-colinear. Let us consider ω a rational 1-form that defines $\mathcal{F}_\chi$, *i.e.*, $i_\chi\omega = 0$. If χ

and Y are generically independent, then $\Omega = \frac{\omega}{i_Y\omega}$ is closed and define $\mathcal{F}_\chi$. If χ and Y are not generically independent, then $Y = f\chi$ with f rational and non-constant. Since $[\chi, Y] = 0$ one has $\chi(f) = 0$. As a result $\mathrm{d}f$ defines $\mathcal{F}_\chi$ and is closed. □

REMARK 7.23. The last two statements can be generalized as follows:

THEOREM 7.24 ([**62**]). *Let ϕ_t be a germ of flow in $\mathrm{Bir}_n(\mathbb{P}^2_\mathbb{C})$, and let χ be its infinitesimal generator. Denote by $\mathrm{G}(\chi)$ the abelian maximal algebraic group contained in $\mathrm{Bir}_n(\mathbb{P}^2_\mathbb{C})$ and that contains $\overline{\langle\phi_t\rangle}^Z$. Then*

- *if $\dim \mathrm{G}(\chi) = 1$, then $\mathcal{F}_\chi$ is either a rational fibration or an elliptic fibration;*
- *if $\dim \mathrm{G}(\chi) \geq 2$, then χ has a strong symmetry.*

In both cases $\mathcal{F}_\chi$ is defined by a closed rational 1-form.

THEOREM 7.25 ([**62**]). *Any germ of birational flow in $\mathrm{Bir}_n(\mathbb{P}^2_\mathbb{C})$ preserves a rational fibration.*

7.3.2. A few words about the classification of germs of quadratic birational flows. Let ϕ_t be a germ of flow in $\mathrm{Bir}_2(\mathbb{P}^2_\mathbb{C})$; then ϕ_t preserves a fibration by lines ([**62**, Theorem 2.16]). In other words up to linear conjugacy

$$\phi_t\colon (z_0, z_1) \dashrightarrow \left(\frac{A(z_1,t)z_0 + B(z_1,t)}{C(z_1,t)z_0 + D(z_1,t)}, \nu(z_1,t)\right)$$

with

- $\nu(z_1,t) = z_1$, or $z_1 + t$, or $\mathrm{e}^{\beta t}z_1$;
- A, B, C, D are polynomials in z_1 and $\deg_{z_1} A \leq 1$, $\deg_{z_1} B \leq 2$, $\deg_{z_1} C = 0$, $\deg_{z_1} D \leq 1$,
- $B(z_1,0) = C(z_1,0) = 0$ and $A(z_1,0) = D(z_1,0)$.

The infinitesimal generator $\chi = \frac{\partial\phi_t}{\partial t}\Big|_{t=0}$ of ϕ_t can be written

$$\frac{\alpha z_0^2 + \ell(z_1)z_0 + P(z_1)}{az_1 + b}\frac{\partial}{\partial z_0} + \varepsilon(z_1)\frac{\partial}{\partial z_1}$$

with α, a, $b \in \mathbb{C}$, ℓ, $P \in \mathbb{C}[z_1]$, $\deg \ell = 1$, $\deg P = 2$ and up to linear conjugacy and scalar multiplication $\varepsilon \in \{0,\, 1,\, z_1\}$.

The above vector fields are classified up to automorphisms of $\mathbb{P}^2_\mathbb{C}$ and renormalization in [**62**, Chapter 2, §2]; such vector fields are detected via the following methods:

- compute explicitely the flow by integration;
- or degenerate χ on another vector field χ_0 that is not rationally integrable;
- or show that a birational model of $\mathcal{F}_\chi$ has an isolated degenerate resonnant singular point (one and only one non-zero eigenvalue), and so $\mathcal{F}_\chi$ has no rational first integral. Then prove that there is no strong symmetry; hence χ is not rationally integrable (Theorem 7.22).

7.4. Nilpotent subgroups of the Cremona group

In [**79**] the nilpotent subgroups of the plane Cremona group are described:

THEOREM 7.26 ([**79**]). *Let* N *be a nilpotent subgroup of* $\mathrm{Bir}(\mathbb{P}^2_\mathbb{C})$. *Assume that, up to finite index,* N *is not abelian. Then*

- *either* N *is a torsion group;*

◇ *or* N *is metabelian up to finite index, i.e.,* [N, N] *is abelian up to finite index.*

EXAMPLES 5. Let α and β be two non-zero complex numbers; the group

$$\langle (z_0, z_1) \mapsto (z_0 + \alpha\beta, z_1),\ (z_0, z_1) \mapsto (z_0 + \alpha z_1, z_1),\ (z_0, z_1) \mapsto (z_0, z_1 + \beta)\rangle$$

is a non-abelian, non-finite and nilpotent subgroup of $\mathrm{Bir}(\mathbb{P}^2_\mathbb{C})$.

If a belongs to $\mathbb{C}(z_1)$, then

$$\langle (z_0, z_1) \mapsto (z_0 + 1, z_1),\ (z_0, z_1) \mapsto (z_0 + z_1, z_1),\ (z_0, z_1) \mapsto (z_0 + a(z_1), z_1 - 1)\rangle,$$

is a non-abelian, non-finite and nilpotent subgroup of $\mathrm{Bir}(\mathbb{P}^2_\mathbb{C})$.

COROLLARY 7.27 ([**79**]). *Let* G *be a group. Assume that* G *contains a subgroup* N *such that*

◇ N *is of nilpotent class* > 1,
◇ N *has no torsion,*
◇ N *is not metabelian up to finite index.*

Then there is no faithful representation of G *into* $\mathrm{Bir}(\mathbb{P}^2_\mathbb{C})$.

REMARK 7.28. Let G be a nilpotent group of nilpotent class n. Take f in G, g in $C^{(n-2)}\mathrm{G}$ and consider $h = [f, g] \in C^{(n-1)}\mathrm{G}$. Since G is of nilpotent class n, then $[f, h] = [g, h] = \mathrm{id}$. In other words any nilpotent group contains a distorted element.

According to Remark 7.28 and Lemma 5.18 one has:

PROPOSITION 7.29. *Let* N *be a nilpotent subgroup of the plane Cremona group. It contains a distorted element which is elliptic or parabolic.*

IDEA OF THE PROOF OF THEOREM 7.26. Take $\mathrm{G} \subset \mathrm{Bir}(\mathbb{P}^2_\mathbb{C})$, a nilpotent subgroup of class k which is not up to finite index of nilpotent class $k - 1$. Denote by Σ_G the set of finitely generated nilpotent subgroups of G that are, up to finite index, not abelian. Then

◇ either any element of Σ_G is finite and G is a torsion group;
◇ or Σ_G contains a non-finite element H.

CLAIM 7.30 ([**76**]). *The group* H *preserves a fibration* $\mathcal{F}$ *that is rational or elliptic.*

Any element of $C^{(k-1)}\mathrm{H}$ preserves fiberwise $\mathcal{F}$. Let ϕ be in $C^{(k-1)}\mathrm{H}$. As $[\phi, \mathrm{G}] = \mathrm{id}$, then

(a) either ϕ preserves fiberwise two distinct fibrations;
(b) or G preserves fiberwise $\mathcal{F}$.

If (a) holds, then ϕ is of finite order; if it is the case for any $\phi \in C^{(k-1)}\mathrm{H}$, then H is, up to finite index, of nilpotent class $k - 1$: contradiction.

If (b) holds, then G is, up to finite index, metabelian. Let us detail why when $\mathcal{F}$ is rational. In that case G is, up to conjugacy, a subgroup of the Jonquières group $\mathcal{J}$. Let pr_2 be the projection $\mathcal{J} \to \mathrm{PGL}(2, \mathbb{C})$. A non-finite nilpotent subgroup of $\mathrm{PGL}(2, \Bbbk)$, where $\Bbbk = \mathbb{C}$ or $\mathbb{C}(z_1)$, is up to finite index, abelian. The group $\mathrm{pr}_2(\mathrm{G})$ is thus, up to finite index, abelian. Consequently we can assume that $\mathrm{pr}_2(C^{(i)}\mathrm{G}) = \{\mathrm{id}\}$ for $1 \leq i \leq k$. In particular $C^{(1)}\mathrm{G}$ is a nilpotent subgroup of $\mathrm{PGL}(2, \mathbb{C}(z_1))$ and as a result is, up to finite index, abelian. □

Idea of the proof of the Claim 7.30. Let us recall that H is a non-finite nilpotent subgroup of $\mathrm{Bir}(\mathbb{P}^2_\mathbb{C})$ with the following properties:

- ◇ H is finitely generated,
- ◇ H is nilpotent of class $k > 0$,
- ◇ H is not, up to finite index, of nilpotent class $k - 1$.

Assume $C^{(k-1)}\mathrm{H}$ is not a torsion group. Then H preserves a fibration that is rational or elliptic. According to Lemma 5.18 a non-trivial element of $C^{(k-1)}\mathrm{G}$ either preserves a unique fibration $\mathcal{F}$ that is rational or elliptic, or is an elliptic birational map. We have the following alternative:

(a) either $C^{(k-1)}\mathrm{G}$ contains an element h that preserves a unique fibration $\mathcal{F}$,
(b) or any element of $C^{(k-1)}\mathrm{G} \smallsetminus \{\mathrm{id}\}$ is elliptic.

Let us look at these eventualities:

(a) Since $[h, \mathrm{G}] = \{\mathrm{id}\}$ any element of G preserves $\mathcal{F}$.
(b) The group $C^{(k-1)}\mathrm{G}$ is finitely generated and abelian. Let $\{a_1, a_2, \ldots, a_n\}$ be a generating set of $C^{(k-1)}\mathrm{G}$. The a_i's are elliptic maps, so there exist a surface S_i, a birational map $\eta_i \colon S_i \dashrightarrow \mathbb{P}^2_\mathbb{C}$ and an integer $k_i > 0$ such that $\eta_i^{-1} \circ a_i^{k_i} \circ \eta_i$ belongs to the neutral component $\mathrm{Aut}(S_i)^0$ of $\mathrm{Aut}(S_i)$. In particular the a_i's fix any curve of negative self-intersection; we can thus assume that S_i is a minimal rational surface. A priori all the S_i are distinct. Nevertheless according to Proposition 2.12 there exist a minimal rational surface S, a birational map $\eta \colon S \dashrightarrow \mathbb{P}^2_\mathbb{C}$ and an integer $k > 0$ such that for any $1 \leq i \leq n$ the map $\eta^{-1} \circ a_i^k \circ \eta$ belongs to the neutral component $\mathrm{Aut}(S)^0$ of $\mathrm{Aut}(S)$.

Minimal rational surfaces are $\mathbb{P}^2_\mathbb{C}$, $\mathbb{P}^1_\mathbb{C} \times \mathbb{P}^1_\mathbb{C}$ and Hirzebruch surfaces $\mathbb{F}_n$, $n \geq 2$. Using
- ◇ on the one hand the description of the automorphism groups of minimal rational surfaces (*see* Chapter 3),
- ◇ and on the other hand the fact that if K is an algebraic Lie subgroup of $\mathrm{GL}(n, \mathbb{C})$, then the semi-simple and nilpotent parts of any element of K belong to K,

we prove that G is, up to finite index and up to conjugacy, contained in the Jonquières group $\mathcal{J}$ (*see* [**79**]).

It remains to consider the case "$C^{(k-1)}\mathrm{G}$ is a torsion group"; the ideas are similar (*see* [**79**, Proposition 4.5]). □

7.5. Centralizers in $\mathrm{Bir}(\mathbb{P}^2_\mathbb{C})$

7.5.1. Centralizers of elliptic birational maps. We will focus on the case of birational self maps of $\mathbb{P}^2_\mathbb{C}$ of infinite order. Note for instance that for a birational self map of $\mathbb{P}^2_\mathbb{C}$ of finite order the situation is wild: consider for instance a birational involution ϕ of $\mathbb{P}^2_\mathbb{C}$. If ϕ is conjugate to an automorphism of $\mathbb{P}^2_\mathbb{C}$, then the centralizer of ϕ in $\mathrm{Bir}(\mathbb{P}^2_\mathbb{C})$ is uncountable but if ϕ is conjugate to a Bertini (or a Geiser) involution, then the centralizer is finite ([**39**]).

According to [**33**] an elliptic birational self map of $\mathbb{P}^2_\mathbb{C}$ of infinite order is conjugate to an automorphism of $\mathbb{P}^2_\mathbb{C}$ which restricts to one of the following automorphisms on some open subset isomorphic to $\mathbb{C}^2$:

- ◇ $(z_0,z_1)\mapsto(\alpha z_0,\beta z_1)$ where α, β belong to $\mathbb{C}^*$ and where the kernel of the group homomorphism
$$\mathbb{Z}^2\to\mathbb{C}^2 \qquad\qquad (i,j)\mapsto\alpha^i\beta^j$$
is generated by $(k,0)$ for some $k\in\mathbb{Z}$;
- ◇ $(z_0,z_1)\mapsto(\alpha z_0,z_1+1)$ where $\alpha\in\mathbb{C}^*$.

We can describe the centralizers of such maps; let us start with the centralizer of $(z_0,z_1)\mapsto(\alpha z_0,\beta z_1)$ where α, β belong to $\mathbb{C}^*$ and where the kernel of the group homomorphism
$$\mathbb{Z}^2\to\mathbb{C}^2 \qquad\qquad (i,j)\mapsto\alpha^i\beta^j$$
is generated by $(k,0)$ for some $k\in\mathbb{Z}$. Recall that $\mathrm{PGL}(2,\mathbb{C})$ is the group of automorphisms of $\mathbb{P}^1_\mathbb{C}$ or equivalently the group of Möbius transformations
$$z_0\dashrightarrow\frac{az_0+b}{cz_0+d}.$$
A direct computation implies the following: for any $\alpha\in\mathbb{C}^*$
$$\big\{\eta\in\mathrm{PGL}(2,\mathbb{C})\,|\,\eta(\alpha z_0)=\alpha\eta(z_0)\big\}=\left\{\begin{array}{l}\mathrm{PGL}(2,\mathbb{C})\text{ if }\alpha=1,\\ \big\{z_0\dashrightarrow\gamma z_0^{\pm1}\,|\,\gamma\in\mathbb{C}^*\big\}\text{ if }\alpha=-1,\\ \big\{z_0\mapsto\gamma z_0\,|\,\gamma\in\mathbb{C}^*\big\}\text{ if }\alpha^2\neq1.\end{array}\right.$$

LEMMA 7.31. *Let us consider* $\phi\colon(z_0,z_1)\mapsto(\alpha z_0,\beta z_1)$ *where* α, β *belong to* $\mathbb{C}^*$ *and where the kernel of the group homomorphism*
$$\mathbb{Z}^2\to\mathbb{C}^2 \qquad\qquad (i,j)\mapsto\alpha^i\beta^j$$
is generated by $(k,0)$ *for some* $k\in\mathbb{Z}$.

The centralizer of ϕ *in* $\mathrm{Bir}(\mathbb{P}^2_\mathbb{C})$ *is*
$$\big\{(z_0,z_1)\dashrightarrow(\eta(z_0),z_1a(z_0^k))\,|\,a\in\mathbb{C}(z_0),\,\eta\in\mathrm{PGL}(2,\mathbb{C}),\,\eta(\alpha z_0)=\alpha\eta(z_0)\big\}.$$

PROOF. Let $\psi\colon(z_0,z_1)\dashrightarrow(\psi_0(z_0,z_1),\psi_1(z_0,z_1))$ be a birational self map of $\mathbb{P}^2_\mathbb{C}$ that commutes with ϕ. Then
$$\psi_0(\alpha z_0,\beta z_1)=\alpha\psi_0(z_0,z_1) \tag{7.1}$$
and
$$\psi_1(\alpha z_0,\beta z_1)=\beta\psi_1(z_0,z_1) \tag{7.2}$$
hold. Denote by ϕ^* the linear automorphism of the $\mathbb{C}$-vector space $\mathbb{C}[z_0,z_1]$ given by
$$\phi^*\colon\varphi(z_0,z_1)\mapsto\varphi(\alpha z_0,\beta z_1).$$
Let us write ψ_i as $\frac{P_i}{Q_i}$ for $i=0,\,1$ where P_i, Q_i are polynomials without common factor. Note that P_0, P_1, Q_0, Q_1 are eigenvectors of ϕ^*, *i.e.*, any of the P_i, Q_i is a product of a monomial in z_0, z_1 with an element of $\mathbb{C}[z_0^k]$. Using (7.1) and (7.2) we get that
$$\left\{\begin{array}{l}\psi_0(z_0,z_1)=z_0a_0(z_0^k),\\ \psi_1(z_0,z_1)=z_1a_1(z_0^k).\end{array}\right.$$
But ψ is birational, so ψ_0 belongs to $\mathrm{PGL}(2,\mathbb{C})$. Furthemore ψ_0 satisfies $\psi_0(\alpha z_0)=\alpha\psi_0(z_0)$. □

Let us now deal with the other possibility:

LEMMA 7.32. *Let ϕ be the automorphism of $\mathbb{P}^2_{\mathbb{C}}$ given by*

$$\phi\colon (z_0, z_1) \mapsto (\alpha z_0, z_1 + \beta)$$

where $\alpha \in \mathbb{C}^*$, $\beta \in \mathbb{C}$. The centralizer of ϕ in $\mathrm{Bir}(\mathbb{P}^2_{\mathbb{C}})$ *is*

$$\left\{(z_0, z_1) \dashrightarrow (\eta(z_0), z_1 + a(z_0)) \mid \left\{ \begin{array}{l} \eta \in \mathrm{PGL}(2, \mathbb{C}),\ \eta(\alpha z_0) = \alpha\eta(z_0), \\ a \in \mathbb{C}(z_0),\ a(\alpha z_0) = a(z_0). \end{array} \right. \right\}$$

PROOF. After conjugacy by $(z_0, z_1) \mapsto (z_0, \beta z_1)$ we can assume that $\beta = 1$.

If $\psi\colon (z_0, z_1) \dashrightarrow (\psi_0(z_0, z_1), \psi_1(z_0, z_1))$ is a birational map that commutes with ϕ, then

$$\psi_0(\alpha z_0, z_1 + 1) = \alpha\psi_0(z_0, z_1) \tag{7.3}$$

and

$$\psi_1(\alpha z_0, z_1 + 1) = \psi_1(z_0, z_1) + 1 \tag{7.4}$$

From (7.3) and [**21**] we get that ψ_0 only depends on z_0. Hence ψ_0 belongs to $\mathrm{PGL}(2, \mathbb{C})$ and commutes with $z_0 \mapsto \alpha z_0$. From (7.4) we get

$$\left\{ \begin{array}{l} \frac{\partial\psi_1}{\partial z_1}(\alpha z_0, z_1 + 1) = \frac{\partial\psi_1}{\partial z_1}(z_0, z_1), \\ \frac{\partial\psi_1}{\partial z_0}(\alpha z_0, z_1 + 1) = \frac{1}{\alpha}\frac{\partial\psi_1}{\partial z_0}(z_0, z_1), \end{array} \right.$$

which again means that both $\frac{\partial\psi_1}{\partial z_0}$ and $\frac{\partial\psi_1}{\partial z_1}$ only depend on z_0. Therefore, $\psi_1\colon (z_0, z_1) \mapsto \gamma z_1 + b(z_0)$ with $\gamma \in \mathbb{C}^*$ and $b \in \mathbb{C}(z_0)$. Then (7.4) can be rewritten:

$$b(\alpha z_0) = b(z_0) + 1 - \gamma$$

which implies that

$$\frac{\partial b}{\partial z_0}(\alpha z_0) = \frac{1}{\alpha}\frac{\partial b}{\partial z_0}(z_0)$$

and that $z_0\frac{\partial b}{\partial z_0}(z_0)$ is invariant under $z_0 \mapsto \alpha z_0$.

If α is not a root of unity, then $\frac{\partial b}{\partial z_0} = \frac{\delta}{z_0}$ for some $\delta \in \mathbb{C}$. As b is rational, δ is zero and b is constant. As a consequence $b(\alpha z_0) = b(z_0) + 1 - \gamma$ implies $\gamma = 1$, that is, $\psi_1\colon (z_0, z_1) \dashrightarrow z_0 + \beta$.

Assume that α is a primitive kth root of unity. The map $\psi\colon (z_0, z_1) \dashrightarrow (\eta(z_0), \gamma z_1 + b(z_0))$ commutes with

$$\phi^k\colon (z_0, z_1) \dashrightarrow (z_0, z_1 + k)$$

if and only if $\gamma(z_1 + k) + b(z_0) = \gamma z_1 + b(z_0) + k$, *i.e.*, if and only if $\gamma = 1$. Then $b(\alpha z_0) = b(z_0) + 1 - 1$ can be rewritten $b(z_0) = b(\alpha z_0)$. □

7.5.2. Centralizers of Jonquières twists. Recall that the subgroup $\mathcal{J}$ of Jonquières maps is isomorphic to $\mathrm{PGL}(2, \mathbb{C}(z_1)) \rtimes \mathrm{PGL}(2, \mathbb{C})$. Let us denote by pr_2 the morphism

$$\mathrm{pr}_2\colon \mathcal{J} \to \mathrm{PGL}(2, \mathbb{C}).$$

Geometrically it corresponds to the action of $\phi \in \mathcal{J}$ on the basis of the invariant fibration $z_1 = \mathrm{cst}$. The kernel of pr_2, *i.e.*, the elements of $\mathcal{J}$ which preserve the fibration $z_1 = \mathrm{cst}$ fiberwise, is a normal subgroup $\mathcal{J}_0 \simeq \mathrm{PGL}(2, \mathbb{C}(z_1))$ of $\mathcal{J}$. Up to a birational conjugacy an element ϕ of $\mathcal{J}_0$ is of one of the following forms ([**77**]):

$$(z_0, z_1) \dashrightarrow (z_0 + a(z_1), z_1), \qquad (z_0, z_1) \dashrightarrow (b(z_1)z_0, z_1),$$

$$(z_0, z_1) \dashrightarrow \left(\frac{c(z_1)z_0 + F(z_1)}{z_0 + c(z_1)}, z_1\right)$$

with $a \in \mathbb{C}(z_1)$, $b \in \mathbb{C}(z_1)^*$, $c \in \mathbb{C}(z_1)$, $F \in \mathbb{C}[z_1]$ and F not a square. Still according to [**77**] the non-finite maximal abelian subgroups of $\mathcal{J}_0$ are

$$\begin{aligned}\mathcal{J}_a &= \big\{(z_0, z_1) \dashrightarrow (z_0 + a(z_1)) \,|\, a \in \mathbb{C}(z_1)\big\},\\ \mathcal{J}_m &= \big\{(z_0, z_1) \dashrightarrow (b(z_1)z_0, z_1) \,|\, a \in \mathbb{C}(z_1)\big\},\\ \mathcal{J}_F &= \big\{(z_0, z_1) \dashrightarrow \left(\frac{c(z_1)z_0 + F(z_1)}{z_0 + c(z_1)}, z_1\right) \,|\, a \in \mathbb{C}(z_1)\big\},\end{aligned}$$

where F denotes an element of $\mathbb{C}[z_1]$ which is not a square. Note that we can assume up to conjugacy that F is a polynomial with roots of multiplicity 1.

If ϕ belongs to $\mathcal{J}_0$, let us denote by $\mathrm{Ab}(\phi)$ the non-finite maximal abelian subgroup of $\mathcal{J}_0$ that contains ϕ. Up to conjugacy

- ◇ either $(z_0, z_1) \dashrightarrow (z_0 + a(z_1), z_1)$ and $\mathrm{Ab}(\phi) = \mathcal{J}_a$;
- ◇ or $(z_0, z_1) \dashrightarrow (b(z_1)z_0, z_1)$ and $\mathrm{Ab}(\phi) = \mathcal{J}_m$;
- ◇ or $(z_0, z_1) \dashrightarrow \left(\frac{c(z_1)z_0+F(z_1)}{z_0+c(z_1)}, z_1\right)$ and $\mathrm{Ab}(\phi) = \mathcal{J}_F$.

PROPOSITION 7.33 ([**61**]). *Let ϕ be an element of $\mathcal{J}_0$ that is a Jonquières twist. Then the centralizer of ϕ in $\mathrm{Bir}(\mathbb{P}^2_\mathbb{C})$ is contained in $\mathcal{J}$.*

PROOF. Consider a birational self map $\varphi\colon (z_0, z_1) \dashrightarrow (\varphi_0(z_0, z_1), \varphi_1(z_0, z_1))$ of $\mathbb{P}^2_\mathbb{C}$ that commutes with ϕ. If φ does not belong to $\mathcal{J}$, then $\varphi_1 = \mathrm{cst}$ is a fibration invariant by ϕ distinct from $z_1 = \mathrm{cst}$. Then ϕ is of finite order (Lemma 8.17): contradiction with the fact that ϕ is a Jonquières twist. □

7.5.2.1. *Centralizers of elements of $\mathcal{J}_a$.* Note that elements of $\mathcal{J}_a$ are not Jonquières twists but elliptic maps. Hence their centralizers are described in §7.5.1. Let us give some details. Let $\phi\colon (z_0, z_1) \dashrightarrow (z_0 + a(z_1), z_1)$ be a non-trivial element of $\mathcal{J}_a$ (*i.e.*, $a \not\equiv 0$). Up to conjugacy by $(z_0, z_1) \dashrightarrow (a(z_1)z_0, z_1)$ one can assume that $a \equiv 1$. The centralizer of $(z_0, z_1) \dashrightarrow (z_0 + 1, z_1)$ is isomorphic to $\mathcal{J}_a \rtimes \mathrm{PGL}(2, \mathbb{C})$ (*see* §7.5.1). Hence

COROLLARY 7.34. *The centralizer of a non-trivial element of $\mathcal{J}_a$ is isomorphic to $\mathcal{J}_a \rtimes \mathrm{PGL}(2, \mathbb{C})$.*

7.5.2.2. *Centralizers of twists of $\mathcal{J}_m$.* An element ϕ of $\mathcal{J}_m$ is a Jonquières twist if and only if up to birational conjugacy

$$(z_0, z_1) \dashrightarrow (a(z_1)z_0, z_1)$$

with $a \in \mathbb{C}(z_1) \smallsetminus \mathbb{C}^*$.

Note that if a belongs to $\mathbb{C}^*$, then $(z_0, z_1) \dashrightarrow (az_0, z_1)$ is an elliptic map whose centralizer is described in §7.5.1. Assume now that $\phi \in \mathcal{J}_m$ is a Jonquières twist. Let $a \in \mathbb{C}(z_1) \smallsetminus \mathbb{C}^*$. Denote by

$$\mathrm{Stab}(a) = \big\{\nu \in \mathrm{PGL}(2, \mathbb{C}) \,|\, a(\nu(z_1)) = a(z_1)^{\pm 1}\big\}$$

the subgroup of $\mathrm{PGL}(2, \mathbb{C})$ and by

$$\mathrm{stab}(a) = \big\{\nu \in \mathrm{PGL}(2, \mathbb{C}) \,|\, a(\nu(z_1)) = a(z_1)\big\}$$

the normal subgroup of $\mathrm{Stab}(a)$. Consider also

$$\overline{\mathrm{stab}(a)} = \big\{(z_0, \nu(z_1)) \,|\, \nu \in \mathrm{stab}(a)\big\}$$

and $\overline{\mathrm{Stab}(a)}$ the group generated by $\overline{\mathrm{stab}(a)}$ and the elements

$$(z_0, z_1) \dashrightarrow \left(\frac{1}{z_0}, \nu(z_1)\right)$$

with $\nu \in \mathrm{Stab}(a) \smallsetminus \mathrm{stab}(a)$.

PROPOSITION 7.35 ([**61**]). *Let ϕ be a Jonquières twist in $\mathcal{J}_m$. The centralizer of ϕ in $\mathrm{Bir}(\mathbb{P}^2_{\mathbb{C}})$ is $\mathcal{J}_m \rtimes \overline{\mathrm{Stab}(a)}$; in particular it is a finite extension of $\mathrm{Ab}(\phi) = \mathcal{J}_m$.*

REMARK 7.36. One can write ϕ as $(z_0, z_1) \dashrightarrow (a(z_1)z_0, z_1)$ with $a \in \mathbb{C}(z_1) \smallsetminus \mathbb{C}^*$. For generic a the group $\overline{\mathrm{Stab}(a)}$ is trivial, so for generic $\phi \in \mathcal{J}_m$ the centralizer of ϕ in $\mathrm{Bir}(\mathbb{P}^2_{\mathbb{C}})$ coincides with $\mathcal{J}_m = \mathrm{Ab}(\phi)$.

PROOF. Write ϕ as $(z_0, z_1) \dashrightarrow (a(z_1)z_0, z_1)$ with $a \in \mathbb{C}(z_1) \smallsetminus \mathbb{C}^*$. If ψ commutes with ϕ, then ψ preserves the fibration $z_1 = \mathrm{cst}$ (Proposition 7.33), *i.e.*,

$$\psi\colon (z_0, z_1) \dashrightarrow \left(\frac{A(z_1)z_0 + B(z_1)}{C(z_1)z_0 + D(z_1)}, \nu(z_1)\right)$$

with $\begin{pmatrix} A & B \\ C & D \end{pmatrix} \in \mathrm{PGL}(2, \mathbb{C}(z_1))$ and $\nu \in \mathrm{PGL}(2, \mathbb{C})$. Since ψ and ϕ commute, the following hold:

$$\begin{cases} A(z_1)C(z_1)\big(1 - a(\nu(z_1))\big) = 0, \\ B(z_1)D(z_1)\big(1 - a(\nu(z_1))\big) = 0. \end{cases}$$

Therefore, $AC \equiv 0$ and $BD \equiv 0$, *i.e.*, $B = C = 0$ or $A = D = 0$.

Assume first that $B = C = 0$, *i.e.*, that

$$\psi\colon (z_0, z_1) \dashrightarrow (A(z_1)z_0, \nu(z_1)).$$

The condition $\phi \circ \psi = \psi \circ \phi$ implies $a(\nu(z_1)) = a(z_1)$. As $\overline{\mathrm{stab}(a)}$ is contained in the centralizer of ϕ in $\mathrm{Bir}(\mathbb{P}^2_{\mathbb{C}})$ the map ϕ belongs to $\mathcal{J}_m \rtimes \overline{\mathrm{stab}(a)}$.

Suppose now that $A = D = 0$, *i.e.*, that $\psi\colon (z_0, z_1) \dashrightarrow \left(\frac{B(z_1)}{z_0}, \nu(z_1)\right)$. The equality $\psi \circ \varphi = \varphi \circ \psi$ implies that $a(\nu(z_1)) = a(z_1)^{-1}$. But $\overline{\mathrm{Stab}(a)}$ is contained in the centralizer of ϕ in $\mathrm{Bir}(\mathbb{P}^2_{\mathbb{C}})$, so ψ belongs to $\mathcal{J}_m \rtimes \overline{\mathrm{Stab}(a)}$. □

7.5.2.3. *Centralizers of elements of $\mathcal{J}_F$.* Let ϕ be a twist in $\mathcal{J}_F$. Let us write ϕ as

$$(z_0, z_1) \dashrightarrow \left(\frac{c(z_1)z_0 + F(z_1)}{z_0 + c(z_1)}, z_1\right)$$

with $c \in \mathbb{C}(z_1)^*$ and $F \in \mathbb{C}[z_1]$ whose roots have multiplicity 1. The curve $\mathcal{C}$ of fixed points of ϕ is given by $z_0^2 = F(z_1)$. Since F has simple roots one has

$$\begin{cases} \mathcal{C} \text{ is rational when } 1 \leq \deg F \leq 2; \\ \text{the genus of } \mathcal{C} \text{ is } 1 \text{ when } 3 \leq \deg F \leq 4; \\ \text{the genus of } \mathcal{C} \text{ is } \geq 2 \text{ when } \deg F \geq 5. \end{cases}$$

◇ Assume first that the genus of $\mathcal{C}$ is positive.

LEMMA 7.37 ([**61**]). *Let*

$$\phi\colon (z_0, z_1) \dashrightarrow \left(\frac{c(z_1)z_0 + F(z_1)}{z_0 + c(z_1)}, z_1\right) \qquad c \in \mathbb{C}(z_1)^*,\ F \in \mathbb{C}[z_1]$$

be a twist in $\mathcal{J}_F$. The curve $z_0^2 = F(z_1)$ and the fibers $z_1 = $cst are invariant and there is no other invariant curve.

PROOF. The map ϕ has two fixed points on a generic fiber which correspond to the intersection of the fiber with the curve $z_0^2 = F(z_1)$. Assume by contradiction that there is another invariant curve $\mathcal{C}$. The curve $\mathcal{C}$ intersects a generic fiber in a finite number of points that are invariant by ϕ. But a Möbius transformation that preserves more than three points is periodic: contradiction with the fact that ϕ is a Jonquières twist, so of infinite order. □

PROPOSITION 7.38 ([**61**]). *Let*

$$\phi\colon (z_0, z_1) \dashrightarrow \left(\frac{c(z_1)z_0 + F(z_1)}{z_0 + c(z_1)}, z_1\right) \qquad c \in \mathbb{C}(z_1)^*,\ F \in \mathbb{C}[z_1]$$

be a twist in $\mathcal{J}_F$. *Assume that* F *has only simple roots and* $\deg F \geq 3$, *i.e., the curve* $z_0^2 = F(z_1)$ *has genus* ≥ 1. *Then the centralizer of* ϕ *in* $\mathrm{Bir}(\mathbb{P}^2_\mathbb{C})$ *is a finite extension of* $\mathrm{Ab}(\phi) = \mathcal{J}_F$.

PROOF. Take $\alpha \in \mathbb{C}$ such that $F(\alpha) \neq 0$. The restriction $\phi_{|z_1=\alpha}$ of ϕ on the fiber $z_1 = \alpha$ has two fixed points: $(\pm\sqrt{F(\alpha)}, \alpha)$. Note that the centralizer $\mathrm{Cent}(\phi)$ of ϕ in $\mathrm{Bir}(\mathbb{P}^2_\mathbb{C})$ is contained in $\mathcal{J}$ (Proposition 7.33). We will focus on elements ψ of $\mathrm{Cent}(\phi)$ that preserve the fibration $z_1 =$cst fiberwise, *i.e.*, on the kernel of

$$\mathrm{pr}_{2|\mathrm{Cent}(\phi)}\colon \mathrm{Cent}(\phi) \to \mathrm{PGL}(2, \mathbb{C}).$$

Note that any $\psi \in \mathrm{Cent}(\phi)$ preserves $\mathcal{C}$ and that the automorphism $\psi_{|\mathcal{C}}$ of $\mathcal{C}$ preserves $\{(\pm\sqrt{F(\alpha)}, \alpha)\}$. Hence either $\psi_{|\mathcal{C}} = \mathrm{id}$, that is, $\psi \in \mathcal{J}_F$, or $\psi_{|\mathcal{C}}$ is the involution $(z_0, z_1) \mapsto (-z_0, z_1)$ of $\mathcal{C}$. Note that the restriction of $\tau\colon (z_0, z_1) \dashrightarrow \left(-\frac{F(z_1)}{z_0}, z_1\right)$ to $\mathcal{C}$ is $\tau_{|\mathcal{C}}\colon (z_0, z_1) \dashrightarrow (-z_0, z_1)$. Therefore, any birational self map of $\mathbb{P}^2_\mathbb{C}$ that preserves both $\mathcal{C}$ and the fibration $z_1 =$cst fiberwise belongs either to $\mathcal{J}_F$ or to $\tau \circ \mathcal{J}_F$. But $\tau \circ \phi \circ \tau^{-1} = \tau \circ \phi \circ \tau = \phi^{-1}$, so τ does not belong to $\mathrm{Cent}(\phi)$. As a result $\ker \mathrm{pr}_{2|\mathrm{Cent}(\phi)} = \mathcal{J}_F$. Any $\varphi \in \mathrm{Cent}(\phi)$ has to preserve $\mathcal{C}$ and the fibration $z_1 =$cst; the restriction $\varphi_{|\mathcal{C}}$ of φ to $\mathcal{C}$ is an automorphism of $\mathcal{C}$ that commutes with the involution $\tau_{|\mathcal{C}}$. The group $\mathrm{Aut}_\tau(\mathcal{C})$ of such automorphisms is a finite group (more precisely if F is generic, then $\mathrm{Aut}_\tau(\mathcal{C}) = \{\mathrm{id}, \tau_{|\mathcal{C}}\}$). □

◇ Assume that $\mathcal{C}$ is rational.

LEMMA 7.39 ([**61**]). *Let* $\phi \in \mathcal{J}_F$ *be a Jonquières twist such that the curve* $\mathcal{C}$ *of fixed points of* ϕ *is rational. Any element that commutes with* ϕ *belongs to* $\mathcal{J}$ *and preserves* $\mathcal{C}$.

PROOF. The curve of fixed points of ϕ is given by $z_0^2 = F(z_1)$. Let ψ be a birational self map of $\mathbb{P}^2_\mathbb{C}$ such that $\phi \circ \psi = \psi \circ \phi$. According to Proposition 7.33 the map ψ preserves the fibration $z_1 =$cst. Either ψ contracts $\mathcal{C}$ or ψ preserves $\mathcal{C}$. But $\mathcal{C}$ is transverse to the fibration $z_1 =$cst, so ψ cannot contract $\mathcal{C}$. As a result φ is an element of $\mathcal{J}$ that preserves $\mathcal{C}$. □

Note that the case $\deg F \geq 3$ has already been studied, so let us assume that $\deg F \leq 2$. Note that if

$$\phi\colon (z_0, z_1) \dashrightarrow \left(\frac{c(z_1)z_0 + z_1}{z_0 + c(z_1)}, z_1\right)$$

and if

$$\varphi\colon (z_0, z_1) \dashrightarrow \left(\frac{z_0}{\gamma z_1 + \delta}, \frac{\alpha z_1 + \beta}{\gamma z_1 + \delta}\right)$$

then $\varphi^{-1} \circ \phi \circ \varphi$ is of the following type:

$$(z_0, z_1) \dashrightarrow \left(\frac{\widetilde{c}(z_1)z_0 + (\alpha z_1 + \beta)(\gamma z_1 + \delta)}{z_0 + \widetilde{c}(z_1)}, z_1\right).$$

In other words thanks to

$$(z_0, z_1) \dashrightarrow \left(\frac{c(z_1)z_0 + z_1}{z_0 + c(z_1)}, z_1\right)$$

we obtain all polynomials $(\alpha z_1 + \beta)(\gamma z_1 + \delta)$ of degree 2 with simple roots. So one can suppose that $\deg F = 1$. Note that if $\deg F = 1$, *i.e.*, $F(z_1) = \alpha z_1 + \beta$, then up to conjugacy by $(z_0, z_1) \mapsto \left(z_0, \frac{z_1 - \beta}{\alpha}\right)$ one can assume that $F\colon z_1 \mapsto z_1$.

LEMMA 7.40 ([**61**]). *Consider the birational self map of* $\mathbb{P}^2_{\mathbb{C}}$ *given by*

$$\phi\colon (z_0, z_1) \dashrightarrow \left(\frac{c(z_1)z_0 + z_1}{z_0 + c(z_1)}, z_1\right)$$

with $c \in \mathbb{C}(z_1)^*$. *If* ψ *is a birational self map of* $\mathbb{P}^2_{\mathbb{C}}$ *that commutes with* ϕ, *then*

- ◇ *either* $\mathrm{pr}_2(\psi) = \frac{\alpha}{z_1}$ *with* $\alpha \in \mathbb{C}^*$;
- ◇ *or* $\mathrm{pr}_2(\psi) = \zeta z_1$ *with* ζ *a root of unity.*

Furthermore $\mathrm{pr}_2(\psi)$ *belongs to the finite group* $\mathrm{stab}\left(\frac{4c^2(z_1)}{c^2(z_1) - z_1}\right)$.

For any α non-zero consider the dihedral group

$$\mathrm{D}_\infty(\alpha) = \langle z_1 \mapsto \frac{\alpha}{z_1},\ z_1 \mapsto \zeta z_1 \mid \zeta \text{ root of unity}\rangle$$

Note that all the $\mathrm{D}_\infty(\alpha)$ are conjugate to $\mathrm{D}_\infty(1)$.

PROPOSITION 7.41 ([**61**]). *Let* $\phi \in \mathcal{J}_F$ *be a Jonquières twist such that the fixed curve of* ϕ *is rational. Up to conjugacy we can assume that*

$$\phi\colon (z_0, z_1) \dashrightarrow \left(\frac{c(z_1)z_0 + z_1}{z_0 + c(z_1)}, z_1\right)$$

with $c \in \mathbb{C}(z_1) \smallsetminus \mathbb{C}$. *The centralizer of* ϕ *in* $\mathrm{Bir}(\mathbb{P}^2_{\mathbb{C}})$ *is*

$$\mathcal{J}_{z_1} \rtimes \left(\mathrm{stab}\left(\frac{4c^2(z_1)}{c^2(z_1) - z_1}\right) \cap \mathrm{D}_\infty(\alpha)\right)$$

for some $\alpha \in \mathbb{C}^*$.

PROOF. Denote by $\mathrm{Cent}(\phi)$ the centralizer of ϕ in $\mathrm{Bir}(\mathbb{P}^2_{\mathbb{C}})$, and by $\mathcal{C}$ the fixed curve of ϕ.

- ◇ Let us first assume that any element of $\mathrm{Cent}(\phi)$ preserves the fibration $z_1 = $cst fiberwise. Then $\mathrm{Cent}(\phi) = \mathcal{J}_{z_1}$.

◇ Assume now that there exists an element ψ in $\mathrm{Cent}(\phi)$ that does not preserve the fibration z_1 =cst fiberwise. According to Lemma 7.40 either $\mathrm{pr}_2(\psi) = \zeta z_1$ with ζ a root of unity, or $\mathrm{pr}_2(\psi) = \frac{\alpha}{z_1}$ with α in $\mathbb{C}^*$.
If $\mathrm{pr}_2(\psi) = \zeta z_1$ with ζ a root of unity, then

$$\frac{4c^2(\zeta z_1)}{c^2(\zeta z_1) - \zeta z_1} = \frac{4c^2(z_1)}{c^2(z_1) - z_1}$$

i.e., $c^2(\zeta z_1) = \zeta c^2(z_1)$. There exists υ such that $\upsilon^2 = \zeta$ and $c(\upsilon^2 z_1) = \upsilon c(z_1)$. Note that $\varphi\colon (z_0, z_1) \mapsto (\upsilon z_0, \upsilon^2 z_1)$ belongs to $\mathrm{Cent}(\phi)$. Note that $\mathrm{pr}_2(\psi \circ \varphi^{-1}) = \mathrm{id}$, so $\psi \circ \varphi^{-1}$ belongs to $\mathcal{J}_{z_1}$. If $\mathrm{pr}_2(\psi) = \frac{\alpha}{z_1}$, then

$$\frac{4c^2\left(\frac{\alpha}{z_1}\right)}{c^2\left(\frac{\alpha}{z_1}\right) - z_1} = \frac{4c^2(z_1)}{c^2(z_1) - z_1}$$

i.e., $c^2\left(\frac{\alpha}{z_1}\right) = \frac{\alpha}{z_1^2} c^2(z_1)$. There exists β in $\mathbb{C}$ such that $\beta^2 = \alpha$ and $c\left(\frac{\beta^2}{z_1}\right) = \frac{\beta}{z_1} c(z_1)$. Note that the map $(z_0, z_1) \mapsto \left(\frac{\beta z_0}{z_1}, \frac{\beta^2}{z_1}\right)$ commutes with ϕ. The map $\psi \circ \varphi^{-1}$ belongs to $\mathrm{Cent}(\phi)$ and preserves the fibration z_1 =cst fiberwise; hence $\psi \circ \varphi^{-1}$ belongs to $\mathcal{J}_{z_1}$. □

We thus have established:

PROPOSITION 7.42 ([**61**]). *The centralizer of a Jonquières twist ϕ that preserves fiberwise the fibration in the plane Cremona group is a finite extension of* $\mathrm{Ab}(\phi)$.

7.5.2.4. *Centralizers of elements of $\mathcal{J} \smallsetminus \mathcal{J}_0$.* The description of the centralizers of elements of $\mathcal{J}_0$ (Proposition 7.42) allows us to describe, up to finite index, the centralizer of elements of $\mathcal{J}$. Generically these maps have a trivial centralizer ([**61**]). A consequence of the study of the centralizers of elements of $\mathcal{J}$ is:

COROLLARY 7.43 ([**61**]). *The centralizer of a Jonquières twist is virtually solvable.*

Zhao has refined this statement:

PROPOSITION 7.44 ([**184**]). *The centralizer of a Jonquières twist whose action on the basis of the rational fibration is of infinite order is virtually abelian.*

7.5.3. What about the others?

7.5.3.1. Let ϕ be a Halphen twist. Up to birational conjugacy one can assume that ϕ is an element of a rational surface S with an elliptic fibration and that this fibration is ϕ-invariant (§2.3). Furthermore we can assume that there is no smooth curve of self-intersection -1 in the fibers, *i.e.*, that the fibration is minimal, and so that ϕ is an automorphism. The elliptic fibration is the unique ϕ-invariant fibration ([**85**]). As a result the fibration is invariant by all elements that commute with ϕ, and the centralizer of ϕ is contained in $\mathrm{Aut}(S)$.

Since the fibration is minimal, the surface S is obtained by blowing up the complex projective plane in the nine base-points of a Halphen pencil and the rank of its Néron-Severi group is equal to 10. The group $\mathrm{Aut}(S)$ can be embedded in the endomorphisms of $H^2(S, \mathbb{Z})$ for the intersection form and preserves the class $[K_S]$ of the canonical divisor, that is, the class of the fibration. The dimension of the orthogonal hyperplane to $[K_S]$ is 9, and the restriction of the intersection form on its hyperplane is semi-negative: its kernel coincides with $\mathbb{Z}[K_S]$. As a consequence $\mathrm{Aut}(S)$ contains an abelian group of finite index with rank ≤ 8. We can thus state:

PROPOSITION 7.45 ([**101**]). *Let ϕ be a Halphen twist. The centralizer of ϕ in $\mathrm{Bir}(\mathbb{P}^2_\mathbb{C})$ contains a subgroup of finite index which is abelian, free and of rank ≤ 8.*

7.5.3.2. We finish the description of the centralizers of birational maps with the case of loxodromic maps in §8.1.2.

CHAPTER 8

Consequences of the action of the Cremona group on an infinite-dimensional hyperbolic space

As we will see in this chapter one of the main techniques to better understand infinite subgroups of $\mathrm{Bir}(\mathbb{P}^2_\mathbb{C})$ is the construction of the action by isometries of the plane Cremona group on an infinite-dimensional hyperbolic space detailed in Chapter 2 and the use of results from hyperbolic geometry and group theory.

In the first section we recall results of Demazure and Beauville that suggest that the plane Cremona group behaves like a rank 2 group. We give an outline of the proof of the description of the centralizer of a loxodromic element of $\mathrm{Bir}(\mathbb{P}^2_\mathbb{C})$. On the one hand it finishes the description of the centralizer of the elements of $\mathrm{Bir}(\mathbb{P}^2_\mathbb{C})$, on the other hand it suggests that $\mathrm{Bir}(\mathbb{P}^2_\mathbb{C})$ behaves as a group of rank 1. We end this section by recalling the description of the morphisms from a countable group with Kazhdan property (T) into $\mathrm{Bir}(\mathbb{P}^2_\mathbb{C})$ which also insinuates that $\mathrm{Bir}(\mathbb{P}^2_\mathbb{C})$ behaves as a group of rank 1.

In the second section we give an outline of the proofs of the description of elliptic subgroups of $\mathrm{Bir}(\mathbb{P}^2_\mathbb{C})$, *i.e.*, the subgroups of $\mathrm{Bir}(\mathbb{P}^2_\mathbb{C})$ all of whose elements are elliptic: if G is such a group, either G is a bounded subgroup of $\mathrm{Bir}(\mathbb{P}^2_\mathbb{C})$, or G is a torsion subgroup ([**172**]). It is thus natural to describe torsion subgroups of $\mathrm{Bir}(\mathbb{P}^2_\mathbb{C})$. In the third section we give an outline of the proof of the fact that if G is a torsion subgroup of $\mathrm{Bir}(\mathbb{P}^2_\mathbb{C})$, then G is isomorphic to a bounded subgroup of $\mathrm{Bir}(\mathbb{P}^2_\mathbb{C})$; furthermore it is isomorphic to a subgroup of $\mathrm{GL}(48,\mathbb{C})$. Let us mention the surprising fact that the proof uses model theory as Malcev already did in [**135**].

The fourth section deals with the Tits alternative and the Burnside problem. We recall the Ping Pong Lemma and give a sketch of the proof of the Tits alternative for the Cremona group, *i.e.*, the proof of

THEOREM 8.1 ([**52, 172**]). *Every subgroup of* $\mathrm{Bir}(\mathbb{P}^2_\mathbb{C})$ *either is virtually solvable, or contains a non-abelian free group.*

One consequence is a positive answer to the Burnside problem for the Cremona group: every finitely generated torsion subgroup of $\mathrm{Bir}(\mathbb{P}^2_\mathbb{C})$ is finite.

The study of solvable groups is a very old problem. For instance let us recall the Lie-Kolchin theorem: any linear solvable subgroup is up to finite index triangularizable ([**124**]). Note that the assumption "up to finite index" is essential: for instance the subgroup

$$\langle\begin{pmatrix} 1 & 0 \\ 1 & -1 \end{pmatrix}, \begin{pmatrix} -1 & 1 \\ 0 & 1 \end{pmatrix}\rangle$$

of $\mathrm{PGL}(2,\mathbb{C})$ is isomorphic to $\mathfrak{S}_3$, so is solvable but is not triangularizable. The fifth section is dedicated to a sketch of the proof of the characterization of the solvable subgroups of the plane Cremona group ([**80, 172**]).

Let us recall a very old question, already asked in 1895 in [**92**]:

> "Tuttavia altre questioni d'indole gruppale relative al gruppo Cremona nel piano (ed a più forte ragione in S_n, $n > 2$) rimangono ancora insolute; ad esempio l'importante questione se il gruppo Cremona contenga alcun sottogruppo invariante (questione alla quale sembra probabile si debba rispondere negativamente)".

In 2013 Cantat and Lamy established that $\mathrm{Bir}(\mathbb{P}^2_\Bbbk)$ is not simple as soon as $\Bbbk$ is algebraically closed ([**59**]). Then in 2016 Lonjou proved that $\mathrm{Bir}(\mathbb{P}^2_\Bbbk)$ is not simple over any field ([**131**]). The sixth section is devoted to normal subgroups of $\mathrm{Bir}(\mathbb{P}^2_\mathbb{C})$ and the non-simplicity of $\mathrm{Bir}(\mathbb{P}^2_\mathbb{C})$. Strategies of [**59**] and [**131**] are evoked. A consequence of one result of [**131**] is the following property: the Cremona group contains infinitely many characteristic subgroups ([**53**]).

Taking the results of the sixth section as a starting point Urech gives a classification of all simple groups that act non-trivially by birational maps on complex compact Kähler surfaces. In particular he gets the two following statements:

THEOREM 8.2 ([**171**]). *A simple group* G *acts non-trivially by birational maps on a rational complex projective surface if and only if* G *is isomorphic to a subgroup of* $\mathrm{PGL}(3,\mathbb{C})$.

THEOREM 8.3 ([**171**]). *Let* G *be a simple subgroup of* $\mathrm{Bir}(\mathbb{P}^2_\mathbb{C})$. *Then*

- ◇ G *does not contain loxodromic elements;*
- ◇ *if* G *contains a parabolic element, then* G *is conjugate to a subgroup of* $\mathcal{J}$*;*
- ◇ *if* G *is an elliptic subgroup, then* G *is either a simple subgroup of an algebraic subgroup of* $\mathrm{Bir}(\mathbb{P}^2_\mathbb{C})$, *or conjugate to a subgroup of* G.

In the last section we give a sketch of the proof of these results.

8.1. A group of rank 1.5

8.1.1. Rank 2 phenomenon. Let $\Bbbk$ be a field. Consider a connected semisimple algebraic group G defined over $\Bbbk$. Let $\Psi\colon \mathrm{G} \to \mathrm{Aut}(\mathrm{G})$ be the mapping $g \mapsto \Psi_g$ where Ψ_g denotes the inner automorphism given by

$$\Psi_g\colon \mathrm{G} \to \mathrm{G}, \qquad\qquad h \mapsto ghg^{-1}.$$

For each g in G one can define Ad_g to be the derivative of Ψ_g at the origin

$$\mathrm{Ad}_g = (D\Psi_g)_{\mathrm{id}}\colon \mathfrak{g} \to \mathfrak{g}$$

where D is the differential and $\mathfrak{g} = T_{\mathrm{id}}\mathrm{G}$ is the tangent space of G at the identity element of G. The map

$$\mathrm{Ad}\colon \mathrm{G} \to \mathrm{Aut}(\mathfrak{g}), \qquad\qquad g \mapsto \mathrm{Ad}_g$$

is a group representation called the *adjoint representation* of G. The $\Bbbk$*-rank* of G is the maximal dimension of a connected algebraic subgroup of G which is diagonalizable over $\Bbbk$ in $\mathrm{GL}(\mathfrak{g})$. Such a maximal diagonalizable subgroup is a *maximal torus*.

THEOREM 8.4 ([**74**, **91**]). *Let $\mathbb{G}_m$ be the multiplicative group over $\mathbb{C}$. Let r be an integer.*

If $\mathbb{G}_m^r$ embeds as an algebraic subgroup in $\mathrm{Bir}(\mathbb{P}^n_\mathbb{C})$, then $r \leq n$. If $r = n$, then the embedding is conjugate to an embedding into the group of diagonal matrices in $\mathrm{PGL}(n+1, \mathbb{C})$.

REMARK 8.5. Theorem 8.4 not only holds for $\mathbb{C}$ but also for any algebraically closed field $\Bbbk$.

In other words the group of diagonal automorphisms D_n plays the role of a maximal torus in $\mathrm{Bir}(\mathbb{P}^n_\mathbb{C})$ and the Cremona group "looks like" a group of rank n.

Furthermore Beauville has shown a finite version of Theorem 8.4 in dimension 2:

THEOREM 8.6 ([**8**]). *Let $p \geq 5$ be a prime.*

If the abelian group $\left(\mathbb{Z}/p\mathbb{Z}\right)^r$ embeds into $\mathrm{Bir}(\mathbb{P}^2_\mathbb{C})$, then $r \leq 2$. Moreover if $r = 2$, then the image of $\left(\mathbb{Z}/p\mathbb{Z}\right)^r$ is conjugate to a subgroup of the group D_2 of diagonal automorphisms of $\mathbb{P}^2_\mathbb{C}$.

REMARK 8.7. This statement not only holds for $\mathbb{C}$ but also for any algebraically closed field $\Bbbk$.

Let us give an idea of the proof. Consider a finite group G of $\mathrm{Bir}(\mathbb{P}^2_\mathbb{C})$. It can be realized as a group of automorphisms of a rational surface S (*see for instance* [**70**]). Moreover one can assume that every birational G-equivariant morphism of S onto a smooth surface with a G-action is an isomorphism. Then according to [**136**]

- ◇ either G preserves a fibration $\pi\colon S \to \mathbb{P}^1$ with rational fibers,
- ◇ or $\mathrm{Pic}(S)^\mathrm{G}$ has rank 1.

In the first case G embeds in the group of automorphisms of the generic fibre $\mathbb{P}^1_{\mathbb{C}(t)}$ of π and Beauville classified the p-elementary subgroups of $\mathrm{Aut}(\mathbb{P}^1_{\mathbb{C}(t)})$.

In the last case S is a del Pezzo surface and the group $\mathrm{Aut}(S)$ is well known. Beauville also classified the p-elementary subgroups of such groups.

Combining this result of those recalled in Chapter 7, §7.5 Zhao gets:

THEOREM 8.8. *Let $\phi \in \mathrm{Bir}(\mathbb{P}^2_\mathbb{C})$ be an element of infinite order. If the centralizer of ϕ is not virtually abelian, then either ϕ is an elliptic map, or a power of ϕ is conjugate to an automorphism of $\mathbb{C}^2$ of the form $(z_0, z_1) \mapsto (z_0, z_1 + 1)$ or $(z_0, z_1) \mapsto (z_0, \beta z_1)$ with $\beta \in \mathbb{C}^*$.*

REMARK 8.9. This statement also holds for $\mathrm{Bir}(\mathbb{P}^2_\Bbbk)$ where $\Bbbk$ is an algebraically closed field ([**184**]).

8.1.2. Rank 1 phenomenon. Generic elements of degree ≥ 2 of $\mathrm{Bir}(\mathbb{P}^2_\mathbb{C})$ are loxodromic and hence cannot be conjugate to elements of the maximal torus D_2. The description of their centralizer is given by:

THEOREM 8.10 ([**31**, **52**]). *Let ϕ be a loxodromic element of $\mathrm{Bir}(\mathbb{P}^2_\mathbb{C})$.*

The infinite cyclic subgroup of $\mathrm{Bir}(\mathbb{P}^2_\mathbb{C})$ generated by ϕ has finite index in the centralizer

$$\mathrm{Cent}(\phi) = \left\{\psi \in \mathrm{Bir}(\mathbb{P}^2_\mathbb{C}) \,|\, \psi \circ \phi = \phi \circ \psi\right\}$$

of ϕ.

REMARK 8.11. Theorem 8.10 holds for any field $\Bbbk$.

The centralizer of a generic element of $\mathrm{SL}(n+1,\mathbb{C})$ is isomorphic to $(\mathbb{C}^*)^n$; Theorem 8.10 suggests that $\mathrm{Bir}(\mathbb{P}^2_{\mathbb{C}})$ behaves as a group of rank 1.

SKETCH OF THE PROOF. If ψ commutes to ϕ, then the isometry ψ_* of $\mathbb{H}^\infty$ preserves the axis $\mathrm{Ax}(\phi)$ of ϕ_* and its two endpoints. Consider the morphism Θ which maps $\mathrm{Cent}(\phi)$ to the group of isometries of $\mathrm{Ax}(\phi)$. View it as a morphism into the group of translations $\mathbb{R}$ of the line. On the one hand the translation lengths are bounded from below by $\log(\lambda_L)$ where λ_L is the Lehmer number, *i.e.*, the unique root >1 of the irreducible polynomial $x^{10}+x^9-x^7-x^6-x^5-x^4-x^3+x+1$ (*see* [**31**]). On the other hand every discrete subgroup of $\mathbb{R}$ is trivial or cyclic. As a result the image of Θ is a cyclic group. Its kernel is made of elliptic elements of $\mathrm{Cent}(\phi)$ fixing all points of $\mathrm{Ax}(\phi)$. Denote by $\mathbf{e}_\phi$ the projection of $\mathbf{e}_0$ on $\mathrm{Ax}(\phi)$. Since $\ker\Theta$ fixes $\mathbf{e}_\phi$, the inequality

$$\mathrm{dist}(\psi_*\mathbf{e}_0,\mathbf{e}_0)\leq 2\mathrm{dist}(\mathbf{e}_0,\mathbf{e}_\phi)$$

holds. As a consequence $\ker\Theta$ is a group of birational maps of bounded degree. From [**34**] the Zariski closure of $\ker\Theta$ in $\mathrm{Bir}(\mathbb{P}^2_{\mathbb{C}})$ is an algebraic subgroup of $\mathrm{Bir}(\mathbb{P}^2_{\mathbb{C}})$. Let us denote by G the connected component of the identity in this group. Assume that $\ker\Theta$ is infinite. Then $\dim\mathrm{G}$ is positive and G is contained, after conjugacy, in the group of automorphisms of a minimal, rational surface ([**26, 91**]). Therefore, G contains a Zariski closed abelian subgroup whose orbits have dimension 1. Those orbits are organised in a pencil of curves that is invariant under the action of ϕ: contradiction with the fact that ϕ_* is loxodromic. As a result $\ker\Theta$ is finite. □

8.1.3. Rank 1 phenomenon. To generalize Margulis's work on linear representations of lattices of simple real Lie groups to non-linear representations, Zimmer proposed to study the actions of lattices on compact varieties ([**185–188**]). One of the main conjectures of the program drawn by Zimmer is: let G be a connex, simple, real Lie group and let Γ be a lattice of G. If there exists a morphism from Γ into the diffeomorphisms group of a compact variety V with infinite image, then the real rank of G is less than or equal to the dimension of V.

In the context of birational self maps one has the following statement that can be seen as another rank one phenomenon:

THEOREM 8.12 ([**52, 76**]). *Let S be a complex projective surface. Let Γ be a countable group with Kazhdan property (T).*

If $\upsilon\colon\Gamma\to\mathrm{Bir}(\mathbb{P}^2_{\mathbb{C}})$ is a morphism with infinite image, then υ is conjugate to a morphism into $\mathrm{PGL}(3,\mathbb{C})$.

REMARK 8.13. Theorem 8.12 indeed holds for any algebraically closed field $\mathbb{k}$.

SKETCH OF THE PROOF. The first step is based on a fixed point property: since Γ has Kazhdan property (T), then $\upsilon(\Gamma)$ acts by isometries on $\mathbb{H}^\infty$ and $(\upsilon(\Gamma))_*$ has a fixed point. Then according to [**72**] all its orbits have bounded diameter. Hence $\rho(\Gamma)$ has bounded degree. There thus exists a birational map $\pi\colon X\dashrightarrow\mathbb{P}^2_{\mathbb{C}}$ such that

- ◇ $\Gamma_S=\pi^{-1}\circ\Gamma\circ\pi$ is a subgroup of $\mathrm{Aut}(S)$;
- ◇ $\mathrm{Aut}(S)^0\cap\Gamma_S$ has finite index in Γ_S.

The classification of algebraic groups of maps of surfaces and the fact that every subgroup of $\mathrm{SL}(2,\mathbb{C})$ having Kazhdan property (T) is finite allow us to prove that:

since $\mathrm{Aut}(S)^0$ contains an infinite group with Kazhdan property (T) the surface S is isomorphic to the projective plane $\mathbb{P}^2_\mathbb{C}$. □

8.2. Subgroups of elliptic elements of $\mathrm{Bir}(\mathbb{P}^2_\mathbb{C})$

A subgroup G of the plane Cremona group is *elliptic* if any element of G is an elliptic birational map.

Let us give an example: a bounded subgroup of $\mathrm{Bir}(\mathbb{P}^2_\mathbb{C})$ is elliptic. But not all elliptic subgroups are bounded; indeed for instance

◇ all elements of $\{(z_0, z_1 + a(z_0)) \,|\, a \in \mathbb{C}(z_0)\}$ are elliptic but $\{(z_0, z_1 + a(z_0)) \,|\, a \in \mathbb{C}(z_0)\}$ contains elements of arbitrarily high degrees;

◇ Wright gives examples of subgroups of $\mathrm{Bir}(\mathbb{P}^2_\mathbb{C})$ isomorphic to a subgroup of roots of unity of $\mathbb{C}^*$ that are not bounded ([**179**]). Let us be more precise. Set $\psi_0 \colon (z_0, z_1) \mapsto (-z_0, -z_1)$ and for any $k \geq 1$,

$$\alpha_k = \exp\left(\frac{\mathbf{i}\pi}{2^k}\right), \quad \phi_k \colon (z_0, z_1) \mapsto (z_1, c_k z_1^{2^k+1} + z_0), \quad \varphi_k = \phi_k^2 \circ \phi_{k-1}^2 \circ \ldots \circ \phi_1^2$$

where c_k denotes an element of $\mathbb{C}^*$. Consider

$$\psi_k = \varphi_k^{-1} \circ \big((z_0, z_1) \mapsto (\alpha_k z_0, \alpha_k^p z_1)\big) \circ \varphi_k$$

where p is an odd integer. The group

$$\mathrm{G} = \bigcup_{k \geq 0} \langle \psi_k \rangle$$

is an abelian group obtained as a growing union of finite cyclic groups that does not preserve any fibration ([**127**]).

This gives all the possibilities for elliptic subgroups of $\mathrm{Bir}(\mathbb{P}^2_\mathbb{C})$:

THEOREM 8.14 ([**172**]). *Let* G *be an elliptic subgroup of the plane Cremona group. Then one of the following holds:*

◇ G *is a bounded subgroup;*

◇ G *preserves a rational fibration;*

◇ G *is a torsion group.*

Furthermore he characterizes torsion subgroups of $\mathrm{Bir}(\mathbb{P}^2_\mathbb{C})$:

THEOREM 8.15 ([**172**]). *Let* $\mathrm{G} \subset \mathrm{Bir}(\mathbb{P}^2_\mathbb{C})$ *be a torsion group. Then* G *is isomorphic to a bounded subgroup of* $\mathrm{Bir}(\mathbb{P}^2_\mathbb{C})$.

Furthermore G *is isomorphic to a subgroup of* $\mathrm{GL}(48, \mathbb{C})$.

As a consequence he gets an analogue of the Theorem of Jordan and Schur:

COROLLARY 8.16 ([**172**]). *There exists a constant* γ *such that every torsion subgroup of* $\mathrm{Bir}(\mathbb{P}^2_\mathbb{C})$ *contains a commutative normal subgroup of index* $\leq \gamma$.

Theorems 8.14 and 8.15 allow us to refine

◇ the result of Cantat about the Tits alternative (§8.4);

◇ the description of solvable subgroups of $\mathrm{Bir}(\mathbb{P}^2_\mathbb{C})$ (*see* §8.5).

The aim of the section is to prove Theorem 8.14. We need the following technical lemmas.

LEMMA 8.17 ([**61**]). *Let* ϕ *be a birational self map of the complex projective plane that fixes pointwise two different rational fibrations. Then* ϕ *is of finite order.*

PROOF. The intersections of the generic fibres of these two fibrations are finite, uniformly bounded. But these intersections are invariant by ϕ so ϕ is of finite order. □

LEMMA 8.18 ([**172**]). *An algebraic subgroup* G *of* $\mathrm{Bir}(\mathbb{P}^2_\mathbb{C})$ *of dimension* ≤ 9 *preserves a unique rational fibration.*

PROOF. According to Theorem 3.46 the group G is conjugate to a subgroup of $\mathrm{Aut}(\mathbb{F}_n)$ for some Hirzebruch surface $\mathbb{F}_n$, $n \geq 2$. As a consequence G preserves a rational fibration $\pi\colon \mathbb{F}_n \to \mathbb{P}^1_\mathbb{C}$. The fibres of π are permuted by G; this yields to a homomorphism

$$f\colon \mathrm{G} \to \mathrm{PGL}(2,\mathbb{C})$$

such that $\dim \ker f \geq 6$.

Assume by contradiction that there exists a second rational fibration $\pi'\colon \mathbb{F}_n \to \mathbb{P}^1_\mathbb{C}$ preserved by G; this yields to a second homomorphism

$$g\colon \mathrm{G} \to \mathrm{PGL}(2,\mathbb{C}).$$

One has $\dim \ker \pi'_{|\ker \pi} > 0$; therefore, $\dim(\ker f \cap \ker g) > 0$. In particular $\ker f \cap \ker g$ contains an element of infinite order: contradiction with Lemma 8.17. □

LEMMA 8.19 ([**172**]). *Let* $\mathrm{G} \subset \mathrm{Bir}(\mathbb{P}^2_\mathbb{C})$ *be an algebraic subgroup isomorphic as an algebraic group to* $\mathbb{C}^*$.

There exists a constant $K(\mathrm{G})$ *such that any elliptic element of*

$$\mathrm{Cent}(\mathrm{G}) = \big\{\varphi \in \mathrm{Bir}(\mathbb{P}^2_\mathbb{C}) \,|\, \varphi\circ\psi = \psi\circ\varphi \quad \forall \psi \in \mathrm{G}\big\}$$

has degree $\leq K(\mathrm{G})$.

PROOF. Up to conjugacy by an element $\psi \in \mathrm{Bir}(\mathbb{P}^2_\mathbb{C})$ one can assume that

$$\mathrm{G} = \big\{(z_0,z_1) \mapsto (\alpha z_0, z_1) \,|\, \alpha \in \mathbb{C}^*\big\}.$$

An elliptic element of Cent(G) is of the following form:

$$\varphi\colon (z_0,z_1) \dashrightarrow (z_0\varphi_1(z_1), \varphi_2(z_1))$$

where φ_1, φ_2 are rational functions. Since $(\deg \varphi^n)_n$ is bounded, φ_1 is constant, and so $\varphi_2\colon z_1 \mapsto \frac{az_1+b}{cz_1+d}$ for some matrix $\begin{pmatrix} a & b \\ c & d \end{pmatrix}$ of $\mathrm{PGL}(2,\mathbb{C})$. In particular $\deg \varphi \leq 2$. The constant $K(\mathrm{G})$ thus only depends on the degree of ψ. □

LEMMA 8.20 ([**172**]). *A group* $\mathrm{G} \subset \mathrm{Bir}(\mathbb{P}^2_\mathbb{C})$ *of monomial elliptic elements is bounded.*

PROOF. The group G is contained in $\mathrm{GL}(2,\mathbb{Z}) \ltimes (\mathbb{C}^*)^2$. Consider the projection $\pi\colon \mathrm{G} \to \mathrm{GL}(2,\mathbb{Z})$. On the one hand $\ker \pi$ is bounded, on the other hand all elements of $\pi(\mathrm{G})$ are bounded. All elliptic elements in $\mathrm{GL}(2,\mathbb{Z}) \subset \mathrm{Bir}(\mathbb{P}^2_\mathbb{C})$ are of finite order, so $\pi(\mathrm{G})$ is a torsion subgroup of $\mathrm{GL}(2,\mathbb{Z})$. Since there are only finitely many conjugacy classes of finite subgroups in $\mathrm{GL}(2,\mathbb{Z})$ the group $\pi(\mathrm{G})$ is finite. Therefore, G is a finite extension of a bounded subgroup; hence G is bounded. □

LEMMA 8.21 ([**172**]). *Let* H *be a semi-simple algebraic subgroup of* $\mathrm{Bir}(\mathbb{P}^2_\mathbb{C})$. *Let* $\mathrm{G} \subset \mathrm{Bir}(\mathbb{P}^2_\mathbb{C})$ *be a group of elliptic elements that normalizes* H. *Then* G *is bounded.*

PROOF. The group H is semi-simple; in particular its group of inner automorphisms has finite index in its group of algebraic automorphisms. As a result there exists $N \in \mathbb{Z}$ such that for any ϕ in G, conjugation by ϕ^N induces an inner automorphism of H. Hence, there exists an element ψ in H such that $\phi^N \circ \psi$ centralizes H. By assumption H is semi-simple, so H contains a closed subgroup D isomorphic as an algebraic group to $\mathbb{C}^*$ and this group is centralized by $\phi^N \circ \psi$. From Lemma 8.19 we get that $\deg(\phi^N \circ \psi)$ is bounded by a constant that depends neither on ϕ, nor on N. As H is an algebraic group, both $\deg \psi$ and $\deg \phi$ are also bounded independently of ϕ and N. Finally G is bounded. □

LEMMA 8.22 ([**172**]). *Let* G *be a subgroup of* $\mathrm{Bir}(\mathbb{P}^2_\mathbb{C})$ *that fixes a point of* $\mathbb{H}^\infty$. *Then*

- ◇ *the degree of all elements in* G *is uniformly bounded;*
- ◇ *there exist a smooth projective surface* S *and a birational map* $\varphi\colon \mathbb{P}^2_\mathbb{C} \dashrightarrow S$ *such that* $\varphi \circ \mathrm{G} \circ \varphi^{-1} \subset \mathrm{Aut}(S)$.

PROOF. Denote by $p \in \mathbb{H}^\infty$ the fixed point of G, and by $\mathbf{e}_0 \in \mathbb{H}^\infty$ the class of a line in $\mathbb{P}^2_\mathbb{C}$. Take an element ψ of G. The action of G on $\mathbb{H}^\infty$ is isometric; hence $d(\psi(\mathbf{e}_0), p) = d(\mathbf{e}_0, p)$, and so $d(\psi(\mathbf{e}_0), p) \leq 2d(\mathbf{e}_0, p)$. This implies

$$\langle \psi(\mathbf{e}_0), \mathbf{e}_0 \rangle \leq \cosh(2d(\mathbf{e}_0, p)) \qquad \forall\, \psi \in \mathrm{G}.$$

Since $\langle \psi(\mathbf{e}_0), \mathbf{e}_0 \rangle = \deg \psi$ the previous inequality can be rewritten as follows:

$$\deg \psi \leq \cosh(2d(\mathbf{e}_0, p)) \qquad \forall\, \psi \in \mathrm{G},$$

i.e., the degrees of all elements in G are uniformly bounded.

According to Weil, G can be regularized (§3.5). □

Let us recall the following statement due to Cantat:

PROPOSITION 8.23 ([**52**]). *Let* Γ *be a finitely generated subgroup of* $\mathrm{Bir}(\mathbb{P}^2_\mathbb{C})$ *of elliptic elements. Then*

- ◇ *either* Γ *is bounded,*
- ◇ *or* Γ *preserves a rational fibration, i.e.,* $\Gamma \subset \mathcal{J}$ *up to birational conjugacy.*

LEMMA 8.24 ([**172**]). *Let* $\mathrm{G} \subset \mathrm{Bir}(\mathbb{P}^2_\mathbb{C})$ *be a group of elliptic elements. Then one of the following holds:*

- ◇ G *preserves a fibration, and so up to birational conjugacy either* $\mathrm{G} \subset \mathcal{J}$, *or* $\mathrm{G} \subset \mathrm{Aut}(S)$ *where* S *is a Halphen surface;*
- ◇ *every finitely generated subgroup of* G *is bounded.*

Furthermore if G *fixes a point* $p \in \partial\mathbb{H}^\infty$ *that does not correspond to the class of a rational fibration, then the second assertion holds.*

PROOF. The group G fixes a point $p \in \mathbb{H}^\infty \cup \partial\mathbb{H}^\infty$ (Theorem 8.44).

If p belongs to $\mathbb{H}^\infty$, then G is bounded.

Let us now assume that p belongs to $\partial\mathbb{H}^\infty$. Then either p corresponds to the class of a general fibre of some fibration, or not.

- ◇ If p corresponds to the class of a general fibre of some fibration $\pi\colon Y \to \mathbb{P}^1_\mathbb{C}$ where Y is a rational surface, then G preserves this fibration and is thus conjugate to a subgroup of $\mathcal{J}$ (if the fibration is rational), or to a subgroup of $\mathrm{Aut}(S)$ where S is a Halphen surface (if the fibration consists of curves of genus 1).

- Suppose now that p does not correspond to the class of a fibration. Let Γ be a finitely generated subgroup of G. Then either Γ is bounded, or Γ preserves a rational fibration (Proposition 8.23). If Γ preserves a rational fibration $\mathcal{F}$, then Γ fixes a point $q\in\partial\mathbb{H}^\infty$ that corresponds to the class of $\mathcal{F}$. Hence p and q are two distinct points preserved by G and G fixes the geodesic line through p and q. In particular G fixes a point in $\mathbb{H}^\infty$ and according to Lemma 8.22 the degrees of all elements in G are uniformly bounded. □

Proof of Theorem 8.14. Consider a subgroup G of $\mathrm{Bir}(\mathbb{P}^2_{\mathbb{C}})$ of elliptic elements. According to Lemma 8.22 either G preserves a rational fibration, or any finitely generated subgroup of G is bounded.

Assume that any finitely generated subgroup of G is bounded. Set

$$n:=\sup\big\{\dim\overline{\Gamma}\,|\,\Gamma\subset \mathrm{G}\text{ finitely generated}\big\}.$$

- If $n=0$, then G is a torsion group.
- If $n=+\infty$, then take Γ a finitely generated subgroup of G such that $\dim\overline{\Gamma}\geq 9$. By Lemma 8.18 the group Γ preserves a unique fibration and this fibration is, again by Lemma 8.18, preserved as well by $\langle\Gamma,\,\phi\rangle$ for any ϕ in G.
- Assume now $n\in\mathbb{N}^*$. Let Γ be a finitely generated subgroup of G such that $\dim\overline{\Gamma}\ =\ n$. Let $\overline{\Gamma}^0$ be the neutral component of Γ. For any $\varphi\in\mathrm{G}$ the group $\langle\overline{\Gamma}^0,\varphi\circ\overline{\Gamma}^0\circ\varphi^{-1}\rangle$ is connected and contained in $\overline{\langle\Gamma,\varphi\circ\Gamma\circ\varphi^{-1}\rangle}$ which is finitely generated and thus of dimension less or equal to n. As a consequence $\langle\overline{\Gamma}^0,\varphi\circ\overline{\Gamma}^0\circ\varphi^{-1}\rangle=\overline{\Gamma}^0$ for any $\varphi\in\mathrm{G}$ and $\overline{\Gamma}^0$ is normalized by G. If $\overline{\Gamma}^0$ is semi-simple, Lemma 8.21 allows us to conclude. Assume that $\overline{\Gamma}^0$ is not semi-simple. Denote by R the radical of $\overline{\Gamma}^0$, *i.e.*, R is the maximal connected normal solvable subgroup of $\overline{\Gamma}^0$. Since $\overline{\Gamma}^0$ is semi-simple the inequality $\dim R>0$ holds. The radical is unique; hence preserved by $\mathrm{Aut}(\overline{\Gamma}^0)$ and in particular normalized by G. Denote by

$$R^{(\ell+1)}=\{\mathrm{id}\}\subsetneq R^{(\ell)}\subset\ldots\subset R^{(2)}\subset R^{(1)}\subset R^{(0)}=R$$

the derived series of R (*i.e.*, $R^{(k+1)}=[R^{(k)},R^{(k)}]$). Note that $\dim R^{(\ell)}>0$ and $R^{(\ell)}$ is abelian. This series is invariant under $\mathrm{Aut}(\overline{\Gamma}^0)$, and so invariant under conjugation by elements of G. In particular G normalizes $R^{(\ell)}$. Since $R^{(\ell)}$ is bounded, $R^{(\ell)}$ is conjugate to one of the groups of Theorem 3.46; in particular $R^{(\ell)}$ can be regularized. In other words, up to birational conjugacy, G is a subgroup of $\mathrm{Bir}(S)$ for some smooth projective surface S on which $R^{(\ell)}$ acts regularly. If all the orbits of $R^{(\ell)}$ have dimension ≤ 1, then G preserves a rational fibration. Assume that $R^{(\ell)}$ has an open orbit $\mathcal{O}$. The group G normalizes $R^{(\ell)}$, so G acts regularly on $\mathcal{O}$. The action of $R^{(\ell)}$ is faithful; as a result $\dim R^{(\ell)}=1$ and $R^{(\ell)}\simeq\mathbb{C}^2$, or $R^{(\ell)}\simeq\mathbb{C}^*\times\mathbb{C}$, or $R^{(\ell)}\simeq\mathbb{C}^*\times\mathbb{C}^*$. If $R^{(\ell)}\simeq\mathbb{C}^2$, then $\mathcal{O}$ is isomorphic to the affine plane, and the action of $R^{(\ell)}$ on $\mathcal{O}$ is given by translations. But the normalizer of $\mathbb{C}^2$ in $\mathrm{Aut}(\mathbb{A}^2_{\mathbb{C}})$ is the group of affine maps $\mathrm{GL}(2,\mathbb{C})\ltimes\mathbb{C}^2$; hence G is bounded. If $R^{(\ell)}\simeq\mathbb{C}^*\times\mathbb{C}$, then we similarly get the inclusion $\mathrm{G}\subset\mathrm{Aut}(\mathbb{C}^*\times\mathbb{C})$. The $\mathbb{C}$-fibration of $\mathbb{C}^*\times\mathbb{C}$ is given by the invertible functions; it is thus preserved by $\mathrm{Aut}(\mathbb{C}^*\times\mathbb{C})$. In particular G preserves a rational fibration. If $R^{(\ell)} \simeq \mathbb{C}^* \times \mathbb{C}^*$, then elements of G are monomial maps, and Lemma 8.20 allows us to conclude. □

8.3. Torsion subgroups of the Cremona group

As we have seen at the beginning of §8.2 some torsion groups can be embedded into $\mathrm{Bir}(\mathbb{P}^2_{\mathbb{C}})$ in such a way that they neither are bounded, nor preserve any fibration. However the group structure of torsion subgroups can be specified:

THEOREM 8.25 ([**172**]). *A torsion subgroup* G *of* $\mathrm{Bir}(\mathbb{P}^2_{\mathbb{C}})$ *is isomorphic to a bounded subgroup of* $\mathrm{Bir}(\mathbb{P}^2_{\mathbb{C}})$.

Furthermore G *is isomorphic to a subgroup of* $\mathrm{GL}(48, \mathbb{C})$.

Malcev used model theory to prove that if for a given group G every finitely generated subgroup can be embedded into $\mathrm{GL}(n, \Bbbk)$ for some field $\Bbbk$, then there exists a field $\Bbbk'$ such that G can be embedded into $\mathrm{GL}(n, \Bbbk')$. Let us briefly introduce the compactness theory from model theory; it states that a set of first-order sentences has a model if and only if any of its finite subsets has a model.

DEFINITION. Let $\{x_i\}_{i\in I}$ be a set of variables. A *condition* is an expression of the form

$$P(x_{i_1}, x_{i_2}, \ldots, x_{i_k}) = 0$$

or an expression of the form

$$\big(P_1(x_{i_1}, x_{i_2}, \ldots, x_{i_k}) \neq 0\big) \vee \big(P_2(x_{i_1}, x_{i_2}, \ldots, x_{i_k}) \neq 0\big) \vee \ldots \\ \vee \big(P_\ell(x_{i_1}, x_{i_2}, \ldots, x_{i_k}) \neq 0\big)$$

where P and the P_i's are polynomials with integer coefficients.

DEFINITION. A *mixed system* is a set of conditions.

DEFINITION. A mixed system S is *compatible* if there exists a field $\Bbbk$ which contains values $\{y_i\}_{i\in I}$ that satisfy S.

THEOREM 8.26 ([**135**]). *If every finite subset of a mixed system S is compatible, then S is compatible.*

Let us now explain the proof of Theorem 8.25. Let G be a torsion subgroup of $\mathrm{Bir}(\mathbb{P}^2_{\mathbb{C}})$. If G is finite, then G is bounded; we can thus assume that G is infinite. Following Theorem 3.46 we will deal with different cases.

- ◇ First assume that every finitely generated subgroup of G is isomorphic to a subgroup of $\mathrm{PGL}(3, \mathbb{C})$. Consider the closed embedding ρ of $\mathrm{PGL}(3, \mathbb{C})$ into $\mathrm{GL}(8, \mathbb{C})$ given by the adjoint representation. Let P_1, P_2, $\ldots$, P_n be polynomials in the set of variables $\{x_{ij}\}_{1\leq i,\, j\leq 8}$ such that $\rho(\mathrm{PGL}(3, \mathbb{C})) \subset \mathrm{GL}(8, \mathbb{C})$ is the zero set of P_1, P_2, $\ldots$, P_n. To any element $g \in$ G we associate an 8×8 matrix of variables (x^g_{ij}). Consider the following mixed system S defined by
  (1) the equations $(x^f_{ij})(x^g_{ij}) = (x^h_{ij})$ for all f, g, $h \in$ G such that $f \circ g = h$;
  (2) the conditions $\big(\bigvee_i x^g_{ii} - 1 \neq 0\big) \vee \big(\bigvee_{i\neq j} x^g_{ij} - 1 \neq 0\big)$;
  (3) $x^{\mathrm{id}}_{ii} - 1 = 0$ and $x^{\mathrm{id}}_{ij} = 0$ for all $1 \leq i \neq j \leq N$;
  (4) $P_k(\{x_{ij}\}) = 0$ for all $1 \leq k \leq n$, for all $g \in$ G;
  (5) $p \neq 0$ for all $p \in \mathbb{Z}^+$ primes.

LEMMA 8.27 ([**172**]). *The system S is compatible.*

PROOF. According to Theorem 8.26 it suffices to show that every finite subset of S is compatible. Let c_1, c_2, $\ldots$, $c_n \in S$ be finitely many conditions. Only finitely many variables x^g_{ij} appear in c_1, c_2, $\ldots$, c_n. Let $\{g_1, g_2, \ldots, g_\ell\} \subset \mathrm{G}$ be the finite set of all elements $g \in \mathrm{G}$ such that for some $1 \leq i, j \leq 8$ the variable x^g_{ij} appears in one of the conditions c_1, c_2, $\ldots$, c_n.

Consider the finitely generated subgroup $\Gamma = \langle g_1, g_2, \ldots, g_\ell\rangle$ of G. By Theorem 8.48 the group Γ is finite. Therefore, by assumption, Γ has a faithful representation to $\mathrm{PGL}(3,\mathbb{C})$. This representation implies that $\mathbb{C}$ contains values that satisfy the conditions c_1, c_2, $\ldots$, c_n. In other words S is compatible. □

As a result there exists a field $\Bbbk$ such that $\Bbbk$ contains values y^g_{ij} for all $1 \leq i, j \leq 8$ and all $g \in \mathrm{G}$ satisfying conditions (1) to (5). Condition (5) asserts that the characteristic of $\Bbbk$ is 0. The group G has at most the cardinality of the continuum since $\mathrm{G} \subset \mathrm{Bir}(\mathbb{P}^2_\mathbb{C})$; the values $\{y^g_{ij}\}$ are thus contained in a subfield $\Bbbk'$ of $\Bbbk$ that has the same cardinality as $\mathbb{C}$. Hence $\Bbbk'$ can be embedded into $\mathbb{C}$ as a subfield. Hence we may suppose that $\Bbbk = \mathbb{C}$. Consider the map

$$\varphi\colon \mathrm{G} \to \mathrm{PGL}(3,\mathbb{C}), \qquad\qquad g \mapsto (y^g_{ij})_{i,j}.$$

Note that

- conditions (1) imply that the image of any element of G is an invertible matrix and that φ is a group automorphism;
- conditions (2) show that this automorphism is injective;
- conditions (3) imply $\varphi(\mathrm{id}) = \mathrm{id}$;
- conditions (4) show that $\varphi(\mathrm{G}) \subset \mathrm{PGL}(3,\mathbb{C}) \subset \mathrm{GL}(8,\mathbb{C})$.

◇ Denote by S_6 the del Pezzo surface of degree 6. If any finitely generated subgroup of G can be embedded into $\mathrm{Aut}(S_6) \simeq \mathrm{D}_2 \rtimes \left(\mathbb{Z}/2\mathbb{Z} \times \mathfrak{S}_3\right)$, a similar reasoning leads to: G is isomorphic to a subgroup of $\mathrm{Aut}(S_6)$.

◇ If any finitely generated subgroup of G can be embedded into

$$\mathrm{Aut}(\mathbb{P}^1_\mathbb{C} \times \mathbb{P}^1_\mathbb{C}) \simeq \big(\mathrm{PGL}(2,\mathbb{C}) \times \mathrm{PGL}(2,\mathbb{C})\big) \rtimes \mathbb{Z}/2\mathbb{Z},$$

then G is isomorphic to a subgroup of $\mathrm{Aut}(\mathbb{P}^1_\mathbb{C} \times \mathbb{P}^1_\mathbb{C})$.

◇ If any finitely generated subgroup of G can be embedded into

$$\mathrm{Aut}(\mathbb{F}_{2n}) \simeq \mathbb{C}[z_0,z_1]_{2n} \rtimes \mathrm{GL}(2,\mathbb{C})/\mu_{2n}$$

for some $n > 0$ (and not necessarily the same for all finitely generated subgroups of G), then G is isomorphic to a subgroup of $\mathrm{GL}(2,\mathbb{C})$ and thus can be embedded in $\mathrm{PGL}(3,\mathbb{C})$.

◇ It remains to consider the case where G contains

- a finitely generated subgroup Γ_1 that cannot be embedded into $\mathrm{Aut}(\mathbb{P}^2_\mathbb{C})$,
- a finitely generated subgroup Γ_2 that cannot be embedded into $\mathrm{Aut}(S_6)$,
- a finitely generated subgroup Γ_3 that cannot be embedded into $\mathrm{Aut}(\mathbb{P}^1_\mathbb{C} \times \mathbb{P}^1_\mathbb{C})$,
- a finitely generated subgroup Γ_4 that cannot be embedded into $\mathrm{Aut}(\mathbb{F}_{2n})$ for all $n > 0$.

The finitely generated subgroup $\Gamma = \langle \Gamma_1, \Gamma_2, \Gamma_3, \Gamma_4\rangle$ is not isomorphic to any subgroup of infinite automorphism group of a del Pezzo surface. Adding finitely many elements if needed we may assume that Γ has order > 648; as a consequence Γ is isomorphic neither to any subgroup of an automorphism group of a del Pezzo surface (Theorem 3.39), nor to a subgroup of $\mathrm{Aut}(\mathbb{F}_{2n})$ for all $n > 0$. Consider a finitely generated subgroup H of G. The finitely generated subgroup $\langle \Gamma, \mathrm{H}\rangle$, and in particular H, is isomorphic to a subgroup of (Theorem 3.46):

- either $\mathrm{Aut}(S,\pi)$ where $\pi\colon S \to \mathbb{P}^1_{\mathbb{C}}$ is an exceptional conic bundle,
- or $\mathrm{Aut}(S,\pi)$ where (S,π) is a $\left(\mathbb{Z}/2\mathbb{Z}\right)^2$-conic bundle and S is not a del Pezzo surface,
- or $\mathrm{Aut}(\mathbb{F}_{2n+1})$ for some $n > 0$.

According to Lemmas 3.42, 3.43 and 3.44 the group H is isomorphic to a subgroup of $\mathrm{PGL}(2,\mathbb{C}) \times \mathrm{PGL}(2,\mathbb{C})$. Therefore, every finitely generated subgroup of G is isomorphic to a subgroup of $\mathrm{PGL}(2,\mathbb{C}) \times \mathrm{PGL}(2,\mathbb{C})$. The group G is thus isomorphic to a subgroup of $\mathrm{PGL}(2,\mathbb{C}) \times \mathrm{PGL}(2,\mathbb{C})$ (Theorem 8.26) and hence to a subgroup of $\mathrm{Aut}(\mathbb{P}^1_{\mathbb{C}} \times \mathbb{P}^1_{\mathbb{C}})$.

LEMMA 8.28 ([**172**]). *Every torsion subgroup of* $\mathrm{Bir}(\mathbb{P}^2_{\mathbb{C}})$ *is isomorphic to a subgroup of* $\mathrm{GL}(48,\mathbb{C})$.

PROOF. Let G be a torsion group of $\mathrm{Bir}(\mathbb{P}^2_{\mathbb{C}})$.

◇ Assume that G is infinite. As we just see, G is isomorphic to a subgroup of $\mathrm{Aut}(\mathbb{P}^2_{\mathbb{C}})$, $\mathrm{Aut}(\mathbb{P}^1_{\mathbb{C}} \times \mathbb{P}^1_{\mathbb{C}})$, $\mathrm{Aut}(S_6)$ or $\mathrm{Aut}(\mathbb{F}_n)$ for some $n \geq 2$. According to the structure of $\mathrm{Aut}(\mathbb{F}_n)$ and Lemma 3.42 all torsion subgroups of $\mathrm{Aut}(\mathbb{F}_n)$ are isomorphic to a subgroup of $\mathrm{GL}(2,\mathbb{C})$ or $\mathrm{PGL}(2,\mathbb{C}) \times \mathbb{C}^*$. But $\mathrm{PGL}(2,\mathbb{C})$ can be embedded into $\mathrm{GL}(3,\mathbb{C})$ and $\mathrm{PGL}(3,\mathbb{C})$ into $\mathrm{GL}(8,\mathbb{C})$, and $\mathrm{Aut}(S_6)$ into $\mathrm{GL}(6,\mathbb{C})$ (Lemma 3.41); the group G is thus isomorphic to a subgroup of $\mathrm{GL}(8,\mathbb{C})$.

◇ Suppose that G is finite and not contained in an infinite bounded subgroup. Then G is contained in the automorphism group (Theorem 3.46)
- either of a del Pezzo surface,
- or of an exceptional fibration,
- or of a $(\mathbb{Z}/2\mathbb{Z})^2$-fibration.

In the first case we get from Lemma 3.47 that G is isomorphic to a subgroup of $\mathrm{GL}(8,\mathbb{C})$.

In the second case the group G can be embedded into $\mathrm{PGL}(2,\mathbb{C}) \times \mathrm{PGL}(2,\mathbb{C})$ (Lemma 3.43).

In the last case G is isomorphic to a subgroup of $\mathrm{GL}(48,\mathbb{C})$ according to [**170**, Lemma 6.2.12]. □

8.4. Tits alternative and Burnside problem

A group G is *virtually solvable* if G contains a finite index solvable subgroup.

A group G *satisfies the Tits alternative* if every subgroup of G either is virtually solvable or contains a non-abelian free subgroup.

A group G *satisfies the Tits alternative for finitely generated subgroups* if every finitely generated subgroup of G either is virtually solvable or contains a non-abelian free subgroup.

Tits showed that linear groups over fields of characteristic zero satisfy the Tits alternative and that linear groups over fields of positive characteristic satisfy the Tits alternative for finitely generated subgroups ([**165**]). Other well-known examples of groups that satisfy the Tits alternative include mapping class groups of surfaces ([**117**]), the outer automorphisms group of the free group of finite rank n ([**17**]), or hyperbolic groups in the sense of Gromov ([**105**]). Lamy studied the group $\mathrm{Aut}(\mathbb{A}^2_{\mathbb{C}})$; in particular using its amalgamated product structure he showed that the Tits alternative holds for $\mathrm{Aut}(\mathbb{A}^2_{\mathbb{C}})$ (*see* [**128**]). In [**52**] Cantat established that $\mathrm{Bir}(\mathbb{P}^2_{\mathbb{C}})$ satisfies the Tits alternative for finitely generated subgroups. Then Urech proved that $\mathrm{Bir}(\mathbb{P}^2_{\mathbb{C}})$ satisfies the Tits alternative ([**171**]).

On the contrary, the group of $\mathcal{C}^\infty$-diffeomorphisms of the circle does not satisfy the Tits alternative ([**45, 99**]).

Note that since solvable subgroups have either polynomial or exponential growth, if G satisfies the Tits alternative, G does not contain groups with intermediate growth.

The main technique to prove that a group contains a non-abelian free group is the Ping Pong Lemma (*for instance* [**71**]):

Lemma 8.29. *Let S be a set. Let g_1 and g_2 be two bijections of S. Assume that S contains two non-empty disjoint subsets S_1 and S_2 such that*

$$g_1^m(S_2)\subset S_1 \qquad g_2^m(S_1)\subset S_2 \qquad \forall\, m\in\mathbb{Z}\smallsetminus\{0\}.$$

Then $\langle g_1,\,g_2\rangle$ is a free group on two generators.

Sketch of the Proof. Let $w=w(a,b)$ be a reduced word that represents a non-trivial element in the free group $\mathbb{F}_2=\langle a,\,b\rangle$. Let us prove that $w(g_1,g_2)$ is a non-trivial map of S. Up to conjugacy by a power of g_1 assume that $w(g_1,g_2)$ starts and ends with a power of g_1:

$$w(g_1,g_2)=g_1^{\ell_n}g_2^{m_n}\dots g_2^{m_1}g_1^{\ell_0}.$$

One checks that $g_1^{\ell_0}$ maps S_2 into S_1, then $g_2^{m_1}g_1^{\ell_0}$ maps S_2 into S_2, $\dots$ and w maps S_2 into S_1. As S_2 is disjoint from S_1 one gets that $w(g_1,g_2)$ is non-trivial. □

Consider a group Γ that acts on a hyperbolic space $\mathbb{H}^\infty$ and that contains two loxodromic isometries ψ_1 and ψ_2 whose fixed points in $\partial\mathbb{H}^\infty$ form two disjoint pairs. Let us take disjoint neighborhoods $S_i\subset\overline{\mathbb{H}^\infty}$ of the fixed point sets of ψ_i, $i=1,\,2$. Then Lemma 8.29 applied to sufficiently high powers ψ_1^n and ψ_2^n of ψ_1 and ψ_2, respectively, produces a free subgroup of Γ. This strategy can be used for the Cremona group acting by isometries on $\mathbb{H}^\infty(\mathbb{P}^2_{\mathbb{C}})$. More precisely Cantat obtained the following result:

Theorem 8.30 ([**52**]). *Let S be a projective surface S over a field $\Bbbk$. The group* $\mathrm{Bir}(S)$ *satisfies the Tits alternative for finitely generated subgroups.*

Then Urech proves:

Theorem 8.31 ([**172**]). *Let S be a complex Kähler surface. Then* $\mathrm{Bir}(S)$ *satisfies the Tits alternative.*

Let us now give a sketch of the proof of this result in the case $S=\mathbb{P}^2_{\mathbb{C}}$.

8.4.1. Subgroups of $\mathrm{Bir}(\mathbb{P}^2_\mathbb{C})$ that contain a loxodromic element. Recall that

- ◇ the subgroup of diagonal automorphisms
$$\mathrm{D}_2 = \{(z_0, z_1) \mapsto (\alpha z_0, \beta z_1) \,|\, \alpha,\, \beta \in \mathbb{C}^*\} \subset \mathrm{PGL}(3, \mathbb{C}) = \mathrm{Aut}(\mathbb{P}^2_\mathbb{C})$$
is a real torus of rank 2;
- ◇ a matrix $A = (a_{ij}) \in \mathrm{GL}(2, \mathbb{Z})$ determines a birational map of $\mathbb{P}^2_\mathbb{C}$:
$$(z_0, z_1) \dashrightarrow \big(z_0^{a_{00}} z_1^{a_{01}}, z_0^{a_{10}} z_1^{a_{11}}\big).$$

The normalizer of D_2 in $\mathrm{Bir}(\mathbb{P}^2_\mathbb{C})$ is the semidirect product
$$\mathrm{Norm}\big(\mathrm{D}_2, \mathrm{Bir}(\mathbb{P}^2_\mathbb{C})\big) = \{\phi \in \mathrm{Bir}(\mathbb{P}^2_\mathbb{C}) \,|\, \phi \circ \mathrm{D}_2 \circ \phi^{-1} = \mathrm{D}_2\} = \mathrm{GL}(2, \mathbb{Z}) \ltimes \mathrm{D}_2.$$
If $M \in \mathrm{GL}(2, \mathbb{Z})$ has spectral radius strictly larger than 1, the associated birational map is loxodromic. In particular there exist loxodromic elements that normalize an infinite elliptic subgroup. Up to conjugacy these are the only examples with this property:

THEOREM 8.32 ([**73**]). *Let* G *be a subgroup of* $\mathrm{Bir}(\mathbb{P}^2_\mathbb{C})$ *containing at least one loxodromic element. Assume that there exists a short exact sequence*
$$1 \longrightarrow \mathrm{A} \longrightarrow \mathrm{G} \longrightarrow \mathrm{B} \longrightarrow 1$$
where A *is infinite and of bounded degree.*

Then G *is conjugate to a subgroup of* $\mathrm{GL}(2, \mathbb{Z}) \ltimes \mathrm{D}_2$.

Urech generalizes this result to the case where A is an infinite group of elliptic elements ([**172**]):

THEOREM 8.33 ([**172**]). *Let* G *be a subgroup of* $\mathrm{Bir}(\mathbb{P}^2_\mathbb{C})$ *containing at least one loxodromic element. Suppose that there exists a short sequence*
$$1 \longrightarrow \mathrm{A} \longrightarrow \mathrm{G} \longrightarrow \mathrm{B} \longrightarrow 1$$
where A *is an infinite group of elliptic elements.*

Then G *is conjugate to a subgroup of* $\mathrm{GL}(2, \mathbb{Z}) \ltimes \mathrm{D}_2$.

In order to give the proof of Theorem 8.33 we need to establish some results.

LEMMA 8.34 ([**172**]). *Let ϕ be a loxodromic monomial map of the complex projective plane. Let Δ_2 be an infinite subgroup of* D_2 *normalized by ϕ.*

Then Δ_2 is dense in D_2 *with respect to the Zariski topology.*

PROOF. Denote by $\overline{\Delta_2}^0$ the neutral component of the Zariski closure of Δ_2.

If $\overline{\Delta_2}^0$ has a dense orbit on $\mathbb{P}^2_\mathbb{C}$, then Δ_2 is dense in D_2. Otherwise the dimension of the generic orbits of $\overline{\Delta_2}^0$ is 1. But ϕ normalizes $\overline{\Delta_2}^0$, so preserves its orbits. In particular ϕ thus preserves a rational fibration: contradiction with the fact that ϕ is loxodromic. □

In [**161**] the classification of tight elements of $\mathrm{Bir}(\mathbb{P}^2_\mathbb{C})$ is given:

THEOREM 8.35 ([**161**]). *Every loxodromic element of the plane Cremona group is rigid.*

Let ϕ be a loxodromic birational self map of the complex projective plane; then

- ◇ *if ϕ is conjugate to a monomial map, no power of ϕ is tight;*
- ◇ *otherwise ϕ^n is tight for some integer n.*

Consider a subgroup G of $\mathrm{Bir}(\mathbb{P}^2_\mathbb{C})$. Let $\phi\in \mathrm{G}$ be a rigid element; then ϕ is also a rigid element in G. The same holds for tight elements but the converse does not: there exist loxodromic maps $\phi\in\mathrm{G}$ such that ϕ is tight in G but not in $\mathrm{Bir}(\mathbb{P}^2_\mathbb{C})$.

Proofs of Theorem 8.35 and Lemma 8.34 imply the following:

THEOREM 8.36 ([**172**]). *Let* G *be a subgroup of* $\mathrm{Bir}(\mathbb{P}^2_\mathbb{C})$. *Let* ϕ *be a loxodromic element. The following assertions are equivalent:*

- ◇ *no power of* ϕ *is tight in* G*;*
- ◇ *there is a subgroup* $\Delta_2\subset \mathrm{G}$ *that is normalized by* ϕ *and a birational self map* ψ *of* $\mathbb{P}^2_\mathbb{C}$ *such that* $\psi\circ\Delta_2\circ\psi^{-1}$ *is a dense subgroup of* D_2 *and* $\psi\circ\varphi\circ\psi^{-1}$ *belongs to* $\mathrm{GL}(2,\mathbb{Z})\ltimes \mathrm{D}_2$.

PROOF OF THEOREM 8.33. The group A fixes a point $p\in\partial\mathbb{H}^\infty\cup\mathbb{H}^\infty$ (Theorem 8.44). Note that if p belongs to $\mathbb{H}^\infty$, then A is bounded and Theorem 8.32 allows us to conclude. Let us assume that p belongs to $\partial\mathbb{H}^\infty$. Note that if A fixes another point q on $\partial\mathbb{H}^\infty$, then A fixes the geodesic between p and q, and so A would be bounded again. Suppose thus that p is the only fixed point of A in $\partial\mathbb{H}^\infty$. Consider a loxodromic map ϕ of N. It normalizes A and so ϕ fixes p. As ϕ is loxodromic, ϕ does not preserve any fibration; consequently p does not correspond to the class of a fibration. From Lemma 8.24 any finitely generated group of elliptic elements that fixes p is bounded. Let G be the subgroup of birational self maps of $\mathbb{P}^2_\mathbb{C}$ that fix p. Denote by L the one-dimensional subspace of $\mathrm{Z}(\mathbb{P}^2_\mathbb{C})$ corresponding to p. The group G fixes p; hence its linear action on $\mathrm{Z}(\mathbb{P}^2_\mathbb{C})$ acts on L by automorphisms preserving the orientation. This implies the existence of a group homomorphism $\rho\colon \mathrm{G}\to\mathbb{R}^*_+$. Note that G does not contain any parabolic element because p does not correspond to the class of a fibration and that loxodromic elements do not fix any vector in $\mathrm{Z}(\mathbb{P}^2_\mathbb{C})$. As a result $\ker\rho$ consists of elliptic elements. But 1 is the only eigenvalue of a map of $\mathrm{Z}(\mathbb{P}^2_\mathbb{C})$ induced by an elliptic birational self map ([**52**]); as a consequence any elliptic birational map of G is contained in $\ker\rho$.

Take a loxodromic map ϕ in G. Let us show by contradiction that no power of ϕ is tight in G. So assume that there exists $n\in\mathbb{Z}$ such that ϕ^n is tight in G. The subgroup N of G is infinite and $\langle\phi\rangle$ has finite index in $\mathrm{Cent}(\phi)$ (Theorem 8.10); there thus exist $\psi\in\mathrm{G}$ that do not commute with ϕ^n. Since all non-trivial elements of $\ll\phi^n\gg$ are loxodromic ([**59**]) the map $\psi\circ\phi^n\circ\psi^{-1}\circ\phi^{-n}$ is loxodromic. But $\rho\big(\psi\circ\phi^n\circ\psi^{-1}\circ\phi^{-n}\big)=1$, *i.e.*, $\psi\circ\phi^n\circ\psi^{-1}\circ\phi^{-n}$ is elliptic: contradiction. Finally no power of ϕ is tight in G. According to Theorem 8.36 there exist $\varphi\in\mathrm{Bir}(\mathbb{P}^2_\mathbb{C})$ and Δ_2 an algebraic subgroup of G such that

- $\varphi\circ\phi\circ\varphi^{-1}$ is monomial;
- $\varphi\circ\Delta_2\circ\varphi^{-1}=\mathrm{D}_2$.

Consider a finitely generated subgroup Γ of $\ker\rho$. The Zariski closure $\overline{\Gamma}$ of Γ is an algebraic subgroup of G because Γ is bounded. Set

$$d=\sup\{\dim\overline{\Gamma}\,|\,\Gamma\subset\ker\rho\text{ finitely generated}\}.$$

We will distinguish the cases d is finite and d is infinite.

- First consider the case $d<\infty$. Note that $\ker\rho$ contains a subgroup conjugated to D_2, so $d\geq 2$. Take Γ a finitely generated subgroup of $\ker\rho$ such that $\dim\overline{\Gamma}=d$. Let $\overline{\Gamma}^0$ be the neutral component of the algebraic group $\overline{\Gamma}$. Let ϕ be an element of G. The group $\phi\circ\overline{\Gamma}^0\circ\phi^{-1}$ is again an algebraic subgroup and $\langle\overline{\Gamma}^0,\,\phi\circ\overline{\Gamma}^0\circ\phi^{-1}\rangle$ is contained in $\overline{\langle\Gamma,\,\phi\circ\Gamma\circ\phi^{-1}\rangle}$.

According to [**111**] the group $\langle\overline{\Gamma}^0,\, \phi\circ\overline{\Gamma}^0\circ\phi^{-1}\rangle$ is closed and connected. On the one hand $\dim\langle\overline{\Gamma}^0,\, \phi\circ\overline{\Gamma}^0\circ\phi^{-1}\rangle\leq d$ and on the other hand $\overline{\Gamma}^0\subset\langle\overline{\Gamma}^0,\, \phi\circ\overline{\Gamma}^0\circ\phi^{-1}\rangle$. As a consequence $\langle\overline{\Gamma}^0,\, \phi\circ\overline{\Gamma}^0\circ\phi^{-1}\rangle=\overline{\Gamma}^0$. In other words ϕ normalizes $\overline{\Gamma}^0$. But $\Gamma\cap\overline{\Gamma}^0$ is infinite, so there exists a birational self map ψ of $\mathrm{Bir}(\mathbb{P}^2_\mathbb{C})$ such that $\psi\circ\mathrm{G}\circ\psi^{-1}\subset\mathrm{GL}(2,\mathbb{Z})\ltimes\mathrm{D}_2$ (Theorem 8.32) and hence $\psi\circ\mathrm{N}\circ\psi^{-1}\subset\mathrm{GL}(2,\mathbb{Z})\ltimes\mathrm{D}_2$.

- Now assume $d=\infty$. Let Γ be a finitely generated subgroup of $\ker\rho$ such that $\dim\overline{\Gamma}\geq 9$. The closure $\overline{\Gamma}$ of Γ preserves a unique rational fibration given by a rational map $\pi\colon\mathbb{P}^2_\mathbb{C}\dashrightarrow\mathbb{P}^1_\mathbb{C}$ (Lemma 8.18). Consider an element ϕ of $\ker\rho$. The algebraic group $\overline{\langle\Gamma,\,\phi\rangle}$ also preserves a unique rational fibration; since $\overline{\Gamma}\subset\overline{\langle\Gamma,\,\phi\rangle}$ this fibration is given by π. As a result $\ker\rho$ preserves a rational fibration. Hence $\ker\rho$ is bounded and the group $\phi\circ\mathrm{G}\circ\phi^{-1}$ is contained in $\mathrm{GL}(2,\mathbb{Z})\ltimes\mathrm{D}_2$ (Theorem 8.32); in particular $\phi\circ\mathrm{N}\circ\phi^{-1}$ is a subgroup of $\mathrm{GL}(2,\mathbb{Z})\ltimes\mathrm{D}_2$. □

LEMMA 8.37 ([**172**]). *Let ϕ and ψ be two loxodromic elements of* $\mathrm{Bir}(\mathbb{P}^2_\mathbb{C})$ *such that* $\mathrm{Ax}(\phi)\neq\mathrm{Ax}(\psi)$. *Then*

◇ *either ϕ and ψ have no common fixed point on $\partial\mathbb{H}^\infty$,*
◇ *or $\langle\phi,\,\psi\rangle$ contains a subgroup* G *and there exists a birational self map φ of the complex projective plane such that*
 - $\varphi\circ\langle\phi,\,\psi\rangle\circ\varphi^{-1}\subset\mathrm{GL}(2,\mathbb{Z})\ltimes\mathrm{D}_2$,
 - $\varphi\circ\mathrm{G}\circ\varphi^{-1}$ *is a dense subgroup of* D_2.

PROOF. Suppose that ϕ and ψ have a common fixed point $p\in\partial\mathbb{H}^\infty$. Denote by L the one-dimensional subspace of $\mathrm{Z}(\mathbb{P}^2_\mathbb{C})$ corresponding to p. The group $\langle\phi,\,\psi\rangle$ generated by ϕ and ψ fixes p, so its linear action on $\mathrm{Z}(\mathbb{P}^2_\mathbb{C})$ acts on L by automorphisms preserving the orientation. A reasoning analogous to that of the proof of Theorem 8.33 implies the existence of a group homomorphism

$$\rho\colon\langle\phi,\,\psi\rangle\to\mathbb{R}^*_+$$

whose kernel consists of elliptic birational maps (*see* Proof of Theorem 8.33).

Assume that ϕ^n is tight for some n. Since $\mathrm{Ax}(\phi)\neq\mathrm{Ax}(\psi)$ the maps ϕ^n and ψ do not commute. According to [**59**] any non-trivial element of $\ll\phi^n\gg$ is loxodromic. Therefore, on the one hand $\psi\circ\phi^n\circ\psi^{-1}\circ\phi^{-n}$ is loxodromic, and on the other hand $\rho(\psi\circ\varphi^n\circ\psi^{-1})=1$; hence $\psi\circ\varphi^n\circ\psi^{-1}$ is elliptic: contradiction. As a result for any k the map ϕ^k is not tight in $\langle\phi,\,\psi\rangle$. Theorem 8.36 implies that there exist a birational self map φ of $\mathbb{P}^2_\mathbb{C}$ and a bounded subgroup $\Delta_2\subset\langle\phi,\,\psi\rangle$ such that

◇ $\varphi\circ\phi\circ\varphi^{-1}$ is monomial;
◇ $\varphi\circ\Delta_2\circ\varphi^{-1}$ is a dense subgroup of D_2.

In particular $\ker\rho\supset\Delta_2$ is thus infinite. Theorem 8.33 allows us to conclude. □

LEMMA 8.38 ([**172**]). *Let ϕ and ψ be two loxodromic elements of* $\mathrm{Bir}(\mathbb{P}^2_\mathbb{C})$ *such that* $\mathrm{Ax}(\phi)\neq\mathrm{Ax}(\psi)$. *Then ϕ and ψ have no common fixed point on $\partial\mathbb{H}^\infty$.*

PROOF. Assume by contradiction that ϕ and ψ have no common fixed point on $\partial\mathbb{H}^\infty$. Lemma 8.37 thus implies that up to birational conjugacy

◇ $\varphi\circ\langle\phi,\,\psi\rangle\circ\varphi^{-1}\subset\mathrm{GL}(2,\mathbb{Z})\ltimes\mathrm{D}_2$,
◇ $\varphi\circ\mathrm{G}\circ\varphi^{-1}\subset\mathrm{D}_2$ is a dense subgroup.

Let us write

$$\phi = d_1 \circ m_1, \qquad\qquad \psi = d_2 \circ m_2,$$

with $d_i \in \mathrm{D}_2$ and $m_i \in \mathrm{GL}(2,\mathbb{Z})$. The group D_2 fixes the axes of all the monomial loxodromic elements; in particular m_1 and m_2 have the same fixed points on $\partial\mathbb{H}^\infty$ as ϕ and ψ. But the group $\langle m_1, m_2\rangle$ does not contain any infinite abelian group, so according to Lemma 8.37 the birational maps m_1 and m_2 have no common fixed point on $\partial\mathbb{H}^\infty$: contradiction. □

LEMMA 8.39 ([**172**]). *Let* G *be a subgroup of the plane Cremona group that contains a loxodromic element. Then one of the following holds:*

- ◇ G *is conjugate to a subgroup of* $\mathrm{GL}(2,\mathbb{Z}) \ltimes \mathrm{D}_2$*;*
- ◇ G *contains a subgroup of index at most* 2 *that is isomorphic to* $\mathbb{Z} \ltimes \mathrm{H}$ *where* H *is a finite group;*
- ◇ G *contains a non-abelian free subgroup.*

PROOF. Let ϕ be a loxodromic map of G.

- ◇ Assume first that all elements in G preserve the axis $\mathrm{Ax}(\phi)$ of ϕ. The group G contains a subgroup H of index at most 2 with the following property: H preserves the orientation of the axis. As a result any element $\psi \in \mathrm{H}$ translates the points on $\mathrm{Ax}(\phi)$ by a constant c_ψ. This yields a group morphism

$$\pi\colon \mathrm{H} \to \mathbb{R}, \qquad\qquad \psi \mapsto c_\psi$$

  such that $\ker\pi$ is a bounded group. From Theorem 8.32 either $\ker\pi$ is finite, or G is conjugate to a subgroup of $\mathrm{GL}(2,\mathbb{Z}) \ltimes \mathrm{D}_2$.
- ◇ Suppose that there is an element $\psi \in \mathrm{G}$ that does not preserve $\mathrm{Ax}(\phi)$. Denote by $\alpha(\phi)$ and $\omega(\phi)$ (resp. $\alpha(\psi)$ and $\omega(\psi)$) the attractive and repulsive fixed points of $\phi_\bullet$ (resp. $\psi_\bullet$). Let $\mathcal{U}_1^+$ (resp. $\mathcal{U}_1^-$, resp. $\mathcal{U}_2^+$, resp. $\mathcal{U}_2^-$) be a small neighborhood of $\alpha(\phi)$ (resp. $\omega(\phi)$, resp. $\alpha(\psi)$, resp. $\omega(\psi)$) in $\partial\mathbb{H}^\infty$. We can assume that $\mathcal{U}_1^+$, $\mathcal{U}_1^-$, $\mathcal{U}_2^+$ and $\mathcal{U}_2^-$ are pairwise disjoint. Set $\mathcal{U}_1 = \mathcal{U}_1^+ \cup \mathcal{U}_1^-$ and $\mathcal{U}_2 = \mathcal{U}_2^+ \cup \mathcal{U}_2^-$. There exist n_1, n_2, n_3, n_4 some positive integers such that

$$\phi^{n_1}(\mathcal{U}_2) \subset \mathcal{U}_1^+, \qquad \phi^{-n_2}(\mathcal{U}_2) \subset \mathcal{U}_1^-, \qquad \psi^{n_3}(\mathcal{U}_1) \subset \mathcal{U}_2^+, \qquad \psi^{-n_4}(\mathcal{U}_1) \subset \mathcal{U}_2^-.$$

  Set $n = \max(n_1, n_2, n_3, n_4)$. Since

$$\phi(\mathcal{U}_1^+) \subset \mathcal{U}_1^+, \qquad \phi^{-1}(\mathcal{U}_1^-) \subset \mathcal{U}_1^-, \qquad \psi(\mathcal{U}_2^+) \subset \mathcal{U}_2^+, \qquad \psi^{-1}(\mathcal{U}_2^-) \subset \mathcal{U}_2^-,$$

  one gets that for any $k \leq n$,

$$\phi^k(\mathcal{U}_2) \subset \mathcal{U}_1, \qquad \phi^{-k}(\mathcal{U}_2) \subset \mathcal{U}_1, \qquad \psi^k(\mathcal{U}_1) \subset \mathcal{U}_2, \qquad \psi^{-k}(\mathcal{U}_1) \subset \mathcal{U}_2.$$

According to the Ping Pong Lemma applied to ϕ^n, ψ^n together with $\mathcal{U}_1$ and $\mathcal{U}_2$ we get that $\langle \psi^n, \phi^n\rangle$ generates a non-abelian free subgroup of G. □

8.4.2. Tits alternative for finitely generated subgroups for automorphism groups and the Jonquières group.

LEMMA 8.40 ([**52**]). *Let* G *be a finitely generated group. Assume that* G *is an extension of a virtually solvable group* R *of length* r *by an other virtually solvable group* Q *of length* q*:*

$$1 \longrightarrow \mathrm{R} \longrightarrow \mathrm{G} \longrightarrow \mathrm{Q} \longrightarrow 1.$$

Then G *is virtually solvable of length* $\leq q + r + 1$.

Hence one has the following statement:

PROPOSITION 8.41 ([**52**]). *Let* G_1 *and* G_2 *be two groups that satisfy the Tits alternative.*

If G *is an extension of* G_1 *by* G_2*, then* G *satisfies the Tits alternative.*

PROOF. Let Γ be a subgroup of G that does not contain a non-abelian free subgroup. For $i \in \{1, 2\}$ denote by $\mathrm{pr}_i\colon \mathrm{G} \to \mathrm{G}_i$ the canonical projection. Since $\mathrm{pr}_i(\mathrm{G})$ does not contain a non-abelian free subgroup, $\mathrm{pr}_i(\Gamma) = \Gamma \cap \mathrm{G}_i$ is virtually solvable (G_i satisfies the Tits alternative). Then according to Lemma 8.40 the group Γ is virtually solvable. □

A first consequence of this result is the following one:

THEOREM 8.42 ([**52**]). *Let* V *be a Kähler compact manifold. Its automorphism group satisfies the Tits alternative.*

PROOF. The group $\mathrm{Aut}(V)$ acts on the cohomology of V. This yields to a morphism ρ from $\mathrm{Aut}(V)$ to $\mathrm{GL}(H^*(V,\mathbb{Z}))$ where $H^*(V,\mathbb{Z})$ denotes the direct sum of the cohomology groups of V. According to [**130**]

- ◇ the kernel of ρ is a complex Lie group with a finite number of connected components;
- ◇ its connected component $\mathrm{Aut}^0(V)$ is an extension of a compact complex torus by a complex algebraic group. We get the result from Proposition 8.41 and the classical Tits alternative. □

A direct consequence of Proposition 8.41 and the Tits alternative for linear groups is:

PROPOSITION 8.43 ([**52**]). *The Jonquières group*

$$\mathcal{J} \simeq \mathrm{PGL}(2,\mathbb{C}) \rtimes \mathrm{PGL}(2,\mathbb{C}(z_0))$$

satisfies the Tits alternative.

8.4.3. "Weak alternative" for isometries of $\mathbb{H}^\infty$. Let us recall some notations and definitions introduced in Chapter 2.

Let $\mathcal{H}$ be a seperable Hilbert space. Let us fix a Hilbert basis $\mathcal{B} = (\mathbf{e}_i)_i$ on $\mathcal{H}$. Consider the scalar product defined on $\mathcal{H}$ by

$$\langle v, v\rangle = v_0^2 - \sum_{i=1}^{\infty} v_i^2$$

where the coordinates v_i are the coordinates of v in $\mathcal{B}$. The *light cone of* $\mathcal{H}$ is the set

$$\mathcal{L}(\mathcal{H}) = \big\{v \in \mathcal{H} \,|\, \langle v, v\rangle = 0\big\}.$$

Let $\mathbb{H}^\infty$ be the connected component of the hyperboloid

$$\big\{v \in \mathcal{H} \,|\, \langle v, v\rangle = 1\big\}$$

that contains $\mathbf{e}_0$. Consider the metric defined on $\mathbb{H}^\infty$ by

$$d(u,v) := \operatorname{arccos}(\langle u, v\rangle).$$

The space $\mathbb{H}^\infty$ is a complete CAT(-1) space, so is hyperbolic (Chapter 2). Its boundary $\partial\mathbb{H}^\infty$ can be identified to $\mathbb{P}(\mathcal{L}(\mathcal{H}))$.

THEOREM 8.44 ([**52**]). *Let* Γ *be a subgroup of* $\mathrm{O}(1,\infty)$.

(1) *If Γ contains a loxodromic isometry ψ, then one of the following properties holds:*
   - ◇ *Γ contains a non-abelian free group,*
   - ◇ *Γ permutes the two fixed points of ψ that lie on $\partial\mathbb{H}^\infty$.*

(2) *If Γ contains no loxodromic isometry, then Γ fixes a point of $\mathbb{H}^\infty \cup \partial\mathbb{H}^\infty$.*

PROOF.

◇ Assume first that Γ contains two loxodromic isometries ϕ and ψ such that the fixed points of ϕ and ψ on $\partial\mathbb{H}^\infty$ are pairwise distinct. According to the Ping Pong Lemma (Lemma 8.29) there are two integers n and m such that ϕ^n and ψ^m generate a subgroup of Γ isomorphic to the free group $\mathbb{F}_2$.

◇ Suppose that Γ contains at least one loxodromic isometry ϕ. Let $\alpha(\phi)$ and $\omega(\phi)$ be the fixed points of ϕ on $\partial\mathbb{H}^\infty$. If Γ contains an element ψ such that

$$\{\alpha(\phi), \omega(\phi)\} \cap \{\alpha(\psi), \omega(\psi)\} = \emptyset$$

then ϕ and $\psi \circ \phi \circ \psi^{-1}$ are two loxodromic isometries to which we can apply the previous argument. Otherwise Γ fixes either $\{\alpha(\phi), \omega(\phi)\}$, or $\{\alpha(\phi)\}$, or $\{\omega(\phi)\}$. Then Γ contains a subgroup of index 2 that fixes $\alpha(\phi)$ of $\omega(\phi)$.

◇ Assume that Γ contains two parabolic isometries ϕ and ψ whose fixed points $\alpha(\phi) \in \partial\mathbb{H}^\infty$ and $\alpha(\psi) \in \partial\mathbb{H}^\infty$ are distinct. Take two elements of $\mathcal{L}(\mathcal{H})$ still denoted $\alpha(\phi)$ and $\alpha(\psi)$ that represent these two points of $\partial\mathbb{H}^\infty$. Let ℓ be a point of $\mathcal{H}$ such that

$$\langle \alpha(\phi), \ell\rangle < 0 \qquad\qquad \langle \alpha(\psi), \ell\rangle > 0.$$

The hyperplane of $\mathcal{H}$ orthogonal to ℓ intersects $\mathbb{H}^\infty$ in a subspace L that "separates" $\alpha(\phi)$ and $\alpha(\psi)$. As a result there exist integers n and m such that $\phi^m(L)$, $\phi^{-m}(L)$, $\psi^n(L)$ and $\psi^{-n}(L)$ don't pairwise intersect. The isometry $\phi^m \circ \psi^n$ has thus two distinct fixed points on $\partial\mathbb{H}^\infty$; hence it is a loxodromic one. Applying the above argument we get that $\langle \phi, \psi\rangle$ contains a free group. Therefore, if Γ contains at least one parabolic isometry, then
   - either Γ contains a non-abelian free group;
   - or Γ fixes a point of $\partial\mathbb{H}^\infty$ that is the unique fixed point of the parabolic isometries of Γ.

◇ Let us finish by assuming that all elements of Γ are elliptic ones. According to [**98**, Chapter 8, Lemma 35 and Corollary 36]
   - either the orbit of any point of $\mathbb{H}^\infty$ is bounded;
   - or the limit set of Γ is a point.

From [**72**, Chapter 2, §b.8] one gets the following alternative: Γ fixes
   - either a point of $\mathbb{H}^\infty$;
   - or a point of $\partial\mathbb{H}^\infty$. □

8.4.4. Proof of Theorem 8.31.

8.4.4.1. *Assume that* G *contains a loxodromic element.* Let G be a subgroup of $\mathrm{Bir}(\mathbb{P}^2_{\mathbb{C}})$ that contains a loxodromic element. According to Lemma 8.39 we have to consider the following three cases:

◇ G is conjugate to a subgroup of $\mathrm{GL}(2,\mathbb{Z}) \ltimes \mathrm{D}_2$ and the Tits alternative holds by Proposition 8.41;

- ◇ G contains a subgroup of index at most 2 that is isomorphic to $\mathbb{Z} \ltimes \mathrm{H}$ where H is a finite group; in other words G is cyclic up to finite index, so the Tits alternative holds;
- ◇ G contains a non-abelian free subgroup, and the Tits alternative holds.

We can thus state:

COROLLARY 8.45 ([**52, 172**]). *Let* G *be a subgroup of* $\mathrm{Bir}(\mathbb{P}^2_\mathbb{C})$ *that contains a loxodromic element. Then* G *satisfies the Tits alternative.*

8.4.4.2. *Assume that* G *contains a parabolic element but no loxodromic element.*

LEMMA 8.46 ([**172**]). *Let* G *be a subgroup of* $\mathrm{Bir}(\mathbb{P}^2_\mathbb{C})$ *that does not contain any loxodromic element but contains a parabolic element. Then* G *is conjugate to a subgroup of* $\mathcal{J}$ *or* $\mathrm{Aut}(S)$, *where* S *is a Halphen surface.*

PROOF. By Theorem 8.44 the group G fixes a point $p \in \mathbb{H}^\infty \cup \partial\mathbb{H}^\infty$. Consider a parabolic element φ of G; then φ has no fixed point in $\mathbb{H}^\infty$ and a unique fixed point q in $\partial\mathbb{H}^\infty$. As a consequence $p = q$. According to Theorem 2.9 there exist a surface S, a birational map $\psi\colon \mathbb{P}^2_\mathbb{C} \dashrightarrow S$, a curve C, and a fibration $\pi\colon S \to C$ such that $\psi \circ \varphi \circ \psi^{-1}$ permutes the fibres of π. In particular $\psi \circ \varphi \circ \psi^{-1}$ preserves the divisor class of a fibre F of π. Since F is a class of a fibre, $F \cdot F = 0$. The point $m \in \mathrm{Z}(\mathbb{P}^2_\mathbb{C})$ corresponding to F, so satisfies $m \cdot m = 0$. Therefore, $q \in \partial\mathbb{H}^\infty$ corresponds to the line passing through the origin and m. It follows that any element in G fixes m, and so preserves the divisor class of F. In other words any element in $\varphi \circ \mathrm{G} \circ \varphi^{-1}$ permutes the fibres of the fibration $\pi\colon S \to C$. If the fibration is rational, then up to birational conjugacy $\mathrm{G} \subset \mathcal{J}$. If the fibration is a fibration of genus 1 curves, there exists a Halphen surface S' such that up to birational conjugacy G is contained in $\mathrm{Bir}(S')$ and preserves the Halphen fibration. By Lemma 2.6 the group G is contained in $\mathrm{Aut}(S')$. □

Assume first that up to birational conjugacy $\mathrm{G} \subset \mathcal{J} \simeq \mathrm{PGL}(2, \mathbb{C}(z_1)) \rtimes \mathrm{PGL}(2, \mathbb{C})$. The Tits alternative for linear groups in characteristic 0 and Proposition 8.41 imply the Tits alternative for G.

Finally suppose that $\mathrm{G} \subset \mathrm{Aut}(S)$ where S is a Halphen surface. The automorphisms groups of Halphen surfaces have been studied ([**56, 101, 104**]). In particular Cantat and Dolgachev prove

THEOREM 8.47 ([**56**]). *Let* S *be a Halphen surface. There exists a homomorphism* $\rho\colon \mathrm{Aut}(S) \to \mathrm{PGL}(2, \mathbb{C})$ *with finite image such that* $\ker \rho$ *is an extension of an abelian group of rank* ≤ 8 *by a cyclic group of order dividing* 24.

In other words the automorphism group of a Halphen surface is virtually abelian; hence G is solvable up to finite index.

8.4.4.3. *Assume that* G *is a group of elliptic elements.* According to Theorems 8.14 and 8.15 one of the following holds:

- ◇ G is isomorphic to a bounded subgroup;
- ◇ G preserves a rational fibration.

Suppose that G is isomorphic to a bounded subgroup; in particular G is isomorphic to a subgroup of linear groups, and so satisfies the Tits alternative.

If G preserves a rational fibration, then G satisfies the Tits alternative (Proposition 8.43).

8.4.5. A consequence of the Tits alternative: the Burnside problem. The Burnside problem posed by Burnside in 1902 asks whether a finitely generated torsion group is finite. Schur showed in 1911 that any finitely generated torsion group that is a subgroup of invertible $n \times n$ complex matrices is finite ([**159**]). One of the tools of the proof is the Jordan-Schur Theorem.

In the 1930's Burnside asked another related question called the restricted Burnside problem: if it is known that a group G with m generators and exponent n is finite, can one conclude that the order of G is bounded by some constant depending only on n and m? In other words are there only finitely many finite groups with m generators of exponent n up to isomorphism?

In 1958 Kostrikin was able to prove that among the finite groups with a given number of generators and a given prime exponent, there exists a largest one: this provides a solution for the restricted Burnside problem for the case of prime exponent ([**125**]).

Later Zelmanov solved the restricted Burnside problem for an arbitrary exponent ([**181, 182**]).

Golod gave a negative answer to the Burnside problem for groups that have a complete system of linear representations ([**103**]).

Later many examples of infinite, finitely generated and torsion groups with even bounded ordres were exhibited ([**118, 134, 148–151**]).

The problem raised by Burnside is still open for homeomorphism (resp. diffeomorphism) groups on closed manifolds. Very few examples are known.

Cantat gave a positive answer to the Burnside problem for the Cremona group:

THEOREM 8.48 ([**52**]). *Every finitely generated torsion subgroup of* $\mathrm{Bir}(\mathbb{P}^2_\mathbb{C})$ *is finite.*

PROOF. Let G be a finitely generated torsion subgroup of $\mathrm{Bir}(\mathbb{P}^2_\mathbb{C})$. From the Tits alternative (Theorem 8.30) G is solvable up to finite index. Since any torsion, solvable, finitely generated group is finite, G is finite. □

8.5. Solvable subgroups of $\mathrm{Bir}(\mathbb{P}^2_\mathbb{C})$

The study of the solvable subgroups of the plane Cremona group starts in [**80**] and goes on in [**172**].

THEOREM 8.49 ([**172**]). *Let* G *be a solvable subgroup of* $\mathrm{Bir}(\mathbb{P}^2_\mathbb{C})$. *Then one of the following holds:*

- ◇ G *is a subgroup of elliptic elements, in particular* G *is isomorphic either to a solvable subgroup of* $\mathcal{J}$, *or to a solvable subgroup of a bounded group;*
- ◇ G *is conjugate to a subgroup of* $\mathcal{J}$*;*
- ◇ G *is conjugate to a subgroup of the automorphism group of a Halphen surface;*
- ◇ G *is conjugate to a subgroup of* $\mathrm{GL}(2,\mathbb{Z}) \ltimes \mathrm{D}_2$ *where*

$$\mathrm{D}_2 = \big\{(z_0, z_1) \mapsto (\alpha z_0, \beta z_1) \,|\, \alpha,\, \beta \in \mathbb{C}^*\big\};$$

- ◇ G *contains a loxodromic element and there exists a finite subgroup* H *of* $\mathrm{Bir}(\mathbb{P}^2_\mathbb{C})$ *such that* $\mathrm{G} = \mathbb{Z} \ltimes H$.

REMARK 8.50. A solvable subgroup of a bounded group is a solvable subgroup from one of the groups that appear in Theorem 3.46.

REMARK 8.51. The centralizer of a birational self map of $\mathbb{P}^2_\mathbb{C}$ that preserves a unique fibration that is rational is virtually solvable (§7.5.2.4); this example illustrates the second case.

Before giving the proof let us state some consequences.

The soluble length of a nilpotent subgroup of $\mathrm{Bir}(\mathbb{P}^2_\mathbb{C})$ can be bounded by the dimension of $\mathbb{P}^2_\mathbb{C}$ as Epstein and Thurston did in the context of Lie algebras and rational vector fields on a connected complex manifold ([**93**]):

COROLLARY 8.52 ([**80**]). *Let* G *be a nilpotent subgroup of* $\mathrm{Bir}(\mathbb{P}^2_\mathbb{C})$ *that is not a torsion group. The soluble length of* G *is bounded by* 2.

Theorem 3.46 allows us to prove:

COROLLARY 8.53 ([**172**]). *The derived length of a bounded solvable subgroup of* $\mathrm{Bir}(\mathbb{P}^2_\mathbb{C})$ *is* ≤ 5.

The derived length of a solvable subgroup of $\mathrm{Bir}(\mathbb{P}^2_\mathbb{C})$ *is at most* 8.

PROOF OF THEOREM 8.49. It decomposes into three parts: G contains a loxodromic element; G does not contain a loxodromic element but G contains a parabolic element; G is a group of elliptic elements.

(1) Assume first that G contains a loxodromic element. Then the Tits alternative and Lemma 8.39 imply the following alternatives:
   - ⋄ either G is conjugate to a subgroup of $\mathrm{GL}(2,\mathbb{Z})\ltimes \mathrm{D}_2$,
   - ⋄ or G contains a subgroup of index at most two that is isomorphic to $\mathbb{Z}\ltimes\mathrm{H}$ where H is a finite group.

(2) Suppose now that G does not contain a loxodromic element but G contains a parabolic element ϕ. The map ϕ preserves a unique fibration $\mathcal{F}$ that is elliptic or rational. Let us prove that any element of G preserves $\mathcal{F}$. Denote by $\alpha(\phi)\in\partial\mathbb{H}^\infty$ the fixed point of ϕ_*. Take one element in the light cone
$$\mathcal{L}\mathrm{Z}(\mathbb{P}^2_\mathbb{C})=\big\{d\in\mathrm{Z}(\mathbb{P}^2_\mathbb{C})\,|\,d\cdot d=0\big\}$$
of $\mathrm{Z}(\mathbb{P}^2_\mathbb{C})$ still denoted by $\alpha(\phi)$ that represents $\alpha(\phi)$. Assume by contradiction that there exists φ in G such that $\varphi(\alpha(\phi))\neq\alpha(\phi)$. The map $\psi=\varphi\circ\phi\circ\varphi^{-1}$ is parabolic and fixes the unique element $\alpha(\psi)$ of $\mathcal{L}\mathrm{Z}(\mathbb{P}^2_\mathbb{C})$ proportional to $\varphi(\alpha(\phi))$. If $\varepsilon>0$ let us denote by $\mathcal{U}\big(\alpha,\varepsilon\big)$ the set
$$\mathcal{U}\big(\alpha,\varepsilon\big)=\big\{\ell\in\mathcal{L}\mathrm{Z}(\mathbb{P}^2_\mathbb{C})\,|\,\alpha\cdot\ell<\varepsilon\big\}.$$
Take $\varepsilon>0$ such that $\mathcal{U}\big(\alpha(\phi),\varepsilon\big)\cap\mathcal{U}\big(\alpha(\psi),\varepsilon\big)=\emptyset$. Since ψ_* is parabolic, $\psi_*^n\big(\mathcal{U}\big(\alpha(\phi),\varepsilon\big)\big)$ is contained in $\mathcal{U}\big(\alpha(\psi),\varepsilon\big)$ for n large enough. For m sufficiently large the following inclusions hold:
$$\phi_*^m\circ\psi_*^n\big(\mathcal{U}\big(\alpha(\phi),\varepsilon\big)\big)\subset\mathcal{U}\left(\alpha(\phi),\frac{\varepsilon}{2}\right)\subsetneq\mathcal{U}\big(\alpha(\phi),\varepsilon\big).$$
This implies that $\phi_*^m\circ\psi_*^n$ is loxodromic: contradiction. So $\alpha(\phi_*)=\alpha(\varphi_*)$ for any $\varphi\in\mathrm{G}$. Finally G is a subgroup either of $\mathcal{J}$, or of the automorphism group of a Halphen surface.

(3) If G is a group of elliptic elements, then according to Theorems 8.14 and 8.15 either G is a bounded subgroup, or G preserves a rational fibration. □

8.6. Normal subgroups of the Cremona group

The strategy of Cantat and Lamy to produce strict, non-trivial, normal subgroups of $\mathrm{Bir}(\mathbb{P}^2_\Bbbk)$ is to let $\mathrm{Bir}(\mathbb{P}^2_\Bbbk)$ act on the hyperbolic space $\mathbb{H}^\infty(\mathbb{P}^2_\Bbbk)$. In the first part of their paper they define the notion of a tight element: an element ϕ of $\mathrm{Bir}(\mathbb{P}^2_\Bbbk)$ is *tight* if it satisfies the following three properties:

- ◇ $\phi_* \in \mathrm{Isom}(\mathbb{H}^\infty)$ is hyperbolic;
- ◇ there exists a positive number ε such that: if ψ belongs to $\mathrm{Bir}(\mathbb{P}^2_\Bbbk)$ and $\psi_*(\mathrm{Ax}(\phi))$ contains two points at distance ε which are at distance at most 1 from $\mathrm{Ax}(\phi)$, then $\psi_*(\mathrm{Ax}(\phi)) = \mathrm{Ax}(\phi)$;
- ◇ if ψ belongs to $\mathrm{Bir}(\mathbb{P}^2_\Bbbk)$ and $\psi_*(\mathrm{Ax}(\phi)) = \mathrm{Ax}(\phi)$, then $\psi\circ\phi\circ\psi^{-1} = \phi$ or $\psi\circ\phi\circ\psi^{-1} = \phi^{-1}$.

The second property is a rigidity property of $\mathrm{Ax}(\phi)$ with respect to isometries ψ_* for $\psi \in \mathrm{Bir}(\mathbb{P}^2_\Bbbk)$; we say that $\mathrm{Ax}(\phi)$ is *rigid* under the action of $\mathrm{Bir}(\mathbb{P}^2_\Bbbk)$. The third property means that the stabilizer of $\mathrm{Ax}(\phi)$ coincides with the normalizer of the cyclic group $\langle\phi\rangle$.

Here since there is no confusion we write $\ll \phi \gg$ for $\ll \phi \gg_{\mathrm{Bir}(\mathbb{P}^2_\Bbbk)}$.

Cantat and Lamy established the following statement:

THEOREM 8.54 ([**59**]). *Let $\Bbbk$ be an algebraically closed field. If $\phi \in \mathrm{Bir}(\mathbb{P}^2_\Bbbk)$ is tight, then there exists a non-zero integer n such that for any non-trivial element ψ of $\ll \phi^n \gg$*

$$\deg\psi \geq \deg(\phi^n).$$

In particular $\ll \phi^n \gg$ is a proper subgroup of $\mathrm{Bir}(\mathbb{P}^2_\Bbbk)$.

In the second part of their article Cantat and Lamy showed that $\mathrm{Bir}(\mathbb{P}^2_\Bbbk)$ contains tight elements. They distinguished two cases: $\Bbbk = \mathbb{C}$ and $\Bbbk \neq \mathbb{C}$. Let us focus on the case $\Bbbk = \mathbb{C}$. They proved that an element ϕ of $\mathrm{Bir}(\mathbb{P}^2_\mathbb{C})$ of the form $a\circ j$, where a is a general element of $\mathrm{PGL}(3,\mathbb{C})$ and j is a Jonquières twist, is tight. Let us explain what general means in this context: any element of $\mathrm{PGL}(3,\mathbb{C})$ suits after removing a countable number of Zariski closed subsets of $\mathrm{PGL}(3,\mathbb{C})$. More precisely they needed the following two conditions:

- ◇ the base-points of ϕ and ϕ^{-1} belong to $\mathbb{P}^2_\mathbb{C}$;
- ◇ $\mathrm{Base}(\phi^k)\cap\mathrm{Base}(\phi^{-i}) = \emptyset$ for any k, $i > 0$.

In [**131**] Lonjou proved the following statement:

THEOREM 8.55 ([**131**]). *For any field $\Bbbk$ the plane Cremona group $\mathrm{Bir}(\mathbb{P}^2_\Bbbk)$ is not simple.*

She did not use the notion of tight element but uses the WPD (weakly properly discontinuous) property. This property was proposed in the context of the mapping class group in [**18**]. An element g of a group G satisfies the WPD property if for any $\varepsilon \geq 0$ for any point $p \in \mathbb{H}^\infty$ there exists a positive integer N such that the set

$$S(\varepsilon,p;N) = \{h \in \mathrm{G} \,|\, \mathrm{dist}(h(p),p) \leq \varepsilon,\ \mathrm{dist}(h(g^N(p)),g^N(p)) \leq \varepsilon\}$$

is finite. Since the elements studied by Lonjou have an axis, she followed the terminology introduced in [**67**] and said that the group G *acts discretely along the axis of g*.

In [**68**] the authors generalized the small cancellation theory for groups acting by isometries on δ-hyperbolic spaces.

Small cancellation theory and the WPD property are connected:

- ◇ in the normal group generated by a family satisfying the small cancellation property elements have a large translation length ([**106**]);
- ◇ if some element g satisfies the WPD property then the conjugates of $\langle g^n\rangle$ form a family satisfying the small cancellation property.

Combining these two statements the following holds:

THEOREM 8.56 ([**68**]). *Let ε be a positive real number. Let* G *be a group acting by isometries on a δ-hyperbolic space X. Let g be a loxodromic element of* G. *If* G *acts discretely along the axis of g, then there exists $n\in\mathbb{N}$ such that for any $h\in\ll g^n\gg\smallsetminus\{\mathrm{id}\}$ the translation length $L(h)$ of h satisfies $L(h)>\varepsilon$.*

In particular, for n big enough $\ll g^n\gg$ is a proper subgroup of G. *Furthermore this subgroup is free.*

As a result to prove Theorem 8.55 Lonjou needed to exhibit elements satisfying the WPD property:

PROPOSITION 8.57 ([**131**]). *Let $n\geq 2$ and let $\Bbbk$ be a field of characteristic which does not divide n. Consider the action of $\mathrm{Bir}(\mathbb{P}^2_\Bbbk)$ on $\mathbb{H}^\infty(\mathbb{P}^2_{\overline{\Bbbk}})$ where $\overline{\Bbbk}$ is the algebraic closure of $\Bbbk$. The group $\mathrm{Bir}(\mathbb{P}^2_\Bbbk)$ acts discretely along the axis of the loxodromic map*

$$h_n\colon (z_0:z_1:z_2)\dashrightarrow\big(z_1z_2^{n-1}:z_1^n-z_0z_2^{n-1}:z_2^n\big).$$

REMARK 8.58. If $\Bbbk$ is an algebraically closed field of characteristic $p>0$, then for any $\ell\geq 1$ one has ([**62**])

$$\ll h_p^\ell\gg=\mathrm{Bir}(\mathbb{P}^2_\Bbbk).$$

Let us explain why when $\Bbbk=\mathbb{C}$.

Let us first establish that

$$\ll\sigma_2\gg=\mathrm{Bir}(\mathbb{P}^2_\mathbb{C}). \tag{8.1}$$

Let ϕ be a birational self map of the complex projective plane. According to the Noether and Castelnuovo Theorem:

$$\phi=(A_1)\circ\sigma_2\circ A_2\circ\sigma_2\circ A_3\circ\ldots\circ A_n\circ(\sigma_2)$$

where the A_i's belong to $\mathrm{PGL}(3,\mathbb{C})$. The group $\mathrm{PGL}(3,\mathbb{C})$ is simple; as a result any A_i can be written as

$$B_1\circ\big((z_0,z_1)\mapsto(-z_0,-z_1)\big)\circ B_1^{-1}\circ B_2\circ\big((z_0,z_1)\mapsto(-z_0,-z_1)\big)\circ B_2^{-1}\circ\ldots\circ B_n\circ\big((z_0,z_1)\\ \mapsto(-z_0,-z_1)\big)\circ B_n^{-1}$$

with B_i in $\mathrm{PGL}(3,\mathbb{C})$. The involutions $(z_0,z_1)\mapsto(-z_0,-z_1)$ and σ_2 are conjugate; therefore, ϕ can be written as a composition of conjugates of σ_2.

Since $\mathrm{PGL}(3,\mathbb{C})$ is simple, for any non-trivial element A of $\mathrm{PGL}(3,\mathbb{C})$ the involution $\iota\colon(z_0,z_1)\mapsto(-z_0,z_1)$ can be written as a composition of conjugates of A. The involutions ι and σ_2 being conjugate, one has

$$\sigma_2=\varphi_1\circ A\circ\varphi_1^{-1}\circ\varphi_2\circ A\circ\varphi_2^{-1}\circ\ldots\circ\varphi_n\circ A\circ\varphi_n^{-1}$$

where the φ_i's are some elements of $\mathrm{Bir}(\mathbb{P}^2_\mathbb{C})$. As a result $\ll\sigma_2\gg\subset\ll A\gg$. But $\ll\sigma_2\gg=\mathrm{Bir}(\mathbb{P}^2_\mathbb{C})$ (see (8.1)), so

$$\ll A\gg=\mathrm{Bir}(\mathbb{P}^2_\mathbb{C}). \tag{8.2}$$

If ϕ belongs to $\mathrm{PGL}(2,\mathbb{C}(z_1))$, then

$$\ll \phi \gg = \mathrm{Bir}(\mathbb{P}^2_\mathbb{C}) \tag{8.3}$$

Indeed since $\mathrm{PGL}(2,\mathbb{C}(z_1))$ is simple, the involution ι can be written as a composition of conjugates of ϕ. But according to (8.2) one has $\ll \iota \gg = \mathrm{Bir}(\mathbb{P}^2_\mathbb{C})$; hence $\ll \phi \gg = \mathrm{Bir}(\mathbb{P}^2_\mathbb{C})$.

If ϕ belongs to $\mathcal{J}$, then

$$\ll \phi \gg = \mathrm{Bir}(\mathbb{P}^2_\mathbb{C}). \tag{8.4}$$

Indeed up to birational conjugacy $\phi\colon (z_0,z_1) \dashrightarrow \big(\phi_1(z_0,z_1),\gamma(z_1)\big)$ where γ is a homothety or a translation. Consider an element $\psi\colon (z_0,z_1) \dashrightarrow \big(\psi_1(z_0,z_1),z_1\big)$ of $\mathrm{PGL}(2,\mathbb{C}(z_1))$. The map $\varphi = [\phi,\psi]$ belongs to

$$\ll \phi \gg \cap \mathrm{PGL}(2,\mathbb{C}(z_1)).$$

If ψ is well chosen, then φ is non-trivial and from (8.3) one gets

$$\ll \phi \gg = \mathrm{Bir}(\mathbb{P}^2_\mathbb{C}).$$

As a result if ϕ is a birational self map of the complex projective plane such that there exists $\psi \in \mathrm{Bir}(\mathbb{P}^2_\mathbb{C})$ for which $[\phi,\psi]$ preserves a rational fibration, then from (8.4)

$$\ll \phi \gg = \mathrm{Bir}(\mathbb{P}^2_\mathbb{C}). \tag{8.5}$$

Let $\phi\colon (z_0,z_1) \mapsto (z_1, P(z_1)-\delta z_0)$, $\delta \in \mathbb{C}^*$, $P \in \mathbb{C}[z_1]$, $\deg P \geq 2$, be a Hénon map. Then $\ll \phi \gg = \mathrm{Bir}(\mathbb{P}^2_\mathbb{C})$. Indeed if $\psi\colon (z_0,z_1) \mapsto (z_0, 2z_1)$, then $[\phi,\psi]$ preserves the rational fibration $z_0 =$cst; one concludes with (8.5).

More generally over any infinite field of characteristic which does not divide n the map h_n does not satisfy the WPD property: this explains the assumptions of Proposition 8.57.

Let us mention that Lonjou got not only the non-simplicity of the plane Cremona group from [**68**] but also the following result:

THEOREM 8.59 ([**131**]). *Let $\mathbb{k}$ be a field. The plane Cremona group*

- ◇ *contains free normal subgroups;*
- ◇ *is SQ-universal, that is, any countable subgroup embeds in a quotient of* $\mathrm{Bir}(\mathbb{P}^2_\mathbb{k})$.

In [**161**] the author proved that any loxodromic element in the Cremona group over any field $\mathbb{k}$ generates a proper normal subgroup; as a result the group $\mathrm{Bir}(\mathbb{P}^2_\mathbb{k})$ is not a simple group. He also gave a criterion in terms of the translation length of a loxodromic map ϕ to know if ϕ is tight and hence if $\ll \phi^n \gg$ is a proper subgroup of $\mathrm{Bir}(\mathbb{P}^2_\mathbb{k})$ for some n.

REMARK 8.60. Let us give the relationship between a tight element and an element that satisfies the WPD property. When we study the action of the Cremona group on $\mathbb{H}^\infty(\mathbb{P}^2_\mathbb{k})$ the axis of any loxodromic element ϕ is rigid and the stabiliser

$$\mathrm{Stab}(\mathrm{Ax}(\phi)) = \big\{\psi \in \mathrm{Bir}(\mathbb{P}^2_\mathbb{k}) \,|\, \psi(\mathrm{Ax}(\phi)) = \mathrm{Ax}(\phi)\big\}$$

of the axis $\mathrm{Ax}(\phi)$ is virtually cyclic if and only if some positive iterate of ϕ is tight ([**59, 131, 161**]). As a result for N large the set $S(\varepsilon,p;N)$ is contained in

$\mathrm{Stab}(\mathrm{Ax}(\phi))$. The map ϕ thus satisfies the WPD property if and only if some positive iterate of ϕ is tight.

REMARK 8.61. Let us recall that a subgroup H of a group G is called a *characteristic subgroup* of G if for every automorphism φ of G the inclusion $\varphi(\mathrm{H}) \subset \mathrm{H}$ holds.

Recall that the examples of elements having the WPD property given by Lonjou are the Hénon maps

$$h_n\colon (z_0 : z_1 : z_2) \dashrightarrow \left(z_1 z_2^{n-1} : z_1^n - z_0 z_2^{n-1} : z_2^n\right)$$

of degree n which is not divisible by the characteristic of $\Bbbk$. The group of automorphisms of $\mathrm{Bir}(\mathbb{P}^2_\mathbb{C})$ is generated by inner automorphisms and the action of $\mathrm{Aut}(\mathbb{C}, +, \cdot)$ (*see* §7.2). As h_n is defined over $\mathbb{Z}$ the subgroup $\ll h^m \gg$ is a characteristic subgroup of $\mathrm{Bir}(\mathbb{P}^2_\mathbb{C})$. One has the following result:

PROPOSITION 8.62 ([**53**]). *The plane Cremona group contains infinitely many characteristic subgroups.*

8.7. Simple groups of $\mathrm{Bir}(\mathbb{P}^2_\mathbb{C})$

This section is devoted to the classification of simple subgroups of $\mathrm{Bir}(\mathbb{P}^2_\mathbb{C})$ (Theorems 8.2 and 8.3) but also to the proof of the following statement:

THEOREM 8.63 ([**171**]). *Let S be a complex surface.*
If G *is a finitely generated simple subgroup of* $\mathrm{Bir}(S)$, *then* G *is finite.*

8.7.1. Simple subgroups of $\mathrm{Bir}(\mathbb{P}^2_\mathbb{C})$. Let us first prove Theorem 8.2. Consider a simple group acting non-trivially on a rational complex surface. Then according to Theorems 8.3 and 3.46 the group G is isomorphic to a subgroup of $\mathrm{PGL}(3,\mathbb{C})$.

Conversely the group $\mathrm{PGL}(3,\mathbb{C}) = \mathrm{Aut}(\mathbb{P}^2_\mathbb{C})$ acts by birational maps on S.

Let us now deal with the proof of Theorem 8.3. Let G be a simple subgroup of the plane Cremona group. We distinguish three cases:
(i) G contains no loxodromic element but a parabolic one;
(ii) G is an elliptic group;
(iii) G contains a loxodromic element.

(i) Assume that G contains no loxodromic element but a parabolic one.

LEMMA 8.64 ([**171**]). *Consider a simple subgroup* G *of* $\mathrm{Bir}(\mathbb{P}^2_\mathbb{C})$ *that contains no loxodromic element but a parabolic element.*

Then G *is conjugate to a subgroup of* $\mathcal{J}$ *and is isomorphic to a subgroup of* $\mathrm{PGL}(2,\mathbb{C})$.

PROOF. According to Lemma 8.46 one has the following alternative: G is conjugate
- either to a subgroup of the automorphism group of a Halphen surface,
- or to a subgroup of $\mathcal{J}$.

But automorphism groups of Halphen surfaces are finite extensions of abelian subgroups (Theorem 8.47), so do not contain infinite simple subgroups. As a result G is conjugate to a subgroup of $\mathcal{J}$. The short exact sequence from the semi-direct product of $\mathcal{J}$ is

$$1 \longrightarrow \mathrm{PGL}(2,\mathbb{C}(z_1)) \longrightarrow \mathcal{J} \stackrel{f}{\longrightarrow} \mathrm{PGL}(2,\mathbb{C}) \longrightarrow 1.$$

The group G is simple and thus contained in the kernel of the image of f. In both cases G is isomorphic to a subgroup of $\mathrm{PGL}(2,\mathbb{C})$. □

(ii) Suppose that G is an elliptic group.

LEMMA 8.65 ([**171**]). *Let* G *be a simple subgroup of the plane Cremona group of elliptic elements. Then*

- *either* G *is a subgroup of an algebraic group of* $\mathrm{Bir}(\mathbb{P}^2_\mathbb{C})$,
- *or* G *is conjugate to a subgroup of* $\mathcal{J}$.

PROOF. According to Theorems 8.14 and 8.15 one of the following holds:

- G is conjugate to a subgroup of an algebraic group;
- G preserves a rational fibration;
- G is a torsion group and G is isomorphic to a subgroup of an algebraic group.

In the first two cases we are done. Let us assume that we are in the third one. Then G is a linear group and according to the Theorem of Jordan and Schur, G has a normal abelian subgroup of finite index. As a consequence G is finite, and so algebraic. □

(iii) Finally we give a sketch of the proof of

THEOREM 8.66 ([**171**]). *A simple subgroup of* $\mathrm{Bir}(\mathbb{P}^2_\mathbb{C})$ *does not contain any loxodromic element.*

Let G be a simple subgroup of $\mathrm{Bir}(\mathbb{P}^2_\mathbb{C})$. Assume by contradiction that G contains a loxodromic map ϕ. Theorems 8.54 and 8.36 imply that ϕ is a monomial map up to birational conjugacy. Looking at the curves contracted by elements of G, Urech gets that all loxodromic elements of G are contained in $\mathrm{GL}(2,\mathbb{Z})\ltimes \mathrm{D}_2$ ([**172**, Lemmas 3.17 and 3.18]). Consider ψ in G. As $\psi\circ\phi\circ\psi^{-1}$ is loxodromic it is monomial. The axis of $\psi\circ\phi\circ\psi^{-1}$ is fixed pointwise by both $\psi\circ \mathrm{D}_2\circ\psi^{-1}$ and D_2. The group H generated by $\psi\circ\phi\circ\psi^{-1}$ and D_2 is thus bounded and according to Theorem 8.33 conjugate to a subgroup of D_2. Hence $\psi\circ \mathrm{D}_2\circ\psi^{-1}$ is contained in D_2 and ψ belongs to $\mathrm{GL}(2,\mathbb{Z})\ltimes \mathrm{D}_2$. Consequently we have the inclusion $\mathrm{G}\subset \mathrm{GL}(2,\mathbb{Z})\ltimes \mathrm{D}_2$ and get a non-trivial morphism $\upsilon\colon \mathrm{G}\to \mathrm{GL}(2,\mathbb{Z})$. The kernel of υ contains an infinite subgroup of D_2 normalized by ϕ (Lemma 8.34): contradiction with the fact that G is simple.

8.7.2. Finitely generated simple subgroups of $\mathrm{Bir}(\mathbb{P}^2_\mathbb{C})$**.** We finish the chapter by giving a sketch of the proof of the following statement:

THEOREM 8.67 ([**171**]). *Any finitely generated simple subgroup of the plane Cremona group is finite.*

This result and the classification of finite subgroups of $\mathrm{Bir}(\mathbb{P}^2_\mathbb{C})$ (*see* [**89**]) imply:

COROLLARY 8.68 ([**171**]). *A finitely generated simple subgroup of* $\mathrm{Bir}(\mathbb{P}^2_\mathbb{C})$ *is isomorphic to*

- *either* $\mathbb{Z}/_{p\mathbb{Z}}$ *for some prime* p;
- *or* $\mathcal{A}_5$;
- *or* $\mathcal{A}_6$;

◇ *or* $\mathrm{PSL}(2,\mathbb{C})$.

Note that the conjugacy classes of these finite groups are also described in [**89**].

REMARK 8.69. Theorem 8.67 also holds for the group of birational self maps of a surface over a field $\Bbbk$.

Let G be a finitely generated subgroup of $\mathrm{Bir}(\mathbb{P}^2_\mathbb{C})$. Let's first see that G does not contain loxodromic elements:

PROPOSITION 8.70 ([**171**]). *Let* G *be a finitely generated subgroup of* $\mathrm{Bir}(\mathbb{P}^2_\mathbb{C})$. *If* G *contains a loxodromic element, then* G *is not simple.*

To prove it we need the following statement.

PROPOSITION 8.71 ([**171**]). *Let* G *be a finitely generated subgroup of* $\mathrm{Bir}(\mathbb{P}^2_\mathbb{C})$. *There exist a finite field* $\Bbbk$ *and a non-trivial morphism* $\upsilon\colon \mathrm{G}\to\mathrm{Bir}(\mathbb{P}^2_\Bbbk)$ *such that for any* ϕ *in* G *the following inequality holds:* $\deg\upsilon(\phi)\leq\deg\phi$.

PROOF OF PROPOSITION 8.70. Let ϕ be a loxodromic element of G.

If ϕ^n is tight in G for some integer n, then Theorem 8.54 allows us to conclude.

If no power of ϕ is tight, then G contains an infinite subgroup Δ_2 that is normalized by ϕ and that is conjugate either to a subgroup of D_2, or to a subgroup of $\mathbb{C}^2$ (Theorem 8.36). In particular the degrees of the elements of Δ_2 are uniformly bounded by an integer N. According to Proposition 8.71 there exist a finite field $\Bbbk$ and a non-trivial morphism $\upsilon\colon \mathrm{G}\to\mathrm{Bir}(\mathbb{P}^2_\Bbbk)$ such that for all ϕ in G:

$$\deg\upsilon(\phi)\leq\deg\phi.$$

In $\mathrm{Bir}(\mathbb{P}^2_\Bbbk)$ there exist only finitely many elements of degree $\leq N$. As a result $\upsilon(\Delta_2)$ is finite. The morphism υ has thus a proper kernel and G is not simple: contradiction. □

We now have the following alternative

(i) G contains a parabolic element,

(ii) G is an elliptic subgroup.

Let us look at these two possibilities.

(i) If G contains a parabolic element, then G is conjugate either to a subgroup of the automorphism group $\mathrm{Aut}(S)$ of a Halphen surface, or to a subgroup of the Jonquières group $\mathcal{J}$.
  - ◇ Assume first that, up to conjugacy, $\mathrm{G}\subset\mathrm{Aut}(S)$ where S is a Halphen surface. Recall that a group G satisfies the *Malcev property* if every finitely generated subgroup Γ of G is residually finite, *i.e.*, for any $g\in\Gamma$ there exist a finite group H and a morphism $\upsilon\colon\Gamma\to\mathrm{H}$ such that g does not belong to $\ker\upsilon$. Malcev showed that linear groups satisfy this property ([**135**]). In [**5**] the authors proved that automorphism groups of scheme over any commutative ring also satisfy this property. Consequently if G contains a parabolic element, then G is, up to conjugacy, a subgroup of $\mathcal{J}$.
  - ◇ Suppose that $\mathrm{G}\subset\mathcal{J}$ up to birational conjugacy. Then G is finite. Indeed:

    LEMMA 8.72 ([**171**]). *Let* $\mathcal{C}$ *be a curve and let* $\mathrm{G}\subset\mathrm{Bir}(\mathbb{P}^1_\mathbb{C}\times\mathcal{C})$ *be a finitely generated simple subgroup that preserves the* $\mathbb{P}^1_\mathbb{C}$*-fibration given by the projection to* $\mathcal{C}$. *Then* G *is finite.*

PROOF. The group G being simple, G is isomorphic either to a subgroup of $\mathrm{PGL}(2,\mathbb{C})$, or to a subgroup of $\mathrm{Aut}(\mathcal{C})$. But both $\mathrm{PGL}(2,\mathbb{C})$ and $\mathrm{Aut}(\mathcal{C})$ satisfy the Malcev property, so G is finite. □

(ii) It remains to look at G when G is a finitely generated, simple, elliptic subgroup of $\mathrm{Bir}(\mathbb{P}^2_{\mathbb{C}})$. Proposition 8.23 asserts that either G is conjugate to a subgroup of $\mathcal{J}$, or G is contained in an algebraic subgroup of $\mathrm{Bir}(\mathbb{P}^2_{\mathbb{C}})$. In the first case Lemma 8.72 allows us to conclude. Let us focus on the last case: algebraic subgroups of $\mathrm{Bir}(\mathbb{P}^2_{\mathbb{C}})$ are linear; hence G is linear and therefore finite since linear groups satisfy the Malcev property.

CHAPTER 9

Big subgroups of automorphisms "of positive entropy"

In this chapter we will focus on automorphisms of surfaces with positive entropy. Recall that a *K3 surface*[1] is a complex, compact, simply connected surface S with a trivial canonical bundle. Equivalently there exists a holomorphic 2-form ω on S which is never zero; ω is unique modulo multiplication by a scalar. Let S be a K3 surface with a holomorphic involution ι. If ι has no fixed point, the quotient of S by $\langle\iota\rangle$ is an *Enriques surface*, otherwise it is a rational surface. Recall that every non-minimal rational surface can be obtained by repeatedly blowing up a minimal rational surface. The *minimal rational surfaces* are the complex projective plane, $\mathbb{P}^1_\mathbb{C}\times\mathbb{P}^1_\mathbb{C}$ and the Hirzebruch surfaces $\mathbb{F}_n$, $n\geq 2$. If S is a complex, compact surface carrying a biholomorphism of positive topological entropy, then S is either a complex torus, or a K3 surface, or an Enriques surface, or a non-minimal rational surface ([**50**]). Although automorphisms of complex tori are easy to describe, it is rather difficult to construct automorphisms on K3 surfaces or rational surfaces. Constructions and dynamical properties of automorphisms of K3 surfaces can be found in [**51**] and [**139**]. The first examples of rational surfaces endowed with biholomorphisms of positive entropy are due to Coble and Kummer ([**64**]):

- ⋄ the Coble surfaces are obtained by blowing up the ten nodes of a nodal sextic in $\mathbb{P}^2_\mathbb{C}$;
- ⋄ the Kummer surfaces are desingularizations of quotients of complex 2-tori by involutions with fixed points.

Obstructions to the existence of such biholomorphisms on rational surfaces are also known: if ϕ is a biholomorphism of a rational surface S such that $\mathrm{h_{top}}(\phi)>0$, then the representation

$$\mathrm{Aut}(S)\to\mathrm{GL}(\mathrm{Pic}(S)) \qquad\qquad g\mapsto g^*$$

has infinite image. Hence according to [**107**] its kernel is finite so that S has no non-zero holomorphic vector field. A second obstruction follows from [**146**]: the surface S has to be obtained by successive blow-ups from the complex projective plane and the number of blow-ups must be at least ten. The first infinite families of examples have been constructed independently in [**140**] and [**12**] by different methods. Since then many constructions have emerged (*see for instance* [**13, 14, 82, 84, 140, 166**]).

In the first section we give three answers to the question "When is a birational self map of a complex projective surface birationally conjugate to an automorphism?" In the second section we deal with constructions of automorphisms of

[1] "So-named in honor of Kummer, Kähler, Kodaira and of the beautiful mountain K2 in Kashmir" ([**175**]).

rational surfaces with positive entropy. In the last section we explain how $\mathrm{SL}(2,\mathbb{Z})$ is realized as a subgroup of automorphisms of a rational surface with the property that every element of infinite order has positive entropy.

9.1. Birational maps and automorphisms

9.1.1. Definitions. Given a birational map $\phi\colon S \dashrightarrow S$ of a projective complex surface its dynamical degree $\lambda(\phi)$ is a positive real number that measures the complexity of the dynamics of ϕ (*see* §2.3). The neperian logarithm $\log\lambda(\phi)$ provides an upper bound for the topological entropy of $\phi\colon S \dashrightarrow S$ and is equal to it under natural assumptions ([**10, 86**]). Let us give an alternative but equivalent definition to that of §2.3. A birational map $\phi\colon S \dashrightarrow S$ of a projective complex surface determines an endomorphism $\phi_*\colon \mathrm{NS}(S)\to\mathrm{NS}(S)$; the dynamical degree $\lambda(\phi)$ of ϕ is defined as the spectral radius of the sequence of endomorphisms $(\phi^n)_*$ as n goes to infinity:

$$\lambda(\phi) = \lim_{n\to+\infty} ||(\phi^n)_*||^{1/n}$$

where $||\cdot||$ denotes a norm on the real vector space $\mathrm{End}(\mathrm{NS}(S))$. This limit exists and does not depend on the choice of the norm. For any ample divisor $D\subset S$:

$$\lambda(\phi) = \lim_{n\to+\infty} (D\cdot(\phi^n)_*D)^{1/n}.$$

The Néron-Severi group of $\mathbb{P}^2_\mathbb{C}$ coincides with the Picard group of $\mathbb{P}^2_\mathbb{C}$, has rank 1, and is generated by the class $\mathbf{e}_0$ of a line

$$\mathrm{NS}(\mathbb{P}^2_\mathbb{C}) = \mathrm{Pic}(\mathbb{P}^2_\mathbb{C}) = \mathbb{Z}\mathbf{e}_0.$$

A map $\phi\in\mathrm{Bir}(\mathbb{P}^2_\mathbb{C})$ acts on $\mathrm{Pic}(\mathbb{P}^2_\mathbb{C})$ by multiplication by $\deg\phi$.

9.1.2. Pisot and Salem numbers. We will give the definitions of Pisot and Salem numbers, for more details *see* [**15**].

A *Pisot number* is an algebraic integer $\lambda\in]1,+\infty[$ whose other Galois conjugates lie in the unit disk. Let us denote by Pis the set of Pisot numbers. It includes all integers ≥ 2 as well as all reciprocical quadratic integers $\lambda>1$. The set Pis is a closed subset of the real line; its infimum is equal to the unique root $\lambda_P>1$ of the cubic equation $x^3=x+1$. The smallest accumulation point of Pis is the golden mean $\lambda_G=\frac{1+\sqrt{5}}{2}$. Note that all Pisot numbers between λ_P and λ_G have been listed.

A *Salem number* is an algebraic integer $\lambda\in]1,+\infty[$ whose other Galois conjugates are in the closed unit disk with at least one on the boundary. The minimal polynomial of λ has thus at least two complex conjugate roots on the unit circle, its roots are permuted by the involution $z\mapsto\frac{1}{z}$ and has degree at least 4. Let Sal be the set of Salem numbers. The unique root $\lambda_L>1$ of the irreducible polynomial $x^{10}+x^9-x^7-x^6-x^5-x^4-x^3+x+1$ is a Salem number. Conjecturally the infimum of Sal is larger than 1 and should be equal to λ_L.

Note that Pis is contained in the closure of Sal.

9.1.3. Dynamical degrees and Pisot and Salem numbers. Let us recall that a birational map $\phi\colon S\dashrightarrow S$ of a compact complex surface is algebraically stable if $(\phi^*)^n=(\phi^n)^*$ for all $n\geq 0$ (*see* §2.3). If ϕ is algebraically stable, then so is ϕ^{-1} and $\lambda(\phi)$ is an algebraic integer. Any birational map of a compact complex surface is conjugate by a birational morphism to an algebraically stable map (Proposition 2.10). From this fact and the Hodge index theorem according to which the intersection form has signature $(1, r_S - 1)$, where r_S denotes the rank of S, Diller and Favre get the following statement:

THEOREM 9.1 ([**85**]). *Let ϕ be a birational self map of a complex projective surface. If $\lambda(\phi)$ is distinct from 1, i.e., if ϕ is loxodromic, then $\lambda(\phi)$ is a Pisot or a Salem number.*

9.1.4. When is a birational map conjugate to an automorphism? A natural question is the following one: when is a birational self map of a complex projective surface birationally conjugate to an automorphism? There are three answers to this question and we will detail it.

9.1.4.1. *A first answer.* Diller and Favre give the first characterization of loxodromic birational maps which are conjugate to an automorphism of a projective surface:

THEOREM 9.2 ([**85**]). *Let $\phi \in \mathrm{Bir}(\mathbb{P}^2_\mathbb{C})$ be a loxodromic map. Assume that ϕ is algebraically stable. The action of ϕ on $\mathrm{H}^{1,1}(\mathbb{P}^2_\mathbb{C})$ admits the eigenvalue $\lambda(\phi) > 1$ with eigenvector $\Theta(\phi)$.*

The map ϕ is birationally conjugate to an automorphism if and only if $\Theta(\phi) \cdot \Theta(\phi) = 0$.

When ϕ is an automorphism, it is easy to check that $\Theta(\phi) \cdot \Theta(\phi) = 0$. We will thus deal with the reciprocical property. Let ϕ be a birational self map of a complex projective surface S. Assume that ϕ is algebraically stable. Hence $\lambda(\phi)$ is equal to the spectral radius of $\phi_* \in \mathrm{End}(\mathrm{NS}(\mathbb{R}, S))$ but also to the spectral radius of $\phi^* = (\phi^{-1})_*$; indeed these endomorphisms are adjoint for the intersection form:

$$\phi_* C \cdot D = C \cdot \phi^* D$$

for all C, D divisor classes. One can factorize ϕ as $\phi = \eta \circ \pi^{-1}$ where $\eta \colon Z \to S$ and $\pi = \pi_1 \circ \ldots \circ \pi_m \colon Z \to S$ are two sequences of point blow-ups. Denote by $F_j \subset Z$ the total transform of the indeterminacy point of π_j^{-1} under the map $\pi_j \circ \ldots \circ \pi_m$. For $1 \leq j \leq m$ let E_j be the direct image of F_j by η. Each E_j, if not zero, is an effective divisor. According to [**85**] we get the following formula called the push-pull formula

$$\phi_* \phi^* C = C + \sum_{j=1}^{m} (C \cdot E_j) E_j \tag{9.1}$$

for all curves (resp. divisor classes) C in S. Since ϕ_* and ϕ^* are adjoint endomorphisms of $\mathrm{NS}(\mathbb{R}, S)$ for the intersection form we get

$$\phi^* C \cdot \phi^* C = C \cdot C \sum_{j=1}^{m} (E_j \cdot C)^2. \tag{9.2}$$

This formula and the Hodge index theorem imply that $\lambda(\phi)$ is a Pisot number or Salem number.

The endomorphisms ϕ^* and ϕ_* preserve both the pseudo effective and nef cones of $\mathrm{NS}(\mathbb{R}, S)$. Suppose that $\lambda(\phi) > 1$. According to the Perron-Frobenius theorem there exists an eigenvector $\Theta(\phi)$ for ϕ_* in the nef cone of $\mathrm{NS}(S)$ such that

$$\phi^* \Theta(\phi) = \lambda(\phi) \Theta(\phi). \tag{9.3}$$

Note that, furthermore, this vector is unique up to scalar form ([**85**]). Both (9.2) and (9.3) imply that

$$(\lambda(\phi)^2-1)\Theta(\phi)\cdot\Theta(\phi)=\sum_{j=1}^{m}(E_j\cdot\Theta(\phi))^2.$$

As a result for all E_j:

$$\Theta(\phi)\cdot\Theta(\phi)=0 \qquad\Longleftrightarrow\qquad \Theta(\phi)\cdot E_j=0.$$

Assume now that $\Theta(\phi)\cdot\Theta(\phi)=0$; then $\Theta(\phi)\cdot E_j=0$ for all E_j. As the E_j's are effective and $\Theta(\phi)$ is nef the $\mathbb{Q}$-vector subspace of $\mathrm{NS}(\mathbb{Q},S)$ generated by the irreducible components of the divisors E_j is contained in $\Theta(\phi)^\perp$. On the orthogonal complement $\Theta(\phi)^\perp$ of the isotropic vector $\Theta(\phi)$ the intersection form is negative and its kernel is the line generated by $\Theta(\phi)$. Equation (9.1) implies

$$\phi_*^k\Theta(\phi)=\frac{1}{\lambda(\phi)^k}\Theta(\phi).$$

But $\lambda(\phi)>1$ and ϕ_* preserves the lattice $\mathrm{NS}(\mathbb{Z},S)$, so $\Theta(\phi)$ is irrational. Consequently the intersection form is negative definite on the $\mathbb{Q}$-vector space generated by all classes of irreducible components of the divisors E_j. According to the Grauert-Mumford contraction theorem ([**3**]) there exists a birational morphism $\eta\colon S\to Y$ that contracts simultaneously all these components. Set $\varphi=\eta\circ\phi\circ\eta^{-1}$. As $\Theta(\phi)$ does not intersect the curves contracted by η the class $\eta_*\Theta(\phi)\in\mathrm{NS}(\mathbb{R},Y)$ is

- ◇ isotropic, and
- ◇ an eigenvector for φ_* with eigenvalue $\lambda(\phi)$.

Let us iterate this process until φ^{-1} does not contract any curve, that is $\varphi\in\mathrm{Aut}(Y)$. If Y is singular, then consider the minimal desingularization $\widetilde{Y}$ of Y; the automorphism φ lifts to an automorphism $\widetilde{\varphi}$ of $\widetilde{Y}$.

As a result one can state

THEOREM 9.3 ([**85**]). *Let S be a complex projective surface. Let ϕ be a loxodromic birational self map of S. Then*

- ◇ *all divisors E_j are orthogonal to $\Theta(\phi)$ if and only if $\Theta(\phi)$ is an isotropic vector;*
- ◇ *if $\Theta(\phi)$ is an isotropic vector, then there exists a birational morphism $\eta\colon S\to Y$ such that $\eta\circ\phi\circ\eta^{-1}$ is an automorphism of Y.*

Then Diller and Favre prove the following statement:

THEOREM 9.4 ([**85**]). *Let $\phi\in\mathrm{Bir}(S)$ (resp. $\psi\in\mathrm{Bir}(\widetilde{S})$) be an algebraically stable map of a complex projective surface S (resp. $\widetilde{S}$). Assume that ϕ and ψ are conjugate via a proper modification. Suppose that $\lambda(\phi)>1$ (or equivalently that $\lambda(\psi)>1$). Then $\Theta(\phi)\cdot\Theta(\phi)=0$ if and only if $\Theta(\psi)\cdot\Theta(\psi)=0$.*

Theorem 9.2 follows from Theorems 9.3 and 9.4.

9.1.4.2. *A second answer.* The following statement gives another characterization of birational maps conjugate to an automorphism of a smooth projective rational surface:

THEOREM 9.5 ([**31, 85**]). *Let ϕ be a birational map of a complex projective surface S.*

◇ *If* $\lambda(\phi)$ *is a Salem number, then there exists a birational map* $\psi\colon \widetilde{S} \dashrightarrow S$ *that conjugates* ϕ *to an automorphism of* $\widetilde{S}$*;*
◇ *if* ϕ *is conjugate to an automorphism, then* $\lambda(\phi)$ *is a quadratic integer or a Salem number.*

Assume that $\lambda(\phi)$ is a Salem number. Denote by $P(t) \in \mathbb{Z}[t]$ the minimal polynomial of $\lambda(\phi)$. But $\lambda(\phi)$ is a Salem number, so there exists a root of P with modulus 1; denote it α. Hence fix an automorphism κ of the field $\mathbb{C}$ such that $\kappa(\lambda(\phi)) = \alpha$. According to Proposition 2.10 we can suppose that ϕ is algebraically stable up to birational conjugacy. The eigenvector $\Theta(\phi)$ thus corresponds to the eigenvalue $\lambda(\phi)$, and so may be taken in $\mathrm{NS}(L, S)$ where L is the splitting field of P. The automorphism κ acts on $\mathrm{NS}(\mathbb{C}, S)$ preserving $\mathrm{NS}(S)$ pointwise. Since ϕ^* is defined over $\mathbb{Z}$ and $\phi^*\Theta(\phi) = \lambda(\phi)\Theta(\phi)$ one obtains

$$\phi^*(\kappa(\Theta(\phi))) = \kappa(\lambda(\phi))\kappa(\Theta(\phi)) = \alpha\kappa(\Theta(\phi));$$

that is $\phi^*\widetilde{\Theta} = \alpha\widetilde{\Theta}$ where $\widetilde{\Theta} = \kappa(\Theta(\phi))$. The divisor classes of the E_j's belong to $\mathrm{NS}(S)$, so they are κ-invariant. As a consequence (9.1) implies

$$\phi_*\phi^*\widetilde{\Theta} = \widetilde{\Theta} + \sum_{j=1}^{m}(\widetilde{\Theta} \cdot E_j)E_j. \tag{9.4}$$

Denote by $\overline{\widetilde{\Theta}}$ the conjugate of $\widetilde{\Theta}$ and by $\overline{\alpha}$ the conjugate of α; from (9.4) one gets

$$(\alpha\overline{\alpha})\widetilde{\theta} \cdot \overline{\widetilde{\theta}} = \phi^*\widetilde{\Theta} \cdot \phi^*\overline{\widetilde{\Theta}}.$$

As $|\alpha| = \alpha\overline{\alpha} = 1$ one gets that $E_j \cdot \widetilde{\Theta} = 0$ for any $1 \leq j \leq m$ and $E_j \cdot \Theta(\phi) = 0$ for any $1 \leq j \leq m$.

Theorem 9.5 follows from Theorem 9.3.

REMARK 9.6. Theorem 9.5 does not extend to quadratic integers (*see* [**31**]).

9.1.4.3. *A third answer.* As we have seen in §1.1 if S is a projective smooth surface, then every $\phi \in \mathrm{Bir}(S)$ admits a minimal resolution, *i.e.*, there exist $\pi_1\colon Z \to S$, $\pi_2\colon Z \to S$ two sequences of blow-ups such that

◇ no (-1)-curves of Z are contracted by both π_1 and π_2;
◇ $\phi = \pi_2 \circ \pi_1^{-1}$.

Denote by $\mathfrak{b}(\phi)$ the number of base-points of ϕ; note that $\mathfrak{b}(\phi)$ is equal to the difference of the ranks of $\mathrm{Pic}(Z)$ and $\mathrm{Pic}(S)$; thus $\mathfrak{b}(\phi)$ is equal to $\mathfrak{b}(\phi^{-1})$. Let us introduce the *dynamical number of the base-points of* ϕ: it is

$$\mu(\phi) = \lim_{k\to+\infty} \frac{\mathfrak{b}(\phi^k)}{k}.$$

Since $\mathfrak{b}(\phi \circ \psi) \leq \mathfrak{b}(\phi) + \mathfrak{b}(\psi)$ for any ϕ, ψ in $\mathrm{Bir}(S)$, $\mu(\phi)$ is a non-negative real number. As $\mathfrak{b}(\phi) = \mathfrak{b}(\phi^{-1})$ one gets $\mu(\phi^k) = |k\mu(\phi)|$ for any $k \in \mathbb{Z}$. Furthermore if $\psi\colon S \dashrightarrow Z$ is a birational map between smooth projective surfaces and if $\phi \in \mathrm{Bir}(S)$, then for all $n \in \mathbb{Z}$:

$$-2\mathfrak{b}(\psi) + \mathfrak{b}(\phi^n) \leq \mathfrak{b}(\psi \circ \varphi^n \circ \psi^{-1}) \leq 2\mathfrak{b}(\psi) + \mathfrak{b}(\phi^n);$$

hence $\mu(\phi) = \mu(\psi \circ \phi \circ \psi^{-1})$. One can thus state the following result:

LEMMA 9.7 ([**33**]). *The dynamical number of base-points is an invariant of conjugation. In particular if* ϕ *is conjugate to an automorphism of a smooth projective surface, then* $\mu(\phi) = 0$.

A base-point p of ϕ is a *persistent base-point* if there exists an integer N such that for any $k \geq N$:

$$\begin{cases} p \in \mathrm{Base}(\phi^k), \\ p \notin \mathrm{Base}(\phi^{-k}). \end{cases}$$

Let p be a point of S or a point infinitely near S such that $p \notin \mathrm{Base}(\phi)$. Consider a minimal resolution of ϕ:

$$\begin{array}{ccc} & Z & \\ {\scriptstyle \pi_2} \swarrow & & \searrow {\scriptstyle \pi_1} \\ S & \overset{}{\underset{\phi}{\dashrightarrow}} & S \end{array}$$

Because p is not a base-point of ϕ it corresponds via π_1 to a point of Z or infinitely near; using π_2 we view this point on S again maybe infinitely near and denote it $\phi^\bullet(p)$. For instance if $S = \mathbb{P}^2_\mathbb{C}$, $p = (1 : 0 : 0)$ and ϕ is the birational self map of $\mathbb{P}^2_\mathbb{C}$ given by

$$(z_0 : z_1 : z_2) \dashrightarrow (z_1 z_2 + z_0^2 : z_0 z_2 : z_2^2)$$

the point $\phi^\bullet(p)$ is not equal to $p = \phi(p)$ but is infinitely near to it. Note that if ϕ, ψ are two birational self maps of S and p is a point of S such that $p \notin \mathrm{Base}(\phi)$, $\phi(p) \notin \mathrm{Base}(\psi)$, then $(\psi \circ \phi)^\bullet(p) = \psi^\bullet(\phi^\bullet(p))$. One can put an equivalence relation on the set of points of S or infinitely near S: the point p is *equivalent* to the point q if there exists an integer k such that $(\phi^k)^\bullet(p) = q$; in particular $p \notin \mathrm{Base}(\phi^k)$ and $q \notin \mathrm{Base}(\phi^{-k})$. Note that the equivalence class is the generalization of a set of orbits for birational maps.

A base-point is *periodic* if

- either $(\phi^k)^\bullet(q) = q$ for some $k \geq 0$,
- or $q \in \mathrm{Base}(\phi^k)$ for any $k \in \mathbb{Z} \smallsetminus \{0\}$ (in particular $(\phi^k)^\bullet(p)$ is never defined for $k \neq 0$).

Let $\mathcal{P}$ be the set of periodic base-points of ϕ. Denote by $\widehat{\mathcal{P}}$ the finite set of points equivalent to a point of $\mathcal{P}$. Both $\mathfrak{b}(\phi)$ and $\mathfrak{b}(\phi^{-1})$ are finite, so there exists $n \in \mathbb{N}$ such that for any $p \in \mathrm{Base}(\phi)$ non-periodic and for any j, $\ell \geq N$:

$$\begin{cases} p \in \mathrm{Base}(\phi^j) \iff p \in \mathrm{Base}(\phi^\ell), \\ p \in \mathrm{Base}(\phi^{-j}) \iff p \in \mathrm{Base}(\phi^{-\ell}). \end{cases}$$

Let us decompose $\mathrm{Base}(\phi)$ into five disjoint sets:

$$\begin{aligned} \mathcal{B}_{++} &= \big\{p \,|\, p \notin \mathcal{P},\, p \in \mathrm{Base}(\phi^j),\, p \in \mathrm{Base}(\phi^{-j}) \quad \forall\, j \geq N\big\}, \\ \mathcal{B}_{+-} &= \big\{p \,|\, p \notin \mathcal{P},\, p \in \mathrm{Base}(\phi^j),\, p \notin \mathrm{Base}(\phi^{-j}) \quad \forall\, j \geq N\big\}, \\ \mathcal{B}_{-+} &= \big\{p \,|\, p \notin \mathcal{P},\, p \notin \mathrm{Base}(\phi^j),\, p \in \mathrm{Base}(\phi^{-j}) \quad \forall\, j \geq N\big\}, \\ \mathcal{B}_{--} &= \big\{p \,|\, p \notin \mathcal{P},\, p \notin \mathrm{Base}(\phi^j),\, p \notin \mathrm{Base}(\phi^{-j}) \quad \forall\, j \geq N\big\}, \end{aligned}$$

and $\mathcal{P}$.

REMARKS 9.8. Note that:

- ⋄ $\mathcal{B}_{+-}$ is the set of persistent base-points of ϕ;
- ⋄ $\mathcal{B}_{-+}$ is the set of persistent base-points of ϕ^{-1};
- ⋄ two equivalent base-points of ϕ belong to the same subsets of $\mathrm{Base}(\phi)$.

Take $k \geq 2N$ an integer. Let us compute $\mathfrak{b}(\phi^k)$. Any base-point of ϕ^k is equivalent to a base-point of ϕ. Let us thus consider a base-point p of ϕ and determine the number $m_{p,k}$ of base-points of ϕ^k which are equivalent to p.

(a) If p belongs to $\mathcal{P}$, then the number of points equivalent to p is less than $\#\mathcal{P}$ and $m_{p,k} \leq \#\mathcal{P}$.

(b) If p does not belong to $\mathcal{P}$, then any point equivalent to p is equal to $(\phi^i)^\bullet(p)$ for some i; furthermore these points are all distinct. Hence $m_{p,k} = \#I_{p,k}$ where

$$I_{p,k} = \{i \in \mathbb{Z} \,|\, p \notin \mathrm{Base}(\phi^i),\, p \in \mathrm{Base}(\phi^{i+k})\}.$$

(b)(i) Suppose that p belongs to $\mathcal{B}_{++}$. Since p does not belong to $\mathrm{Base}(\phi^i)$, the following inequalities hold: $-N < i < N$, and so $m_{p,k} < 2N$.

(b)(ii) If p belongs to $\mathcal{B}_{--}$, then p belongs to $\mathrm{Base}(\phi^{i+k})$; hence $-N < i+k < N$ and $m_{p,k} < 2N$.

(b)(iii) Assume that p belongs to $\mathcal{B}_{-+}$. As $p \notin \mathrm{Base}(\phi^i)$ (resp. $p \in \mathrm{Base}(\phi^{i+k})$), one has $-N < i$ (resp. $i+k \leq N$). These two conditions imply $-N < i \leq N-k$. But $k > 2N$, so $m_{p,k} = 0$.

(b)(iv) Finally consider a point p in $\mathcal{B}_{+-}$. The fact that $p \notin \mathrm{Base}(\phi^i)$ (resp. $p \in \mathrm{Base}(\phi^{i+k})$) yields $i < N$ (resp. $-N < i+k$). As a result $-N-k < i < N$ and $m_{p,k} \leq 2N+k$. Conversely if $i \leq -N$ and $i+k \geq N$, then $p \notin \mathrm{Base}(\phi^i)$ and $p \in \mathrm{Base}(\phi^{i+k})$, *i.e.*, $i \in I_{p,k}$. As a consequence $m_{p,k} \geq \#[N-k, -N] = k - 2N + 1$. Finally

$$-2N \leq m_{p,k} - k \leq 2N.$$

Consequently there exist two constants α, β (independent on k) such that for all $k \geq 2N$:

$$\nu k + \alpha \leq \mathfrak{b}(\phi^k) \leq \nu k + \beta$$

where ν is the number of equivalence classes of persistent base-points of ϕ (recall that $\mathcal{B}_{+-}$ is the set of persistent base-points of ϕ). But $\mu(\phi) = \lim\limits_{k\to+\infty} \dfrac{\mathfrak{b}(\phi^k)}{k}$, so $\mu(\phi) = \nu$. One can thus state:

PROPOSITION 9.9 ([**33**]). *Let S be a smooth projective surface. Let ϕ be a birational self map of S.*

Then $\mu(\phi)$ coincides with the number of equivalence classes of persistent base-points of ϕ. In particular $\mu(\phi)$ is an integer.

The following statement gives another characterization of birational maps which are conjugate to an automorphism of a projective surface; contrary to the two previous ones it works for all maps of $\mathrm{Bir}(S)$.

THEOREM 9.10 ([**33**]). *Let ϕ be a birational self map of a smooth projective surface. Then ϕ is conjugate to an automorphism of a smooth projective surface if and only if $\mu(\phi) = 0$.*

REMARK 9.11. This characterization was implicitely used in [**12–14, 82**].

Let us give an example of [**82**]. Consider the birational self map of $\mathbb{P}^2_{\mathbb{C}}$ given by

$$\psi\colon (z_0 : z_1 : z_2) \dashrightarrow (z_0z_2^2 + z_1^3 : z_1z_2^2 : z_2^3);$$

it has five base-points: $p = (1:0:0)$ and four points infinitely near. Denote by $\widehat{P_1}$ the collection of these points. Similarly ψ^{-1} has five base-points: $(1:0:0)$ and four points infinitely near; let $\widehat{P_2}$ be the collection of these points. Consider the automorphism A given by

$$A\colon (z_0 : z_1 : z_2) \mapsto \big(\alpha z_0 + 2(1-\alpha)z_1 + (2+\alpha-\alpha^2)z_2 : -z_0 + (\alpha+1)z_2 : z_0 - 2z_1 + (1-\alpha)z_2\big)$$

with $\alpha \in \mathbb{C} \smallsetminus \{0, 1\}$. Then

- ◇ $\widehat{P_1}$, $A(\widehat{P_2})$, and $(A \circ \psi \circ A)(\widehat{P_2})$ have distinct supports;
- ◇ $\widehat{P_1} = (A \circ \psi)^2 \circ A(\widehat{P_2})$.

As a result the base-points of $\phi = A \circ \psi$ are non-persistent, so ϕ is conjugate to an automorphism of a rational surface; this rational surface is $\mathbb{P}^2_\mathbb{C}$ blown up in $\widehat{P_1}$, $A(\widehat{P_2})$, and $(A \circ \psi \circ A)(\widehat{P_2})$. Furthermore $\lambda(A \circ \psi) > 1$.

Proof of Theorem 9.10. Lemma 9.7 shows that if ϕ is conjugate to an automorphism of a smooth projective surface, then $\mu(\phi) = 0$.

Let us prove the converse. Assume that $\mu(\phi) = 0$. One can suppose that by blowing up points ϕ is algebraically stable (Proposition 2.10). Therefore, ϕ has no periodic base-point and $\mathcal{B}_{++} = \emptyset$. Furthermore $\mu(\phi) = 0$ corresponds to $\mathcal{B}_{+-} = \mathcal{B}_{-+} = \emptyset$. All base-points thus belong to $\mathcal{B}_{--}$. Assume that ϕ is not an automorphism of S. Let $\tau\colon Z \to S$ be the blow-up of the base-points of ϕ. The morphism $\chi = \phi \circ \tau\colon Z \to S$ is the blow-up of the base-points of ϕ^{-1}. Consider a (-1)-curve $E \subset Z$ contracted by χ. The image $\chi(E)$ of E is a proper point of S that belongs to $\mathrm{Base}(\phi^{-1})$. Since ϕ is algebraically stable, then for all $k \geq 0$

$$\chi(E) \notin \mathrm{Base}(\phi^k).$$

As a result $\phi^k \circ \chi\colon Z \dashrightarrow S$ is well-defined at any point of E. The curve $C = \tau(E)$ is thus an irreducible curve of S contracted by ϕ^{k+1}; any base-point of ϕ^{k+1} that belongs to C as proper of infinitely near point is also a base-point of ϕ. This finite set of points is contained in $\mathcal{B}_{--}$; so there is $n > 0$ such that no base-point of ϕ^n belongs to C. Since C is blown down by ϕ^n, C is a (-1)-curve of S. Contracting C conjugates ϕ to an algebraically stable birational map all of whose base-points are in $\mathcal{B}_{--}$. The rank of the Picard group of this new surface is strictly less than the rank of $\mathrm{Pic}(S)$. Consequently if we repeat this process, it has to stop. In other words ϕ is conjugate to an automorphism of a smooth projective surface. □

9.2. Constructions of automorphisms with positive entropy

9.2.1. McMullen's idea. In [**140**] McMullen establishes a result similar to Torelli's theorem for K3 surfaces: he constructs automorphisms on some rational surfaces prescribing the action of the automorphisms on cohomological groups of the surface.

The relationship between the Coxeter group and the birational geometry of the plane, used by McMullen, has been discussed since 1895 (*see* [**123**]) and has been much developed since then (*see for instance* [**64, 87, 90, 101, 108**]).

A rational surface S is a *marked blow-up* of $\mathbb{P}^2_\mathbb{C}$ if it is presented as a blow-up $\pi\colon S \to \mathbb{P}^2_\mathbb{C}$ of $\mathbb{P}^2_\mathbb{C}$ at n distinct points p_1, p_2, $\ldots$, p_n. The marking determines the basis for $\mathrm{Pic}(S)$ given by the hyperplane bundle and the classes of the exceptional curves over the p_j. The first step toward finding an automorphism ϕ of S is to construct a plausible candidate for its linear action ϕ^* on the Picard group. Note that candidate actions must preserve the intersection form, the class of the canonical divisor, and the set of effective classes. Let us mention two sorts of involutions on $\mathrm{Pic}(S)$ that satisfy these restrictions:

◇ an abstraction of the involution σ_2,
◇ the involution that swaps the basis elements corresponding to two different exceptional curves.

If we compose such involutions one gets a Coxeter group W_n

◇ that is infinite as soon as $n \geq 9$,
◇ has elements with positive spectral radius when $n \geq 10$.

Furthermore except in some degenerate situations an element $w \in \mathrm{W}_n$ transforms the basis of $\mathrm{Pic}(S)$ corresponding to the given marking into a basis corresponding to some other marking $\varphi' \colon S \to \mathbb{P}^2_{\mathbb{C}}$. If the base-points of the new marking coincide, up to an element of $\mathrm{Aut}(\mathbb{P}^2_{\mathbb{C}})$, with those of the original, then one obtains an automorphism $\phi = \varphi^{-1} \circ \varphi'$ of S with $\phi^* = w$. The main problem with this approach is that it is not easy, given $w \in \mathrm{W}_n$, to see how the base-points of the two markings are related. The problem is easier if the base-points of the original marking lie along an elliptic curve; indeed in that case the new base-points also lie on this elliptic curve. Computations are thus computations on a curve, so are simpler. The best case is the case of a cuspidal cubic as there is a one-parameter subgroup of $\mathrm{Aut}(\mathbb{P}^2_{\mathbb{C}})$ fixing such a curve. McMullen proved

THEOREM 9.12 ([**140**]). *For any $n \geq 10$ the standard element w of W_n may be realized by an automorphism ϕ of a marked blow-up S with an invariant cuspidal anticanonical curve. The entropy of ϕ is the spectral radius of w which is positive.*

The following question "What are the elements of $w \in W_n$ which may be realized by rational surface automorphisms?" was also considered in [**84**] and [**166**]. Diller gave a rather thorough enumeration of the possibilities for quadratic birational maps which have an invariant curve. Such maps are determined by the data consisting of three orbit lengths (n_1, n_2, n_3) and a permutation of $\{1, 2, 3\}$. Diller also showed that not all orbit data, and not all $w \in \mathrm{W}_n$, are realizable by maps with invariant curve. Uehara established the following statement:

THEOREM 9.13 ([**166**]). *For every $w \in \mathrm{W}_n$ with spectral radius > 1 there is a rational surface automorphism ϕ such that the spectral radius of ϕ^* is the same as the spectral radius of w.*

Uehara's method combines elements of McMullen's and Diller's approaches. Given $w \in \mathrm{W}_n$ he prescribed a set of orbit data and proved that these orbit data can be realized by an automorphism ϕ. The induced ϕ^* has the same spectral radius as w, although the two may not be conjugate.

REMARK 9.14. While McMullen's and Diller's constructions involve automorphisms with invariant curves note that in [**12**] the authors showed that rational surface automorphisms of positive entropy do not necessarily possess invariant curves.

9.2.2. Bedford and Kim construction. In [**11**] and [**12**] the authors found automorphisms within a specific two-parameters family of plane birational maps. The initial observation in the two papers is the same: for certain parameter pairs all points of indeterminacy for all iterates of the map in question can be eliminated by performing finitely many point blow-ups. The map then lifts to an automorphism of the resulting rational surface. This idea was "systematized" in [**82**].

In [**12**] the authors prove that essentially all examples of rational surfaces automorphisms associated to Coxeter elements can be found within the two-parameter birational family $(f_{a,b})_{(a,b)}$ given by $f_{a,b}(z_0, z_1) = \left(z_1, \frac{z_1+a}{z_0+b}\right)$.

9.3. Automorphisms are pervasive

9.3.1. Automorphisms of del Pezzo surfaces. Any del Pezzo surface S contains a finite number of (-1)-curves (*i.e.*, smooth curves isomorphic to $\mathbb{P}^1_\mathbb{C}$ and of self-intersection -1). Each of them can be contracted to get another del Pezzo surface of degree $(K_S)^2+1$. There are, moreover, the only reducible curves of S of negative self-intersection. If $S \neq \mathbb{P}^2_\mathbb{C}$, then there is a finite number of conic bundles $S \to \mathbb{P}^1_\mathbb{C}$ up to automorphism of $\mathbb{P}^1_\mathbb{C}$ and each of them has exactly $8-(K_S)^2$ singular fibers.

This latter fact can be found by contracting one component in each singular fiber which is the union of two (-1)-curves, obtaining a line bundle on a del Pezzo surface, isomorphic to $\mathbb{P}^1_\mathbb{C}\times\mathbb{P}^1_\mathbb{C}$ or to the Hirzebruch surface $\mathbb{F}_1$ and having degree 8.

For more details see [**74, 138**].

Automorphisms of del Pezzo surfaces of order 4. Set

$$S = \big\{(z_0 : z_1 : z_2 : z_3) \in \mathbb{P}(2,1,1,1) \,|\, z_0^2 - z_1^4 = z_2z_3(z_2+z_3)(z_2+\mu z_3)\big\}$$

where μ belongs to $\mathbb{C} \smallsetminus \{0, 1\}$. The surface S is a del Pezzo one of degree 2. The automorphism β given by

$$\beta\colon (z_0 : z_1 : z_2 : z_3) \mapsto (z_0 : \mathbf{i}z_1 : z_2 : z_3)$$

fixes pointwise the elliptic curve given by $z_0 = 0$. When μ varies all possible elliptic curves are obtained. Moreover $\operatorname{rk}\operatorname{Pic}(S)^\beta = 1$.

There are other automorphisms β of order 4 of rational surfaces S such that β^2 fixes an elliptic curve but none for which $\operatorname{rk}\operatorname{Pic}(S)^\beta = 1$ (*see* [**28**]).

Automorphisms of del Pezzo surfaces of order 6. Let us give explicit possibilities for automorphisms of order 6.

(i) Set

$$S = \big\{(z_0 : z_1 : z_2 : z_3) \in \mathbb{P}(3,1,1,2) \,|\, z_0^2 = z_3^3 + \mu z_1^4 z_3 + z_1^6 + z_2^6\big\}$$

for some general $\mu \in \mathbb{C}$ such that S is smooth. The surface S is a del Pezzo surface of degree 1. Consider on S:

$$\alpha\colon (z_0 : z_1 : z_2 : z_3) \mapsto (z_0 : z_1 : -\mathbf{j}z_2 : z_3)$$

where $\mathbf{j} = \mathrm{e}^{2\mathbf{i}\pi/3}$.

The automorphism α fixes pointwise the elliptic curve given by $z_2 = 0$. When μ varies all possible elliptic curves are obtained. The equality $\operatorname{rk}\operatorname{Pic}(S)^\alpha = 1$ holds (*see* [**89**, Corollary 6.11]).

(ii) Set

$$S = \big\{(z_0 : z_1 : z_2 : z_3) \in \mathbb{P}^3_\mathbb{C} \,|\, z_0z_1^2 + z_0^3 + z_2^3 + z_3^3 + \mu z_0z_2z_3 = 0\big\}$$

where μ is such that the cubic surface is smooth. The surface is a del Pezzo surface of degree 3. Consider on S the automorphism α given by

$$\alpha\colon (z_0 : z_1 : z_2 : z_3) \mapsto (z_0 : -z_1 : \mathbf{j}z_2 : \mathbf{j}^2z_3).$$

Note that α^3 fixes pointwise the elliptic curve $z_1 = 0$ and α acts on it via a translation of order 3. When μ varies all possible elliptic curves are obtained. The equality $\mathrm{rk}\,\mathrm{Pic}(S)^\alpha = 1$ holds ([**89**]).

(iii) Set

$$S = \big\{(z_0 : z_1 : z_2 : z_3) \in \mathbb{P}^3_\mathbb{C} \,|\, z_0^3 + z_1^3 + z_2^3 + (z_1 + \mu z_2)z_3^2 = 0\big\}$$

where $\mu \in \mathbb{C}$ is such that the cubic surface is smooth. It is a del Pezzo surface of degree 3. Consider α defined by $\alpha\colon (z_0 : z_1 : z_2 : z_3) \mapsto (\mathbf{j}z_0 : z_1 : z_2 : z_3)$. The automorphism α^3 fixes pointwise the elliptic curve $z_3 = 0$ and α acts on it via an automorphism of order 3 with three fixed points. When μ varies the birational class of α changes but not the isomorphism class of the elliptic curve fixed by α^3.

9.3.2. Outline of the construction.

9.3.2.1. *The central involution of* $\mathrm{SL}(2,\mathbb{Z})$ *and its image into* $\mathrm{Bir}(\mathbb{P}^2_\mathbb{C})$. Set $A = \begin{pmatrix} 1 & 1 \\ 0 & 1 \end{pmatrix}$ and $B = \begin{pmatrix} 0 & 1 \\ -1 & 0 \end{pmatrix}$. A presentation of $\mathrm{SL}(2,\mathbb{Z})$ is given by (*see* [**147**])

$$\langle A,\, B \,|\, B^4 = (AB)^3 = 1,\; B^2(AB) = (AB)B^2\rangle.$$

As a result the quotient of $\mathrm{SL}(2,\mathbb{Z})$ by its center is a free product of $\mathbb{Z}/2\mathbb{Z}$ and $\mathbb{Z}/3\mathbb{Z}$ generated by the classes $[B]$ of B and $[AB]$ of AB:

$$\mathrm{PSL}(2,\mathbb{Z}) = \langle [B],\, [AB] \,|\, [B]^2 = [AB]^3 = \mathrm{id}\rangle.$$

Recall that $\mathrm{SL}(2,\mathbb{R})$ acts on the upper half plane

$$\mathbb{H} = \big\{x + \mathbf{i}y \in \mathbb{C} \,|\, x,\, y \in \mathbb{R},\, y > 0\big\}$$

by Möbius transformations

$$\mathrm{SL}(2,\mathbb{R}) \times \mathbb{H} \to \mathbb{H}, \qquad \left(\begin{pmatrix} a & b \\ c & d \end{pmatrix}, z\right) \mapsto \frac{az+b}{cz+d}$$

the hyperbolic structure of $\mathbb{H}$ being preserved. This yields to a natural notion of elliptic, parabolic and loxodromic elements of $\mathrm{SL}(2,\mathbb{R})$. If M belongs to $\mathrm{SL}(2,\mathbb{Z})$ one can be more precise and check the following observations:

- ⋄ M is elliptic if and only if M has finite order;
- ⋄ M is parabolic if and only if M has infinite order and its trace is ± 2;
- ⋄ M is loxodromic if and only if M has infinite order and its trace is $\neq \pm 2$.

Up to conjugacy the elliptic elements of $\mathrm{SL}(2,\mathbb{Z})$ are

$$\begin{pmatrix} -1 & 0 \\ 0 & -1 \end{pmatrix}, \begin{pmatrix} 0 & 1 \\ -1 & -1 \end{pmatrix}, \begin{pmatrix} 0 & 1 \\ -1 & 0 \end{pmatrix}, \begin{pmatrix} 0 & -1 \\ 1 & 0 \end{pmatrix}, \begin{pmatrix} 0 & -1 \\ 1 & 1 \end{pmatrix}.$$

In particular an element of finite order is of order 2, 3, 4 or 6.

A parabolic element of $\mathrm{SL}(2,\mathbb{Z})$ is up to conjugacy one of the following ones

$$\begin{pmatrix} 1 & a \\ 0 & 1 \end{pmatrix} \qquad \begin{pmatrix} -1 & a \\ 0 & -1 \end{pmatrix}$$

with $a \in \mathbb{Z}$.

Since $B^2 \in \mathrm{SL}(2,\mathbb{Z})$ is an involution its image by any embedding $\theta\colon \mathrm{SL}(2,\mathbb{Z}) \to \mathrm{Bir}(\mathbb{P}^2_\mathbb{C})$ is a birational involution. As we have seen in §6.1 an element of order 2 of the Cremona group is up to conjugacy one of the following:

- ⋄ an automorphism of $\mathbb{P}^2_\mathbb{C}$,

◇ a Jonquières involution of degree ≥ 2,
◇ a Bertini involution,
◇ a Geiser involution.

Since B^2 commutes with $\mathrm{SL}(2,\mathbb{Z})$ the group $\theta\big(\mathrm{SL}(2,\mathbb{Z})\big)$ is contained in the centralizer of $\theta(B^2)$. But if $\theta(B^2)$ is a Bertini involution or a Geiser involution, then the centralizer of $\theta(B^2)$ is finite ([**39**]). As a result $\theta(B^2)$ is conjugate either to an automorphism of $\mathbb{P}^2_{\mathbb{C}}$, or to a Jonquières involution. Assume that $\theta(B^2)$ is not linearisable; $\theta(B^2)$ fixes thus pointwise a unique irreducible curve Γ of genus ≥ 1. Denote by G the image of θ. The group G preserves Γ and the action of G on Γ gives the exact sequence

$$1 \longrightarrow \mathrm{G}' \longrightarrow \mathrm{G} \longrightarrow \mathrm{H} \longrightarrow 1$$

where H is a subgroup of $\mathrm{Aut}(\Gamma)$ and G' contains $\theta(B^2)$ and fixes Γ. The genus of Γ is positive; hence H cannot coincide with ${}^{\mathrm{G}}\!/_{\langle\theta(B^2)\rangle}$, a free product of ${}^{\mathbb{Z}}\!/_{2\mathbb{Z}}$ and ${}^{\mathbb{Z}}\!/_{3\mathbb{Z}}$. As a consequence $\mathrm{G}' \lhd \mathrm{G}$ strictly contains $\langle\theta(B^2)\rangle$; thus G' is infinite and not abelian. In particular the group of birational maps fixing pointwise Γ is infinite and not abelian. So according to [**38**] the curve Γ has genus 1. One can now state:

LEMMA 9.15 ([**32**]). *Let θ be an embedding of* $\mathrm{SL}(2,\mathbb{Z})$ *into the plane Cremona group. Then up to birational conjugacy*

◇ *either $\theta(B^2)$ is an automorphism of $\mathbb{P}^2_{\mathbb{C}}$ of order* 2,
◇ *or $\theta(B^2)$ is a Jonquières involution of degree* 3 *fixing (pointwise) an elliptic curve.*

9.3.2.2. *Existence of infinitely many loxodromic embeddings of* $\mathrm{SL}(2,\mathbb{Z})$ *into* $\mathrm{Bir}(\mathbb{P}^2_{\mathbb{C}})$. Let us consider the standard embedding

$$\theta_e\colon \mathrm{SL}(2,\mathbb{Z})\to\mathrm{Bir}(\mathbb{P}^2_{\mathbb{C}}), \qquad \begin{pmatrix} a & b\\ c & d\end{pmatrix}\mapsto\Big((z_0:z_1:z_2)\mapsto(az_0+bz_1:cz_0+dz_1:z_2)\Big).$$

Note that $\theta_e\big(\mathrm{SL}(2,\mathbb{Z})\big)$ is a subgroup of $\mathrm{PGL}(3,\mathbb{C})$ that preserves the line L_{z_2} of equation $z_2=0$ and acts on it via the maps

$$\mathrm{SL}(2,\mathbb{Z}) \to \mathrm{PSL}(2,\mathbb{Z}) \subset \mathrm{PSL}(2,\mathbb{C}) = \mathrm{Aut}(L_{z_2}).$$

Pick $\mu\in\mathbb{C}^*$ such that the point $p=(\mu:1:0)\in L_{z_2}$ has a trivial isotropy group under the action of $\mathrm{PSL}(2,\mathbb{Z})$. Fix an even integer $k>0$; consider ψ the conjugation of

$$\psi'\colon (z_0:z_1:z_2) \dashrightarrow (z_0^k : z_0^{k-1}z_1+z_2^k : z_0^{k-1}z_2)$$

by $(z_0:z_1:z_2)\mapsto(z_0+\mu z_1:z_1:z_2)$. Then define the morphism $\theta_k\colon \mathrm{SL}(2,\mathbb{Z})\to\mathrm{Bir}(\mathbb{P}^2_{\mathbb{C}})$ as follows:

$$\theta_k(B)=\theta_e(B)\colon (z_0:z_1:z_2)\mapsto(z_1:-z_0:z_2), \qquad \theta_k(AB)=\psi\circ\theta_e(AB)\circ\psi^{-1}.$$

The map ψ' restricts to an automorphism of the affine plane where $z_0\neq 0$, commutes with $\theta_k(B^2)=\theta_e(B^2)=(z_0:z_1:-z_2)\in\mathrm{Aut}(\mathbb{P}^2_{\mathbb{C}})$ and acts trivially on L_{z_2}. Since ψ commutes with $\theta_k(B^2)$ the map $\theta_k(AB)$ commutes with $\theta_k(B^2)$. As a result θ_k is a well-defined morphism. As $\psi_{|L_{z_2}\smallsetminus\{p\}}=\mathrm{id}$ the actions of θ_e and θ_k on L_{z_2} are the same; θ_k is thus an embedding.

LEMMA 9.16 ([**32**]). *Let n be a positive integer. Let a_1, $\ldots$, a_n, b_1, $\ldots$, b_n be $2n$ elements in $\{-1,\,1\}$. The birational self map of $\mathbb{P}^2_{\mathbb{C}}$*

$$\theta_k\big(B^{b_n}(AB)^{a_n}B^{b_{n-1}}(AB)^{a_{n-1}}\ldots B^{b_1}(AB)^{a_1}\big)$$

has degree k^{2n} and has exactly $2n$ proper base-points, all lying on L_{z_2}.

More precisely the base-points are

$$p,\ \big((AB)^{a_1}\big)^{-1}(p),\ \big(B^{b_1}(AB)^{a_1}\big)^{-1}(p),$$
$$\big((AB)^{a_2}B^{b_1}(AB)^{a_1}\big)^{-1}(p),\ \ldots,\ \big((AB)^{a_n}B^{b_{n-1}}(AB)^{a_{n-1}}\ldots B^{b_1}(AB)^{a_1}\big)^{-1}(p),$$
$$\big(B^{b_n}(AB)^{a_n}B^{b_{n-1}}(AB)^{a_{n-1}}\ldots B^{b_1}(AB)^{a_1}\big)^{-1}(p).$$

This result implies the existence of infinitely many loxodromic embeddings of $\mathrm{SL}(2,\mathbb{Z})$ into $\mathrm{Bir}(\mathbb{P}^2_{\mathbb{C}})$:

COROLLARY 9.17 ([**32**]). *Let n be a positive integer. Let a_1, a_2, $\ldots$, a_n, b_1, b_2, $\ldots$, b_n be $2n$ elements in $\{-1,\,1\}$. The birational self map of $\mathbb{P}^2_{\mathbb{C}}$*

$$\theta_k\big(B^{b_n}(AB)^{a_n}B^{b_{n-1}}(AB)^{a_{n-1}}\ldots B^{b_1}(AB)^{a_1}\big)$$

has dynamical degree k^{2n}.

In particular, θ_k is a loxodromic embedding and

$$\big\{\lambda(\phi)\,|\,\phi\in\theta_k\big(\mathrm{SL}(2,\mathbb{Z})\big)\big\}=\big\{1,\,k^2,\,k^4,\,k^6,\,\ldots\big\}.$$

PROOF. Let us consider an element of infinite order of $\mathrm{SL}(2,\mathbb{Z})$; it is conjugate to

$$\varphi=B^{b_n}(AB)^{a_n}B^{b_{n-1}}(AB)^{a_{n-1}}\ldots B^{b_1}(AB)^{a_1}$$

where a_1, a_2, $\ldots$, a_n, b_1, b_2, $\ldots$, $b_n\in\{-1,\,1\}$. According to Lemma 9.16 the degree of $\theta_k(\varphi^r)$ is equal to k^{2nr}. As a consequence $\lambda\big(\theta_k(\varphi)\big)=k^{2n}$. □

IDEA OF THE PROOF OF LEMMA 9.16. We proceed by induction on n. Let us detail the case $n=1$. The birational map ψ has degree k and has a unique proper base-point $p=(\mu:1:0)\in L_{z_2}$. The same holds for ψ^{-1}. Moreover $\psi_{|L_{z_2}\smallsetminus\{p\}}=\psi^{-1}_{|L_{z_2}\smallsetminus\{p\}}=\mathrm{id}$. Since $\theta_e(AB)^{a_1}\in\mathrm{Aut}(\mathbb{P}^2_{\mathbb{C}})$ moves the point p onto another point of L_{z_2}, the map $\theta_k\big((AB)^{a_1}\big)$ has degree k^2 and exactly two proper base-points which are p and $\big((AB)^{a_1}\big)^{-1}(p)=(\psi\circ\theta_e)(AB)^{-a_1}$. As $\theta_k(S)$ belongs to $\mathrm{Aut}(\mathbb{P}^2_{\mathbb{C}})$, $\theta_k(B^{b_1}(AB)^{a_1})$ also has degree k^2 and two proper base-points which are p and $\big((AB)^{a_1}\big)^{-1}(p)$. □

9.3.2.3. *Description of loxodromic embeddings for which the central element fixes (pointwise) an elliptic curve.* Let us note that

$$\mathrm{SL}(2,\mathbb{Z})=\langle\alpha,\,\beta\,|\,\beta^4=\mathrm{id},\,\alpha^3=\beta^2\rangle$$

(take the presentation we gave before and set $\alpha^2=AB$, $\beta=B$) and that

$$\mathrm{SL}(2,\mathbb{Z})=\langle\alpha,\,\beta\,|\,\alpha^6=\beta^4=\alpha^3\beta^2=\mathrm{id}\rangle.$$

In this section we will use this last presentation.

We say that a curve is *fixed* by a birational map if it is pointwise fixed, and say that a curve is *invariant* or *preserved* if the map induces a birational action on the curve.

All conjugacy classes of elements of order 4 and 6 in $\mathrm{Bir}(\mathbb{P}^2_{\mathbb{C}})$ have been classified in [**28**]. Many of them can act on del Pezzo surfaces of degree 1, 2, 3 or 4.

Del Pezzo surfaces X, Y of degree ≤ 4 and automorphisms $\alpha \in \mathrm{Aut}(X)$, resp. $\beta \in \mathrm{Aut}(Y)$ of order 6, resp. 4, so that

- ◇ α^3 and β^2 fix pointwise an elliptic curve,
- ◇ and that $\mathrm{Pic}(X)^\alpha$, $\mathrm{Pic}(Y)^\beta$ both have rank 1,

are defined to create the embedding. Contracting (-1)-curves invariant by the involutions α^3 and β^2 (but not by α, β which act minimally on X and Y) we get rational morphisms $X \to X_4$ and $Y \to Y_4$ where X_4, Y_4 are del Pezzo surfaces on which α^3 and β^2 act minimally. Furthermore X_4 and Y_4 are del Pezzo surfaces of degree 4, both $\mathrm{Pic}(X_4)^{\alpha^3}$ and $\mathrm{Pic}(Y_4)^{\beta^2}$ have rank 2 and are generated by the fibers of the two conic bundles on X_4 and Y_4. Choosing a birational map $X_4 \dashrightarrow Y_4$ conjugating α^3 to β^2 (which exists if and only if the elliptic curves are isomorphic), which is general enough, we obtain a loxodromic embedding

$$\mathrm{SL}(2,\mathbb{Z}) \to \mathrm{Bir}(\mathbb{P}^2_\mathbb{C}).$$

To prove that there is no other relation in $\langle \alpha, \beta \rangle$ and that all elements of infinite order are loxodromic, the morphisms $X \to X_4$ and $Y \to Y_4$ and the actions of α and β on $\mathrm{Pic}(X)^{\alpha^3}$ and $\mathrm{Pic}(Y)^{\beta^2}$ are described; furthermore the composition of the elements does what is expected.

Bibliography

[1] Maria Alberich-Carramiñana, *Geometry of the plane Cremona maps*, Lecture Notes in Mathematics, vol. 1769, Springer-Verlag, Berlin, 2002. MR1874328

[2] James W. Alexander, *On the factorization of Cremona plane transformations*, Trans. Amer. Math. Soc. **17** (1916), no. 3, 295–300, DOI 10.2307/1989004. MR1501043

[3] Wolf P. Barth, Klaus Hulek, Chris A. M. Peters, and Antonius Van de Ven, *Compact complex surfaces*, 2nd ed., Ergebnisse der Mathematik und ihrer Grenzgebiete. 3. Folge. A Series of Modern Surveys in Mathematics [Results in Mathematics and Related Areas. 3rd Series. A Series of Modern Surveys in Mathematics], vol. 4, Springer-Verlag, Berlin, 2004. MR2030225

[4] Hyman Bass, Edwin H. Connell, and David Wright, *The Jacobian conjecture: reduction of degree and formal expansion of the inverse*, Bull. Amer. Math. Soc. (N.S.) **7** (1982), no. 2, 287–330, DOI 10.1090/S0273-0979-1982-15032-7. MR663785

[5] Hyman Bass and Alexander Lubotzky, *Automorphisms of groups and of schemes of finite type*, Israel J. Math. **44** (1983), no. 1, 1–22, DOI 10.1007/BF02763168. MR693651

[6] Lionel Bayle and Arnaud Beauville, *Birational involutions of* $\mathbf{P}^2$, Asian J. Math. **4** (2000), no. 1, 11–17, DOI 10.4310/AJM.2000.v4.n1.a2. Kodaira's issue. MR1802909

[7] Arnaud Beauville, *Complex algebraic surfaces*, London Mathematical Society Lecture Note Series, vol. 68, Cambridge University Press, Cambridge, 1983. Translated from the French by R. Barlow, N. I. Shepherd-Barron and M. Reid. MR732439

[8] Arnaud Beauville, *p-elementary subgroups of the Cremona group*, J. Algebra **314** (2007), no. 2, 553–564, DOI 10.1016/j.jalgebra.2005.07.040. MR2344578

[9] Arnaud Beauville and Jérémy Blanc, *On Cremona transformations of prime order* (English, with English and French summaries), C. R. Math. Acad. Sci. Paris **339** (2004), no. 4, 257–259, DOI 10.1016/j.crma.2004.06.015. MR2092009

[10] Eric Bedford and Jeffrey Diller, *Energy and invariant measures for birational surface maps*, Duke Math. J. **128** (2005), no. 2, 331–368, DOI 10.1215/S0012-7094-04-12824-6. MR2140266

[11] Eric Bedford and Kyounghee Kim, *Periodicities in linear fractional recurrences: degree growth of birational surface maps*, Michigan Math. J. **54** (2006), no. 3, 647–670, DOI 10.1307/mmj/1163789919. MR2280499

[12] Eric Bedford and Kyounghee Kim, *Dynamics of rational surface automorphisms: linear fractional recurrences*, J. Geom. Anal. **19** (2009), no. 3, 553–583, DOI 10.1007/s12220-009-9077-8. MR2496566

[13] Eric Bedford and Kyounghee Kim, *Continuous families of rational surface automorphisms with positive entropy*, Math. Ann. **348** (2010), no. 3, 667–688, DOI 10.1007/s00208-010-0498-2. MR2677899

[14] Eric Bedford and Kyounghee Kim, *Dynamics of rational surface automorphisms: rotation domains*, Amer. J. Math. **134** (2012), no. 2, 379–405, DOI 10.1353/ajm.2012.0015. MR2905001

[15] M.-J. Bertin, A. Decomps-Guilloux, M. Grandet-Hugot, M. Pathiaux-Delefosse, and J.-P. Schreiber, *Pisot and Salem numbers*, Birkhäuser Verlag, Basel, 1992. With a preface by David W. Boyd. MR1187044

[16] E. Bertini, *Ricerche sulle trasformazioni univoche involutorie nel piano*, Annali di Mat. **8** (1877), 254–287.

[17] Mladen Bestvina, Mark Feighn, and Michael Handel, *The Tits alternative for* $\mathrm{Out}(F_n)$. *I. Dynamics of exponentially-growing automorphisms*, Ann. of Math. (2) **151** (2000), no. 2, 517–623, DOI 10.2307/121043. MR1765705

[18] Mladen Bestvina and Koji Fujiwara, *Bounded cohomology of subgroups of mapping class groups*, Geom. Topol. **6** (2002), 69–89, DOI 10.2140/gt.2002.6.69. MR1914565

[19] A. Białynicki-Birula, *Remarks on the action of an algebraic torus on k^n* (English, with Russian summary), Bull. Acad. Polon. Sci. Sér. Sci. Math. Astronom. Phys. **14** (1966), 177–181. MR200279

[20] Garrett Birkhoff, *Lie groups simply isomorphic with no linear group*, Bull. Amer. Math. Soc. **42** (1936), no. 12, 883–888, DOI 10.1090/S0002-9904-1936-06457-8. MR1563459

[21] Jérémy Blanc, *Conjugacy classes of affine automorphisms of $\mathbb{K}^n$ and linear automorphisms of $\mathbb{P}^n$ in the Cremona groups*, Manuscripta Math. **119** (2006), no. 2, 225–241, DOI 10.1007/s00229-005-0617-7. MR2215969

[22] Jérémy Blanc, *Finite abelian subgroups of the Cremona group of the plane* (English, with English and French summaries), C. R. Math. Acad. Sci. Paris **344** (2007), no. 1, 21–26, DOI 10.1016/j.crma.2006.11.026. MR2286582

[23] Jérémy Blanc, *Finite abelian subgroups of the Cremona group of the plane* (English, with English and French summaries), C. R. Math. Acad. Sci. Paris **344** (2007), no. 1, 21–26, DOI 10.1016/j.crma.2006.11.026. MR2286582

[24] Jérémy Blanc, *The number of conjugacy classes of elements of the Cremona group of some given finite order* (English, with English and French summaries), Bull. Soc. Math. France **135** (2007), no. 3, 419–434, DOI 10.24033/bsmf.2541. MR2430188

[25] Jérémy Blanc, *Linearisation of finite abelian subgroups of the Cremona group of the plane*, Groups Geom. Dyn. **3** (2009), no. 2, 215–266, DOI 10.4171/GGD/55. MR2486798

[26] Jérémy Blanc, *Sous-groupes algébriques du groupe de Cremona* (French, with English and French summaries), Transform. Groups **14** (2009), no. 2, 249–285, DOI 10.1007/s00031-008-9046-5. MR2504924

[27] Jérémy Blanc, *Groupes de Cremona, connexité et simplicité* (French, with English and French summaries), Ann. Sci. Éc. Norm. Supér. (4) **43** (2010), no. 2, 357–364, DOI 10.24033/asens.2123. MR2662668

[28] Jérémy Blanc, *Elements and cyclic subgroups of finite order of the Cremona group*, Comment. Math. Helv. **86** (2011), no. 2, 469–497, DOI 10.4171/CMH/231. MR2775137

[29] Jérémy Blanc, *Simple relations in the Cremona group*, Proc. Amer. Math. Soc. **140** (2012), no. 5, 1495–1500, DOI 10.1090/S0002-9939-2011-11004-9. MR2869134

[30] Jérémy Blanc, *Algebraic elements of the Cremona groups*, From classical to modern algebraic geometry, Trends Hist. Sci., Birkhäuser/Springer, Cham, 2016, pp. 351–360, DOI 10.1007/978-3-319-32994-9_7. MR3776658

[31] Jérémy Blanc and Serge Cantat, *Dynamical degrees of birational transformations of projective surfaces*, J. Amer. Math. Soc. **29** (2016), no. 2, 415–471, DOI 10.1090/jams831. MR3454379

[32] Jérémy Blanc and Julie Déserti, *Embeddings of* $\mathrm{SL}(2,\mathbb{Z})$ *into the Cremona group*, Transform. Groups **17** (2012), no. 1, 21–50, DOI 10.1007/s00031-012-9174-9. MR2891210

[33] Jérémy Blanc and Julie Déserti, *Degree growth of birational maps of the plane*, Ann. Sc. Norm. Super. Pisa Cl. Sci. (5) **14** (2015), no. 2, 507–533. MR3410471

[34] Jérémy Blanc and Jean-Philippe Furter, *Topologies and structures of the Cremona groups*, Ann. of Math. (2) **178** (2013), no. 3, 1173–1198, DOI 10.4007/annals.2013.178.3.8. MR3092478

[35] Lionel Nguyen Van Thé, *Fixed points in compactifications and combinatorial counterparts* (English, with English and French summaries), Ann. H. Lebesgue **2** (2019), 149–185, DOI 10.5802/ahl.16. MR3978387

[36] Jérémy Blanc and Isac Hedén, *The group of Cremona transformations generated by linear maps and the standard involution* (English, with English and French summaries), Ann. Inst. Fourier (Grenoble) **65** (2015), no. 6, 2641–2680. MR3449593

[37] Jérémy Blanc and Susanna Zimmermann, *Topological simplicity of the Cremona groups*, Amer. J. Math. **140** (2018), no. 5, 1297–1309, DOI 10.1353/ajm.2018.0032. MR3862066

[38] Jérémy Blanc, Ivan Pan, and Thierry Vust, *Sur un théorème de Castelnuovo* (French, with English and French summaries), Bull. Braz. Math. Soc. (N.S.) **39** (2008), no. 1, 61–80, DOI 10.1007/s00574-008-0072-7. MR2390684

[39] Jérémy Blanc, Ivan Pan, and Thierry Vust, *On birational transformations of pairs in the complex plane*, Geom. Dedicata **139** (2009), 57–73, DOI 10.1007/s10711-008-9316-3. MR2481837

[40] Jérémy Blanc and Egor Yasinsky, *Quotients of groups of birational transformations of cubic del Pezzo fibrations* (English, with English and French summaries), J. Éc. polytech. Math. **7** (2020), 1089–1112, DOI 10.5802/jep.136. MR4136435

[41] Jérémy Blanc and Susanna Zimmermann, *Topological simplicity of the Cremona groups*, Amer. J. Math. **140** (2018), no. 5, 1297–1309, DOI 10.1353/ajm.2018.0032. MR3862066

[42] Christian Bonatti, Sylvain Crovisier, and Amie Wilkinson, *The C^1 generic diffeomorphism has trivial centralizer*, Publ. Math. Inst. Hautes Études Sci. **109** (2009), 185–244, DOI 10.1007/s10240-009-0021-z. MR2511588

[43] Armand Borel and Jacques Tits, *Homomorphismes "abstraits" de groupes algébriques simples* (French), Ann. of Math. (2) **97** (1973), 499–571, DOI 10.2307/1970833. MR316587

[44] Nicolas Bourbaki, *General topology. Chapters 5–10*, Elements of Mathematics (Berlin), Springer-Verlag, Berlin, 1998. Translated from the French; Reprint of the 1989 English translation. MR1726872

[45] Matthew G. Brin and Craig C. Squier, *Groups of piecewise linear homeomorphisms of the real line*, Invent. Math. **79** (1985), no. 3, 485–498, DOI 10.1007/BF01388519. MR782231

[46] Michel Brion, *Some basic results on actions of nonaffine algebraic groups*, Symmetry and spaces, Progr. Math., vol. 278, Birkhäuser Boston, Boston, MA, 2010, pp. 1–20, DOI 10.1007/978-0-8176-4875-6_1. MR2562620

[47] Felix E. Browder (ed.), *Mathematical developments arising from Hilbert problems*, Proceedings of Symposia in Pure Mathematics, Vol. XXVIII, American Mathematical Society, Providence, R. I., 1976. MR0419125

[48] Marco Brunella, *Birational geometry of foliations*, IMPA Monographs, vol. 1, Springer, Cham, 2015. MR3328860

[49] Marc Burger, Alessandra Iozzi, and Nicolas Monod, *Equivariant embeddings of trees into hyperbolic spaces*, Int. Math. Res. Not. **22** (2005), 1331–1369, DOI 10.1155/IMRN.2005.1331. MR2152540

[50] Serge Cantat, *Dynamique des automorphismes des surfaces projectives complexes* (French, with English and French summaries), C. R. Acad. Sci. Paris Sér. I Math. **328** (1999), no. 10, 901–906, DOI 10.1016/S0764-4442(99)80294-8. MR1689873

[51] Serge Cantat, *Dynamique des automorphismes des surfaces $K3$* (French), Acta Math. **187** (2001), no. 1, 1–57, DOI 10.1007/BF02392831. MR1864630

[52] Serge Cantat, *Sur les groupes de transformations birationnelles des surfaces* (French, with English and French summaries), Ann. of Math. (2) **174** (2011), no. 1, 299–340, DOI 10.4007/annals.2011.174.1.8. MR2811600

[53] Serge Cantat, *The Cremona group in two variables*, European Congress of Mathematics, Eur. Math. Soc., Zürich, 2013, pp. 211–225. MR3469123

[54] Serge Cantat, *Morphisms between Cremona groups, and characterization of rational varieties*, Compos. Math. **150** (2014), no. 7, 1107–1124, DOI 10.1112/S0010437X13007835. MR3230847

[55] Serge Cantat and Yves de Cornulier, *Distortion in Cremona groups*, Ann. Sc. Norm. Super. Pisa Cl. Sci. (5) **20** (2020), no. 2, 827–858. MR4105920

[56] Serge Cantat and Igor Dolgachev, *Rational surfaces with a large group of automorphisms*, J. Amer. Math. Soc. **25** (2012), no. 3, 863–905, DOI 10.1090/S0894-0347-2012-00732-2. MR2904576

[57] Serge Cantat and Charles Favre, *Symétries birationnelles des surfaces feuilletées* (French, with English summary), J. Reine Angew. Math. **561** (2003), 199–235, DOI 10.1515/crll.2003.066. MR1998612

[58] Serge Cantat and Stéphane Lamy, *Groupes d'automorphismes polynomiaux du plan* (French, with English summary), Geom. Dedicata **123** (2006), 201–221, DOI 10.1007/s10711-007-9126-z. MR2299735

[59] Serge Cantat and Stéphane Lamy, *Normal subgroups in the Cremona group*, Acta Math. **210** (2013), no. 1, 31–94, DOI 10.1007/s11511-013-0090-1. With an appendix by Yves de Cornulier. MR3037611

[60] G. Castelnuovo, *Le trasformationi generatrici del gruppo cremoniano nel piano*, Atti R. Accad. Sci. Torino **36** (1901), 861–874.

[61] Dominique Cerveau and Julie Déserti, *Centralisateurs dans le groupe de Jonquières* (French), Michigan Math. J. **61** (2012), no. 4, 763–783, DOI 10.1307/mmj/1353098512. MR3049289

[62] Dominique Cerveau and Julie Déserti, *Transformations birationnelles de petit degré* (French), Cours Spécialisés [Specialized Courses], vol. 19, Société Mathématique de France, Paris, 2013. MR3155973

[63] Ian Chiswell, *Introduction to Λ-trees*, World Scientific Publishing Co., Inc., River Edge, NJ, 2001. MR1851337

[64] Arthur B. Coble, *Algebraic geometry and theta functions*, Revised printing. American Mathematical Society Colloquium Publication, vol. X, American Mathematical Society, Providence, R.I., 1961. MR0123958

[65] M. Coornaert, T. Delzant, and A. Papadopoulos, *Géométrie et théorie des groupes* (French), Lecture Notes in Mathematics, vol. 1441, Springer-Verlag, Berlin, 1990. Les groupes hyperboliques de Gromov. [Gromov hyperbolic groups]; With an English summary. MR1075994

[66] Y. Cornulier, *Nonlinearity of some subgroups of the planar Cremona group*, Unpublished manuscript, 2013, `http://www.normalesup.org/~cornulier/crelin.pdf`.

[67] Rémi Coulon, *Théorie de la petite simplification: une approche géométrique [d'après F. Dahmani, V. Guirardel, D. Osin et S. Cantat, S. Lamy]* (French), Astérisque **380, Séminaire Bourbaki. Vol. 2014/2015** (2016), Exp. No. 1089, 1–33, DOI 10.1016/j.physletb.2016.08.052. MR3522169

[68] F. Dahmani, V. Guirardel, and D. Osin, *Hyperbolically embedded subgroups and rotating families in groups acting on hyperbolic spaces*, Mem. Amer. Math. Soc. **245** (2017), no. 1156, v+152, DOI 10.1090/memo/1156. MR3589159

[69] Tommaso de Fernex, *On planar Cremona maps of prime order*, Nagoya Math. J. **174** (2004), 1–28, DOI 10.1017/S0027763000008795. MR2066103

[70] Tommaso de Fernex and Lawrence Ein, *Resolution of indeterminacy of pairs*, Algebraic geometry, de Gruyter, Berlin, 2002, pp. 165–177. MR1954063

[71] Pierre de la Harpe, *Topics in geometric group theory*, Chicago Lectures in Mathematics, University of Chicago Press, Chicago, IL, 2000. MR1786869

[72] Pierre de la Harpe and Alain Valette, *La propriété (T) de Kazhdan pour les groupes localement compacts (avec un appendice de Marc Burger)* (French, with English summary), Astérisque **175** (1989), 158. With an appendix by M. Burger. MR1023471

[73] Thomas Delzant and Pierre Py, *Kähler groups, real hyperbolic spaces and the Cremona group*, Compos. Math. **148** (2012), no. 1, 153–184, DOI 10.1112/S0010437X11007068. MR2881312

[74] Michel Demazure, *Sous-groupes algébriques de rang maximum du groupe de Cremona* (French), Ann. Sci. École Norm. Sup. (4) **3** (1970), 507–588. MR284446

[75] Michel Demazure, Henry Charles Pinkham, and Bernard Teissier (eds.), *Séminaire sur les Singularités des Surfaces* (French), Lecture Notes in Mathematics, vol. 777, Springer, Berlin, 1980. Held at the Centre de Mathématiques de l'École Polytechnique, Palaiseau, 1976–1977. MR579026

[76] Julie Déserti, *Groupe de Cremona et dynamique complexe: une approche de la conjecture de Zimmer* (French), Int. Math. Res. Not., posted on 2006, Art. ID 71701, 27, DOI 10.1155/IMRN/2006/71701. MR2233717

[77] Julie Déserti, *Sur les automorphismes du groupe de Cremona* (French, with English summary), Compos. Math. **142** (2006), no. 6, 1459–1478, DOI 10.1112/S0010437X06002478. MR2278755

[78] Julie Déserti, *Le groupe de Cremona est hopfien* (French, with English and French summaries), C. R. Math. Acad. Sci. Paris **344** (2007), no. 3, 153–156, DOI 10.1016/j.crma.2006.12.005. MR2292279

[79] Julie Déserti, *Sur les sous-groupes nilpotents du groupe de Cremona* (French, with English and French summaries), Bull. Braz. Math. Soc. (N.S.) **38** (2007), no. 3, 377–388, DOI 10.1007/s00574-007-0050-5. MR2344204

[80] Julie Déserti, *On solvable subgroups of the Cremona group*, Illinois J. Math. **59** (2015), no. 2, 345–358. MR3499516

[81] Julie Déserti, *Some properties of the group of birational maps generated by the automorphisms of $\mathbb{P}^n_{\mathbb{C}}$ and the standard involution*, Math. Z. **281** (2015), no. 3-4, 893–905, DOI 10.1007/s00209-015-1512-x. MR3421645

[82] Julie Déserti and Julien Grivaux, *Automorphisms of rational surfaces with positive entropy*, Indiana Univ. Math. J. **60** (2011), no. 5, 1589–1622, DOI 10.1512/iumj.2011.60.4427. MR2997001

[83] Jean A. Dieudonné, *La géométrie des groupes classiques* (French), Springer-Verlag, Berlin-New York, 1971. Troisième édition; Ergebnisse der Mathematik und ihrer Grenzgebiete, Band 5. MR0310083

[84] Jeffrey Diller, *Cremona transformations, surface automorphisms, and plane cubics*, Michigan Math. J. **60** (2011), no. 2, 409–440, DOI 10.1307/mmj/1310667983. With an appendix by Igor Dolgachev. MR2825269

[85] J. Diller and C. Favre, *Dynamics of bimeromorphic maps of surfaces*, Amer. J. Math. **123** (2001), no. 6, 1135–1169. MR1867314

[86] Tien-Cuong Dinh and Nessim Sibony, *Une borne supérieure pour l'entropie topologique d'une application rationnelle* (French, with English summary), Ann. of Math. (2) **161** (2005), no. 3, 1637–1644, DOI 10.4007/annals.2005.161.1637. MR2180409

[87] Igor Dolgachev and David Ortland, *Point sets in projective spaces and theta functions* (English, with French summary), Astérisque **165** (1988), 210 pp. (1989). MR1007155

[88] Igor V. Dolgachev, *Classical algebraic geometry*, Cambridge University Press, Cambridge, 2012. A modern view. MR2964027

[89] Igor V. Dolgachev and Vasily A. Iskovskikh, *Finite subgroups of the plane Cremona group*, Algebra, arithmetic, and geometry: in honor of Yu. I. Manin. Vol. I, Progr. Math., vol. 269, Birkhäuser Boston, Boston, MA, 2009, pp. 443–548, DOI 10.1007/978-0-8176-4745-2_11. MR2641179

[90] Igor V. Dolgachev and De-Qi Zhang, *Coble rational surfaces*, Amer. J. Math. **123** (2001), no. 1, 79–114. MR1827278

[91] F. Enriques, *Sui gruppi continui di transformazioni cremoniani nel piano*, Rend. Accad. Lincei **1er sem.** (1893).

[92] F. Enriques, *Conferenze di geometria: fondamenti di una geometria iperspaziale*, Bologna (1895).

[93] D. B. A. Epstein and W. P. Thurston, *Transformation groups and natural bundles*, Proc. London Math. Soc. (3) **38** (1979), no. 2, 219–236, DOI 10.1112/plms/s3-38.2.219. MR531161

[94] R. P. Filipkiewicz, *Isomorphisms between diffeomorphism groups*, Ergodic Theory Dynam. Systems **2** (1982), no. 2, 159–171 (1983), DOI 10.1017/s0143385700001486. MR693972

[95] S. P. Franklin, *Spaces in which sequences suffice*, Fund. Math. **57** (1965), 107–115, DOI 10.4064/fm-57-1-107-115. MR180954

[96] C. F. Geiser, *Ueber zwei geometrische Probleme* (German), J. Reine Angew. Math. **67** (1867), 78–89, DOI 10.1515/crll.1867.67.78. MR1579359

[97] Étienne Ghys, *Sur les groupes engendrés par des difféomorphismes proches de l'identité* (French, with English summary), Bol. Soc. Brasil. Mat. (N.S.) **24** (1993), no. 2, 137–178, DOI 10.1007/BF01237675. MR1254981

[98] Étienne Ghys and Pierre de la Harpe, *La propriété de Markov pour les groupes hyperboliques* (French), Sur les groupes hyperboliques d'après Mikhael Gromov (Bern, 1988), Progr. Math., vol. 83, Birkhäuser Boston, Boston, MA, 1990, pp. 165–187, DOI 10.1007/978-1-4684-9167-8_9. MR1086657

[99] Étienne Ghys and Vlad Sergiescu, *Sur un groupe remarquable de difféomorphismes du cercle* (French), Comment. Math. Helv. **62** (1987), no. 2, 185–239, DOI 10.1007/BF02564445. MR896095

[100] Marat Gizatullin, *On some tensor representations of the Cremona group of the projective plane*, New trends in algebraic geometry (Warwick, 1996), London Math. Soc. Lecture Note Ser., vol. 264, Cambridge Univ. Press, Cambridge, 1999, pp. 111–150, DOI 10.1017/CBO9780511721540.007. MR1714823

[101] M. H. Gizatullin, *Rational G-surfaces* (Russian), Izv. Akad. Nauk SSSR Ser. Mat. **44** (1980), no. 1, 110–144, 239. MR563788

[102] M. Kh. Gizatullin, *Defining relations for the Cremona group of the plane* (Russian), Izv. Akad. Nauk SSSR Ser. Mat. **46** (1982), no. 5, 909–970, 1134. MR675525

[103] E. S. Golod, *On nil-algebras and finitely approximable p-groups* (Russian), Izv. Akad. Nauk SSSR Ser. Mat. **28** (1964), 273–276. MR0161878

[104] Julien Grivaux, *Parabolic automorphisms of projective surfaces (after M. H. Gizatullin)*, Mosc. Math. J. **16** (2016), no. 2, 275–298, DOI 10.17323/1609-4514-2016-16-2-275-298. MR3480704

[105] M. Gromov, *Hyperbolic groups*, Essays in group theory, Math. Sci. Res. Inst. Publ., vol. 8, Springer, New York, 1987, pp. 75–263, DOI 10.1007/978-1-4613-9586-7_3. MR919829

[106] Vincent Guirardel, *Geometric small cancellation*, Geometric group theory, IAS/Park City Math. Ser., vol. 21, Amer. Math. Soc., Providence, RI, 2014, pp. 55–90, DOI 10.1090/pcms/021/03. MR3329725

[107] Brian Harbourne, *Rational surfaces with infinite automorphism group and no antipluricanonical curve*, Proc. Amer. Math. Soc. **99** (1987), no. 3, 409–414, DOI 10.2307/2046337. MR875372

[108] Brian Harbourne, *Iterated blow-ups and moduli for rational surfaces*, Algebraic geometry (Sundance, UT, 1986), Lecture Notes in Math., vol. 1311, Springer, Berlin, 1988, pp. 101–117, DOI 10.1007/BFb0082911. MR951643

[109] Robin Hartshorne, *Algebraic geometry*, Springer-Verlag, New York-Heidelberg, 1977. Graduate Texts in Mathematics, No. 52. MR0463157

[110] J. F. P. Hudson, *On sunny collapsing*, Proc. Cambridge Philos. Soc. **66** (1969), 247–250, DOI 10.1017/s0305004100044935. MR247630

[111] James E. Humphreys, *Linear algebraic groups*, Springer-Verlag, New York-Heidelberg, 1975. Graduate Texts in Mathematics, No. 21. MR0396773

[112] V. A. Iskovskih, *Minimal models of rational surfaces over arbitrary fields* (Russian), Izv. Akad. Nauk SSSR Ser. Mat. **43** (1979), no. 1, 19–43, 237. MR525940

[113] V. A. Iskovskikh, *Generators and relations in a two-dimensional Cremona group* (Russian), Vestnik Moskov. Univ. Ser. I Mat. Mekh. **5** (1983), 43–48. MR722450

[114] V. A. Iskovskikh, *Proof of a theorem on relations in the two-dimensional Cremona group* (Russian), Uspekhi Mat. Nauk **40** (1985), no. 5(245), 255–256. MR810819

[115] V. A. Iskovskikh, *On finite subgroups of the Cremona group*, Vietnam J. Math. **33** (2005), no. Special Issue, 61–80. MR2227704

[116] V. A. Iskovskikh, F. K. Kabdykairov, and S. L. Tregub, *Relations in a two-dimensional Cremona group over a perfect field* (Russian), Izv. Ross. Akad. Nauk Ser. Mat. **57** (1993), no. 3, 3–69, DOI 10.1070/IM1994v042n03ABEH001542; English transl., Russian Acad. Sci. Izv. Math. **42** (1994), no. 3, 427–478. MR1243341

[117] N. V. Ivanov, *Algebraic properties of the Teichmüller modular group* (Russian), Dokl. Akad. Nauk SSSR **275** (1984), no. 4, 786–789. MR745513

[118] Sergei V. Ivanov, *The free Burnside groups of sufficiently large exponents*, Internat. J. Algebra Comput. **4** (1994), no. 1-2, ii+308, DOI 10.1142/S0218196794000026. MR1283947

[119] J. P. Jouanolou, *Équations de Pfaff algébriques* (French), Lecture Notes in Mathematics, vol. 708, Springer, Berlin, 1979. MR537038

[120] Gaston Julia, *Mémoire sur la permutabilité des fractions rationnelles* (French), Ann. Sci. École Norm. Sup. (3) **39** (1922), 131–215. MR1509242

[121] Gaston Julia, *Œuvres de Gaston Julia. Vol. I* (French), Gauthier-Villars, Paris, 1968. Edited by Michel Hervé. MR708273

[122] Heinrich W. E. Jung, *Über ganze birationale Transformationen der Ebene* (German), J. Reine Angew. Math. **184** (1942), 161–174, DOI 10.1515/crll.1942.184.161. MR8915

[123] S. Kantor, *Theorie der eindeutigen periodischen Transformationen in der Ebene* (German), J. Reine Angew. Math. **114** (1895), 50–108, DOI 10.1515/crll.1895.114.50. MR1580369

[124] M. I. Kargapolov and Ju. I. Merzljakov, *Fundamentals of the theory of groups*, Graduate Texts in Mathematics, vol. 62, Springer-Verlag, New York-Berlin, 1979. Translated from the second Russian edition by Robert G. Burns. MR551207

[125] A. I. Kostrikin, *On Burnside's problem* (Russian), Dokl. Akad. Nauk SSSR **119** (1958), 1081–1084. MR0132101

[126] H. Kraft, *Regularization of rational group actions*, arXiv:1808.08729 (2018).

[127] Stéphane Lamy, *Dynamique des groupes paraboliques d'automorphismes polynomiaux de* $\mathbb{C}^2$ (French, with English and French summaries), Bol. Soc. Brasil. Mat. (N.S.) **32** (2001), no. 2, 185–212, DOI 10.1007/BF01243867. MR1860869

[128] Stéphane Lamy, *L'alternative de Tits pour* $\mathrm{Aut}[\mathbb{C}^2]$ (French, with French summary), J. Algebra **239** (2001), no. 2, 413–437, DOI 10.1006/jabr.2000.8701. MR1832900

[129] Stéphane Lamy, *Une preuve géométrique du théorème de Jung* (French, with French summary), Enseign. Math. (2) **48** (2002), no. 3-4, 291–315. MR1955604

[130] David I. Lieberman, *Compactness of the Chow scheme: applications to automorphisms and deformations of Kähler manifolds*, Fonctions de plusieurs variables complexes, III (Sém. François Norguet, 1975), Lecture Notes in Math., vol. 670, Springer, Berlin, 1978, pp. 140–186. MR521918

[131] Anne Lonjou, *Non simplicité du groupe de Cremona sur tout corps* (French, with English and French summaries), Ann. Inst. Fourier (Grenoble) **66** (2016), no. 5, 2021–2046. MR3533276

[132] Anne Lonjou, *Pavage de Voronoï associé au groupe de Cremona* (French, with English summary), Publ. Mat. **63** (2019), no. 2, 521–599, DOI 10.5565/PUBLMAT6321905. MR3980934

[133] Anne Lonjou, *Sur l'hyperbolicité de graphes associés au groupe de Cremona* (French, with English and French summaries), Épijournal Géom. Algébrique **3** (2019), Art. 11, 29. MR4015911

[134] I. G. Lysënok, *Infinite Burnside groups of even period* (Russian, with Russian summary), Izv. Ross. Akad. Nauk Ser. Mat. **60** (1996), no. 3, 3–224, DOI 10.1070/IM1996v060n03ABEH000077; English transl., Izv. Math. **60** (1996), no. 3, 453–654. MR1405529

[135] A. Malcev, *On isomorphic matrix representations of infinite groups* (Russian, with English summary), Rec. Math. [Mat. Sbornik] N.S. **8 (50)** (1940), 405–422. MR0003420

[136] Ju. I. Manin, *Rational surfaces over perfect fields* (Russian, with English summary), Inst. Hautes Études Sci. Publ. Math. **30** (1966), 55–113. MR225780

[137] Ju. I. Manin, *Rational surfaces over perfect fields. II* (Russian), Mat. Sb. (N.S.) **72 (114)** (1967), 161–192. MR0225781

[138] Yu. I. Manin, *Cubic forms*, 2nd ed., North-Holland Mathematical Library, vol. 4, North-Holland Publishing Co., Amsterdam, 1986. Algebra, geometry, arithmetic; Translated from the Russian by M. Hazewinkel. MR833513

[139] Curtis T. McMullen, *Dynamics on K3 surfaces: Salem numbers and Siegel disks*, J. Reine Angew. Math. **545** (2002), 201–233, DOI 10.1515/crll.2002.036. MR1896103

[140] Curtis T. McMullen, *Dynamics on blowups of the projective plane*, Publ. Math. Inst. Hautes Études Sci. **105** (2007), 49–89, DOI 10.1007/s10240-007-0004-x. MR2354205

[141] Michael McQuillan, *Diophantine approximations and foliations*, Inst. Hautes Études Sci. Publ. Math. **87** (1998), 121–174. MR1659270

[142] Luís Gustavo Mendes, *Kodaira dimension of holomorphic singular foliations*, Bol. Soc. Brasil. Mat. (N.S.) **31** (2000), no. 2, 127–143, DOI 10.1007/BF01244239. MR1785264

[143] Ashot Minasyan and Denis Osin, *Acylindrical hyperbolicity of groups acting on trees*, Math. Ann. **362** (2015), no. 3-4, 1055–1105, DOI 10.1007/s00208-014-1138-z. MR3368093

[144] Ashot Minasyan and Denis Osin, *Correction to: Acylindrical hyperbolicity of groups acting on trees [MR3368093]*, Math. Ann. **373** (2019), no. 1-2, 895–900, DOI 10.1007/s00208-018-1699-3. MR3968890

[145] David Mumford, *Hilbert's fourteenth problem–the finite generation of subrings such as rings of invariants*, Mathematical developments arising from Hilbert problems (Proc. Sympos. Pure Math., Vol. XXVIII, Northern Illinois Univ., De Kalb, Ill., 1974), Amer. Math. Soc.., Providence, R. I., 1976, pp. 431–444. MR0435076

[146] Masayoshi Nagata, *On rational surfaces. I. Irreducible curves of arithmetic genus* 0 *or* 1, Mem. Coll. Sci. Univ. Kyoto Ser. A. Math. **32** (1960), 351–370, DOI 10.1215/kjm/1250776405. MR126443

[147] Morris Newman, *Integral matrices*, Academic Press, New York-London, 1972. Pure and Applied Mathematics, Vol. 45. MR0340283

[148] P. S. Novikov and S. I. Adjan, *Infinite periodic groups. I* (Russian), Izv. Akad. Nauk SSSR Ser. Mat. **32** (1968), 212–244. MR0240178

[149] P. S. Novikov and S. I. Adjan, *Infinite periodic groups. II* (Russian), Izv. Akad. Nauk SSSR Ser. Mat. **32** (1968), 251–524. MR0240179

[150] P. S. Novikov and S. I. Adjan, *Infinite periodic groups. III* (Russian), Izv. Akad. Nauk SSSR Ser. Mat. **32** (1968), 709–731. MR0240180

[151] A. Yu. Ol′shanskiĭ, *The Novikov-Adyan theorem* (Russian), Mat. Sb. (N.S.) **118(160)** (1982), no. 2, 203–235, 287. MR658789

[152] J. Palis, *Rigidity of the centralizers of diffeomorphisms and structural stability of suspended foliations*, Differential topology, foliations and Gelfand-Fuks cohomology (Proc. Sympos., Pontifícia Univ. Católica, Rio de Janeiro, 1976), Lecture Notes in Math., vol. 652, Springer, Berlin, 1978, pp. 114–121. MR505654

[153] J. Palis and J.-C. Yoccoz, *Centralizers of Anosov diffeomorphisms on tori*, Ann. Sci. École Norm. Sup. (4) **22** (1989), no. 1, 99–108. MR985856

[154] J. Palis and J.-C. Yoccoz, *Rigidity of centralizers of diffeomorphisms*, Ann. Sci. École Norm. Sup. (4) **22** (1989), no. 1, 81–98. MR985855

[155] Ivan Pan, *Une remarque sur la génération du groupe de Cremona* (French, with English and French summaries), Bol. Soc. Brasil. Mat. (N.S.) **30** (1999), no. 1, 95–98, DOI 10.1007/BF01235676. MR1686984

[156] V. L. Popov, *Tori in the Cremona groups* (Russian, with Russian summary), Izv. Ross. Akad. Nauk Ser. Mat. **77** (2013), no. 4, 103–134, DOI 10.1070/im2013v077n04abeh002659; English transl., Izv. Math. **77** (2013), no. 4, 742–771. MR3135700

[157] J. F. Ritt, *Permutable rational functions*, Trans. Amer. Math. Soc. **25** (1923), no. 3, 399–448, DOI 10.2307/1989297. MR1501252

[158] Maxwell Rosenlicht, *Some basic theorems on algebraic groups*, Amer. J. Math. **78** (1956), 401–443, DOI 10.2307/2372523. MR82183

[159] I. Schur, *Über Gruppen linearer Substitutionen mit Koeffizienten aus einem algebraischen Zahlkörper* (German), Math. Ann. **71** (1911), no. 3, 355–367, DOI 10.1007/BF01456850. MR1511662

[160] Jean-Pierre Serre, *Arbres, amalgames,* SL_2 (French), Société Mathématique de France, Paris, 1977. Avec un sommaire anglais; Rédigé avec la collaboration de Hyman Bass; Astérisque, No. 46. MR0476875

[161] N.I. Shepherd-Barron, *Some effectivity questions for plane Cremona transformations*, arXiv:1311.6608 (2013).

[162] Christophe Soulé, *Groupes opérant sur un complexe simplicial avec domaine fondamental* (French), C. R. Acad. Sci. Paris Sér. A-B **276** (1973), A607–A609. MR317307

[163] Robert Steinberg, *Some consequences of the elementary relations in* SL_n, Finite groups—coming of age (Montreal, Que., 1982), Contemp. Math., vol. 45, Amer. Math. Soc., Providence, RI, 1985, pp. 335–350, DOI 10.1090/conm/045/822247. MR822247

[164] Richard G. Swan, *Generators and relations for certain special linear groups*, Advances in Math. **6** (1971), 1–77 (1971), DOI 10.1016/0001-8708(71)90027-2. MR284516

[165] J. Tits, *Free subgroups in linear groups*, J. Algebra **20** (1972), 250–270, DOI 10.1016/0021-8693(72)90058-0. MR286898

[166] Takato Uehara, *Rational surface automorphisms with positive entropy* (English, with English and French summaries), Ann. Inst. Fourier (Grenoble) **66** (2016), no. 1, 377–432. MR3477879

[167] Hiroshi Umemura, *Sur les sous-groupes algébriques primitifs du groupe de Cremona à trois variables* (French), Nagoya Math. J. **79** (1980), 47–67. MR587409

[168] Hiroshi Umemura, *Maximal algebraic subgroups of the Cremona group of three variables. Imprimitive algebraic subgroups of exceptional type*, Nagoya Math. J. **87** (1982), 59–78. MR676586

[169] Hiroshi Umemura, *On the maximal connected algebraic subgroups of the Cremona group. I*, Nagoya Math. J. **88** (1982), 213–246. MR683251

[170] C. Urech, *Subgroups of Cremona groups*, Ph.D. thesis, Iniversity of Basel/University of Rennes 1, 2017.

[171] Christian Urech, *Simple groups of birational transformations in dimension two*, Comment. Math. Helv. **95** (2020), no. 2, 211–246, DOI 10.4171/CMH/486. MR4115282

[172] Christian Urech, *Subgroups of elliptic elements of the Cremona group*, J. Reine Angew. Math. **770** (2021), 27–57, MR4193461.

[173] Christian Urech and Susanna Zimmermann, *A new presentation of the plane Cremona group*, Proc. Amer. Math. Soc. **147** (2019), no. 7, 2741–2755, DOI 10.1090/proc/14479. MR3973878

[174] André Weil, *On algebraic groups of transformations*, Amer. J. Math. **77** (1955), 355–391, DOI 10.2307/2372535. MR74083

[175] André Weil, *Scientific works. Collected papers. Vol. II (1951–1964)* (French), Springer-Verlag, New York-Heidelberg, 1979. MR537935

[176] Hermann Weyl, *On unitary metrics in projective space*, Ann. of Math. (2) **40** (1939), no. 1, 141–148, DOI 10.2307/1968550. MR1503455

[177] James V. Whittaker, *On isomorphic groups and homeomorphic spaces*, Ann. of Math. (2) **78** (1963), 74–91, DOI 10.2307/1970503. MR150750

[178] A. Wiman, *Zur Theorie der endlichen Gruppen von birationalen Transformationen in der Ebene* (German), Math. Ann. **48** (1896), no. 1-2, 195–240, DOI 10.1007/BF01446342. MR1510931

[179] David Wright, *Abelian subgroups of* $\operatorname{Aut}_k(k[X, Y])$ *and applications to actions on the affine plane*, Illinois J. Math. **23** (1979), no. 4, 579–634. MR540400

[180] David Wright, *Two-dimensional Cremona groups acting on simplicial complexes*, Trans. Amer. Math. Soc. **331** (1992), no. 1, 281–300, DOI 10.2307/2154009. MR1038019

[181] Michael Vaughan-Lee, *On Zel′manov's solution of the restricted Burnside problem*, J. Group Theory **1** (1998), no. 1, 65–94, DOI 10.1515/jgth.1998.003. MR1490159

[182] Michael Vaughan-Lee, *The restricted Burnside problem*, 2nd ed., London Mathematical Society Monographs. New Series, vol. 8, The Clarendon Press, Oxford University Press, New York, 1993. MR1364414

[183] D.-Q. Zhang, *Automorphisms of finite order on rational surfaces*, J. Algebra **238** (2001), no. 2, 560–589, DOI 10.1006/jabr.2000.8673. With an appendix by I. Dolgachev. MR1823774

[184] S. Zhao, *Centralizers of elements of infinite order in plane Cremona groups*, arXiv:1907.11203 (2019).

[185] R. J. Zimmer, *Kazhdan groups acting on compact manifolds*, Invent. Math. **75** (1984), no. 3, 425–436, DOI 10.1007/BF01388637. MR735334

[186] Robert J. Zimmer, *On connection-preserving actions of discrete linear groups*, Ergodic Theory Dynam. Systems **6** (1986), no. 4, 639–644, DOI 10.1017/S0143385700003746. MR873437

[187] Robert J. Zimmer, *Actions of semisimple groups and discrete subgroups*, Proceedings of the International Congress of Mathematicians, Vol. 1, 2 (Berkeley, Calif., 1986), Amer. Math. Soc., Providence, RI, 1987, pp. 1247–1258. MR934329

[188] Robert J. Zimmer, *Lattices in semisimple groups and invariant geometric structures on compact manifolds*, Discrete groups in geometry and analysis (New Haven, Conn., 1984), Progr. Math., vol. 67, Birkhäuser Boston, Boston, MA, 1987, pp. 152–210, DOI 10.1007/978-1-4899-6664-3_6. MR900826

[189] Susanna Zimmermann, *The Cremona group of the plane is compactly presented*, J. Lond. Math. Soc. (2) **93** (2016), no. 1, 25–46, DOI 10.1112/jlms/jdv054. MR3455780

Index

Notation index